Cell Biology: Basics to Breakthroughs

Edited by

K.N. Aruljothi
Department of Genetic Engineering, College of Engineering and Technology SRM Institute of Science and Technology, Kattankulathur, Chengalpattu Tamil Nadu, India

Prakash Gangadaran
Department of Medicine & Department of Nuclear Medicine. School of Medicine, Kyungpook National University. Daegu, Republic of Korea

Krishnan Anand
Department of Chemical Pathology, School of Pathology, Faculty of Health Sciences, University of the Free State, Bloemfontein, South Africa

&

Satish Ramalingam, K. Kumaran & Kruthika Prakash
Department of Genetic Engineering, College of Engineering and Technology, SRM Institute of Science and Technology, Kattankulathur, Chengalpattu, Tamil Nadu, India

Cell Biology: Basics to Breakthroughs

Editors: K.N. Aruljothi, Prakash Gangadaran, Krishnan Anand, Satish Ramalingam, K. Kumaran & Kruthika Prakash

ISBN (Online): 979-8-89881-243-0

ISBN (Print): 979-8-89881-244-7

ISBN (Paperback): 979-8-89881-245-4

© 2026, Bentham Books imprint.

Published by Bentham Science Publishers Pte. Ltd. Singapore, in collaboration with Eureka Conferences, USA. All Rights Reserved.

First published in 2026.

BENTHAM SCIENCE PUBLISHERS LTD.
End User License Agreement (for non-institutional, personal use)

This is an agreement between you and Bentham Science Publishers Ltd. Please read this License Agreement carefully before using the ebook/echapter/ejournal (**"Work"**). Your use of the Work constitutes your agreement to the terms and conditions set forth in this License Agreement. If you do not agree to these terms and conditions then you should not use the Work.

Bentham Science Publishers agrees to grant you a non-exclusive, non-transferable limited license to use the Work subject to and in accordance with the following terms and conditions. This License Agreement is for non-library, personal use only. For a library / institutional / multi user license in respect of the Work, please contact: permission@benthamscience.org.

Usage Rules:

1. All rights reserved: The Work is the subject of copyright and Bentham Science Publishers either owns the Work (and the copyright in it) or is licensed to distribute the Work. You shall not copy, reproduce, modify, remove, delete, augment, add to, publish, transmit, sell, resell, create derivative works from, or in any way exploit the Work or make the Work available for others to do any of the same, in any form or by any means, in whole or in part, in each case without the prior written permission of Bentham Science Publishers, unless stated otherwise in this License Agreement.
2. You may download a copy of the Work on one occasion to one personal computer (including tablet, laptop, desktop, or other such devices). You may make one back-up copy of the Work to avoid losing it.
3. The unauthorised use or distribution of copyrighted or other proprietary content is illegal and could subject you to liability for substantial money damages. You will be liable for any damage resulting from your misuse of the Work or any violation of this License Agreement, including any infringement by you of copyrights or proprietary rights.

Disclaimer:

Bentham Science Publishers does not guarantee that the information in the Work is error-free, or warrant that it will meet your requirements or that access to the Work will be uninterrupted or error-free. The Work is provided "as is" without warranty of any kind, either express or implied or statutory, including, without limitation, implied warranties of merchantability and fitness for a particular purpose. The entire risk as to the results and performance of the Work is assumed by you. No responsibility is assumed by Bentham Science Publishers, its staff, editors and/or authors for any injury and/or damage to persons or property as a matter of products liability, negligence or otherwise, or from any use or operation of any methods, products instruction, advertisements or ideas contained in the Work.

Limitation of Liability:

In no event will Bentham Science Publishers, its staff, editors and/or authors, be liable for any damages, including, without limitation, special, incidental and/or consequential damages and/or damages for lost data and/or profits arising out of (whether directly or indirectly) the use or inability to use the Work. The entire liability of Bentham Science Publishers shall be limited to the amount actually paid by you for the Work.

General:

1. Any dispute or claim arising out of or in connection with this License Agreement or the Work (including non-contractual disputes or claims) will be governed by and construed in accordance with the laws of Singapore. Each party agrees that the courts of the state of Singapore shall have exclusive jurisdiction to settle any dispute or claim arising out of or in connection with this License Agreement or the Work (including non-contractual disputes or claims).
2. Your rights under this License Agreement will automatically terminate without notice and without the

need for a court order if at any point you breach any terms of this License Agreement. In no event will any delay or failure by Bentham Science Publishers in enforcing your compliance with this License Agreement constitute a waiver of any of its rights.
3. You acknowledge that you have read this License Agreement, and agree to be bound by its terms and conditions. To the extent that any other terms and conditions presented on any website of Bentham Science Publishers conflict with, or are inconsistent with, the terms and conditions set out in this License Agreement, you acknowledge that the terms and conditions set out in this License Agreement shall prevail.

Bentham Science Publishers Pte. Ltd.
No. 9 Raffles Place
Office No. 26-01
Singapore 048619
Singapore
Email: subscriptions@benthamscience.net

CONTENTS

FOREWORD i
PREFACE ii
ACKNOWLEDGEMENTS iii
LIST OF CONTRIBUTORS iv

CHAPTER 1 FOUNDATIONS OF LIFE: CELLS AND ORIGIN 1
Meghana A. Shakthi, Harin N. Ganesh, R. Kirubakaran and K.N. Aruljothi
    INTRODUCTION 1
        Theory of Spontaneous Generation 1
        Abiogenesis for the Origin of Life 2
        Cell and its Contents 3
            Cell Organelles 5
        Organelle Biogenesis and the Endosymbiotic Theory 6
            Chemical Composition of Cell 7
            Cytoplasmic Composition of the Cell 8
        Structural and Functional Aspects of the Cell 8
            Cell Shape and How it Impacts Function 8
            Cell Size Variation 9
        Integration of Cells into Tissues 10
        Extracellular Matrix and Connective Tissues 11
        Plant and Animal Cells 12
            Plant Cell 12
            Evolution of Photosynthesis 14
            Animal Cell 14
        Cells Under the Microscope 15
    CONCLUSION 16
    ACKNOWLEDGEMENTS 16
    REFERENCE 16

CHAPTER 2 PROKARYOTES VS. EUKARYOTES: COMPARATIVE STRUCTURAL AND FUNCTIONAL INSIGHTS 20
Kruthika Prakash, Sayantani Chattopadhyay, T. Sasitharan, Soumyadeep Maiti,
Sanjana Dhayalan, K. Kumaran, R. Kirubakaran and K.N. Aruljothi
    INTRODUCTION TO PROKARYOTES 21
        Prokaryotic Diversity 22
        Classification of Prokaryotes 22
        Morphological Diversity 22
        Metabolic Diversity 23
        Ecological Diversity 23
        Cell Envelope 24
        Cytoplasm and External Structures 24
        Prokaryotic Genetics 25
        Genome Structure 25
        Gene Transfer Mechanisms 25
        Mutation and Adaptation 25
        Prokaryotes and Human Health 26
        Introduction to Eukaryotes 26
        Characteristics of Eukaryotes 26

WHITTAKER CLASSIFICATION - A BREAKTHROUGH IN THE STUDY OF EUKARYOTE ANATOMY	27
DEVELOPMENT OF MULTICELLULAR ORGANISMS	28
Biotechnological Applications of Eukaryotes	30
CONCLUSION	30
REFERENCES	31

CHAPTER 3 NUCLEUS: THE CONTROL CENTRE OF THE CELL 34
K. Kumaran, Harin N. Ganesh and K.N. Aruljothi

INTRODUCTION	34
History	35
STRUCTURE AND FUNCTION	36
Structure of Nucleus in Eukaryotes	36
Function of Nucleus in Eukaryotes	38
Structure of Genetic Material in Prokaryotes	38
Function of Genetic Material in Prokaryotes	39
DISORDERS ASSOCIATED WITH NUCLEUS	39
RECENT RESEARCH IN THE NUCLEUS	41
Therapeutics	42
CONCLUSION	43
ACKNOWLEDGEMENTS	44
REFERENCES	44

CHAPTER 4 ENDOPLASMIC RETICULUM: THE CELLULAR FACTORY 47
Shanmuga Priya, K. Kumaran, Sanjana Dhayalan, Krishnan Anand and K.N. Aruljothi

INTRODUCTION	47
FUNCTIONS OF RER	48
The Process of Protein Folding Associated with Molecular Chaperones	49
ER as the Quality Control Center	50
ER Stress	50
Diabetes Type II and ER Stress in Autophagy	52
Therapeutics	53
ER Stress – Therapeutics and Curative Treatment for Acute Spinal Cord Injury	53
Therapeutic for Ulcerative Colitis	54
SMOOTH ENDOPLASMIC RETICULUM	55
Evolutionary Aspects of SER	56
Calcium Storage in Smooth Endoplasmic Reticulum	56
The Mitochondrial-Associated Endoplasmic Reticulum Membrane (MAM)	57
Understanding the Role of MAM in Cardiovascular Disease	57
RECENT TRENDS IN ER RESEARCH	58
CONCLUSION	59
ACKNOWLEDGEMENTS	59
REFERENCES	59

CHAPTER 5 GOLGI APPARATUS: SHAPING AND SHIPPING CELLULAR PROTEINS 61
S. Sahasra, Sanjana Dhayalan, K. Kumaran and K.N. Aruljothi

INTRODUCTION	61
STRUCTURE OF GOLGI APPARATUS	62
CIS Face	62
MEDIAL Face	63
TRANS Face	63

FUNCTION OF THE GOLGI APPARATUS	64
Protein Glycosylation	64
Lipid and Polysaccharide Metabolism	67
Protein Sorting and Export	68
GOLGI APPARATUS DYSFUNCTION AND ASSOCIATED DISEASES	68
Neurodegenerative Diseases	68
Infectious Disease	69
Cancer	69
Genetic Mutation	69
THERAPEUTICS	70
Gaucher Disease (GD)	70
Parkinson's Disease	70
GTPase Rab1	71
LRRK2 Mutations	71
Rab8 and Rab10	71
PLA2G6 Mutations	71
α-Synuclein Aggregates	71
Cancer Therapy	72
Inhibitors of Peripheral Golgi-Associated Proteins - TAS-116	72
Compounds Affecting GA Protein Localization-2-(Substituted Phenyl)-Benzimidazole (2-PB)	72
Nanodelivery Approaches	72
RECENT TRENDS	73
COVID-19	73
Golgi Dynamics and Regulation	73
GOLGI AND DISEASE	73
Golgi and Cell Signalling	73
Golgi and Infectious Diseases	74
CONCLUSION	74
ACKNOWLEDGEMENTS	74
REFERENCES	74
CHAPTER 6 RIBOSOMES: ENGINES OF PROTEIN SYNTHESIS	77
Sanjana Dhayalan, Shalini Roy, K. Kumaran and *K.N. Aruljothi*	
INTRODUCTION	77
Ribosome in Prokaryotes and Eukaryotes	78
Origin and Evolution of the Ribosome	79
Ribosomes Fingerprint	81
STRUCTURE OF RIBOSOME	82
Composition of the Ribosome	84
FUNCTION OF RIBOSOME	85
Ribosome Biogenesis (Ribi)	86
Ribi in Eukaryotic Cells	86
Ribi in Prokaryotic Cells	86
Role of the Ribosome in Translation – Initiation	87
Role of the ribosome in Translation - Elongation	88
Role of the Ribosome in Translation – Termination	89
Ribosome Recycling	90
Regulation of Termination	91
DYSFUNCTION AND DISEASE OF THE RIBOSOME	91
	95

  RP Mutations and Cancer 95
  Onco-Ribosomes and Cancer Development 95
 DIAGNOSTIC AND THERAPEUTIC PROPERTY OF RIBOSOMES 96
  RP as Anti-microbial 96
  Therapeutic Targeting of Ribi and mTORC1 in Cancer 97
 RECENT TRENDS 97
  Advancements in Ribosome Profiling and Cancer 97
  Ribosome Profiling Reveals Non-Canonical ORF Translation in High-Risk Childhood
  Medulloblastoma 98
 CONCLUSION 99
 ACKNOWLEDGEMENTS 99
 REFERENCES 99

CHAPTER 7 LYSOSOMES: THE CELL'S DIGESTIVE SYSTEM 102
B. Surya, Faizaan Khan, Abhinav Roy, Sanjana Dhayalan, K. Kumaran, Krishnan Anand and *K.N. Aruljothi*
 INTRODUCTION 102
  Discovery of Lysosomes 103
  Evolution of Lysosomes 104
   Early Endosomes 105
   Recycling Endosomes 105
   Late Endosomes 105
 FUNCTION 106
  Autophagy (Process of Self-destruction) 106
  Phagocytosis [Function of Specialized Cells (macrophages, Neutrophils)] 106
  Lysosomal Membrane Proteins 107
  Lysosome Reformation 108
  Endocytic Lysosome Reformation (ELR) 109
  Autophagic Lysosome Reformation (ALR) 109
  Phagocytic Lysosome Reformation (PLR) 110
  Key Players of Membrane Fusion 111
 LYSOSOMAL DYSFUNCTION 114
  Lysosome Storage Disorders (LSD) 115
   Parkinson's Disease 116
   Muscular Dystrophy 118
 FUTURE DIRECTIONS FOR LYSOSOME RESEARCH 119
 CONCLUSION 120
 ACKNOWLEDGEMENTS 120
 REFERENCES 120

CHAPTER 8 VACUOLES: THE STORAGE VAULTS OF THE CELL 122
S. Sahasra, S. Aswini, K. Kumaran, Sanjana Dhayalan and *K.N. Aruljothi*
 INTRODUCTION 122
  Types of Vacuoles 123
   Plant Vacuoles 123
   Animal Vacuoles 124
   Fungal Vacuoles 124
  Differences from Plant and Animal Vacuoles 125
   Protist Vacuole 125
  Structure 126
  Tonoplast 126
  Morphological Changes 127

Dynamic Structures	127
Components	128
Inclusion Bodies in the Vacuole	128
Pigments	129
Vacuoles vs. Vesicles	129
Structure	129
Function	129
FUNCTION OF VACUOLES	130
Vacuolar Membrane-collapse System	131
Vacuole-plasma Membrane-fusion System	131
Autophagosome-lysosome Fusion	132
Formation of Autolysosomes	132
DIAGNOSTIC AND THERAPEUTIC APPLICATION OF VACUOLES	134
RECENT TRENDS	134
CONCLUSION	135
ACKNOWLEDGEMENTS	135
REFERENCES	136

CHAPTER 9 PLASMA MEMBRANE: GATEWAY AND SENTINEL OF CELLULAR EXCHANGE 138
S. Aswini, K. Kumaran, Sanjana Dhayalan, Prakash Gangadaran and K.N. Aruljothi

INTRODUCTION	139
STRUCTURE	139
Gorter and Grendel Membrane Theory (1920)	139
Pauci-molecular Theory	140
Fluid Mosaic Model (1972)	140
Peripheral Proteins (Extrinsic Proteins)	141
Integral Proteins (Intrinsic Proteins)	141
Handerson and Unwin's Membrane Theory	142
Evolutionary Aspects of Plasma Membrane	143
FUNCTIONS OF PLASMA MEMBRANE	143
Acting as a Physical Barrier	143
Selective Permeability	143
Endocytosis by the Plasma Membrane	143
Exocytosis	144
Facilitating Communication and Signalling among Cells	144
Providing Shape to the Cytoskeleton and Maintaining Cell Potential	144
Membrane Transport Mechanism	145
Fick's First Law	145
Osmosis	146
Passive Diffusion	146
Facilitated Diffusion	147
Glucose Transporter	148
Potassium Channels	149
Calcium-activated Potassium Channel	149
Inwardly Rectifying Potassium Channel	149
Tandem Pore Domain Potassium Channel	149
Voltage-gated Potassium Channel	150
Sodium Channel	150
Aquaporin	151

  Active Transport 151
  Group Translocation 151
 ABNORMALITIES IN PLASMA MEMBRANE 153
  Hypertension 153
  Sphingolipids-related Disorders 153
   Basis for Sphingolipidoses-neuronal Vulnerability 154
  Neurotrophic Signalling 154
  Insulin Signalling 155
  Impacts of Sphingolipid Accumulation on the use of Cellular Energy 155
 DIAGNOSTIC AND THERAPEUTIC ROLE OF PLASMA MEMBRANE 156
 RECENT TRENDS 156
  CAR-T Therapy 156
  Liposomal Drug Delivery Systems 157
 CONCLUSION 158
 ACKNOWLEDGEMENTS 158
 REFERENCES 158

CHAPTER 10 MITOCHONDRIA AND CHLOROPLASTS: EVOLUTIONARY ENGINES OF THE CELL 161
 .K. Kumaran, Keerthana Ramesh, Sanjana Dhayalan, Gargii Chatterjee and K.N Aruljothi
 MITOCHONDRIA 162
  Introduction 162
   Origin and Evolution of Mitochondria 163
   Mitochondrial Inheritance in Eukaryotes 163
  Structure of Mitochondria 164
  Function of Mitochondria 166
   Electron Transport Chain and Oxidative Phosphorylation 166
   Mitochondria-associated Membranes 168
  Mitochondrial DNA Mutations 168
   Role of Mitochondria in Cancer and Aging 169
  Recent Trends in the Biology of Mitochondria 170
 CHLOROPLAST 173
  Introduction 173
   Origin and Evolution of Chloroplasts 173
  Structure of Chloroplast 174
   Dynamic Thylakoid Architecture 175
   Structural Plasticity 175
   Light Intensity and Thylakoid Structure 175
   Light and Dark Reactions 176
  Recent Trends in Malfunctions in Chloroplasts 177
   Autophagy and Chloroplast Degradation 177
   Sensitivity to Thermal Stress 177
   Production of Reactive Oxygen Species (ROS) 178
   Impact on Mitochondrial Function 178
 CONCLUSION 178
 ACKNOWLEDGEMENTS 179
 REFERENCES 179

CHAPTER 11 CYTOSKELETON: THE CELL'S BACKBONE AND HIGHWAY 182
 Danyal Reyaz, Sparshika Mishra, Ranjini Sengupta, T. Sasitharan, S. Gnanavel and K.N. Aruljothi

INTRODUCTION	182
Types of Cytoskeletal Fibers	183
MICROFILAMENTS	184
Structure	184
Organisation	185
Treadmilling	185
Organization of Actin Filaments	185
Functions	186
Protrusions of the Cell Surface	186
Muscle Contraction	186
Contractile Assemblies of Actin and Myosin in Non-muscle Cells	188
Cytokinesis	188
Locomotion	189
INTERMEDIATE FILAMENTS	189
Intermediate Filaments in the Cell	189
Composition	190
Types of Intermediate Filaments	191
TYPE I: Acidic Keratins	191
TYPE II: Basic Keratins	191
TYPE V: Nuclear Lamins	194
Structure And Assembly of Intermediate Filaments	194
Intermediate Filament: Organization And Intracellular Localization	195
Intermediate Filaments and Cell Signalling	196
MICROTUBULES	196
Microtubules Assembly	197
Role of MAPs in the Organization of Microtubules in Cells	197
Microtubule Motors	198
Cargo Transport by Microtubules	199
Diseases Associated with Microtubules	199
CONCLUSION	201
ACKNOWLEDGEMENTS	201
REFERENCES	201
CHAPTER 12 SIGNAL TRANSDUCTION PATHWAYS ORCHESTRATE CELLULAR COMMUNICATION: A NARRATIVE REVIEW	206
Kruthika Prakash, Raksa Arun, Srisri Satishkartik, Sayantani Chattopadhyay, Mashira Rahman, Prakash Gangadaran and *K.N. Aruljothi*	
INTRODUCTION	206
TYPES OF CELL SIGNALING	207
Autocrine Signaling	207
Paracrine Signaling	207
Endocrine Signaling	208
Juxtacrine Signaling	208
Role of Signal Transduction in Cellular Communication	208
Receptors	208
Signaling Molecules	208
Signal Transduction Proteins	208
Effector Proteins	209
CELL-CELL SIGNALING	209
Modes of Cell-cell Communication	209
Direct Cell-cell Contact	209

Types of Signaling Molecules	212
Proteins and Peptides	212
Steroids	212
Small Molecules	212
Importance in Tissue Homeostasis	212
Immune Response	212
Growth Regulation	213
Tissue Repair	213
PATHWAYS OF INTRACELLULAR SIGNAL TRANSDUCTION	213
Phases of Signal Transduction	214
Reception	214
Transduction	214
Response	214
Amplification	215
Termination	215
CELL SURFACE RECEPTORS	215
Types of Cell Surface Receptors	215
GPCRs	216
RTKs	217
Ligand-gated Ion Channels	217
Cytokine Receptors	217
Receptor–ligand Interaction	218
GPCR PATHWAYS	219
Mechanism of GPCR Activation	219
Signaling by GPCRs	220
cAMP Pathways	220
IP3/DAG Pathway	221
Physiological and Pathological Roles	222
MAPK PATHWAYS	223
Components of MAPK Pathways	223
MAPK Pathways	224
Extracellular Signal-regulated Kinase (ERK1/2) Pathway	225
JNK Signaling Pathway	226
p38 MAPK Pathway	227
Dysregulated MAPK Signaling in Disease	228
CROSS-TALK BETWEEN GPCRS AND THE MAPK PATHWAY	229
REGULATION AND TERMINATION OF SIGNALING PATHWAYS	230
Feedback Mechanisms	231
Role of Phosphatases and Ubiquitination	231
Implications for Drug Development	232
CONCLUSION	232
CONSENT FOR PUBLICATION	233
CONFLICT OF INTEREST	233
ACKNOWLEDGEMENTS	233
REFERENCES	233
CHAPTER 13 CELL DEATH: MECHANISMS AND MYSTERIES BEYOND APOPTOSIS	245

Shambhavi Jha, Rohan Vyas, S. Manvi, Vasanth Kanth T.L., Keerthivasu Ramasamy, K.N. Aruljothi and *Ramya Lakshmi Rajendran*

INTRODUCTION	246
The Historical Tapestry of Cell Death Research	247

    Exploring Diverse Cell Death Pathways 248
PROGRAMMED CELL DEATH 249
    Delving into Apoptosis 250
    Extrinsic and Intrinsic Pathways in Apoptosis: Dual Routes to Cellular Demise 251
        Extrinsic Pathway 251
        Intrinsic Pathway 252
    Convergence and Caspase Activation 253
        Cell Shrinkage 253
        Membrane Blebbing 253
        DNA Fragmentation 253
        Recent Advancements 255
    Anoikis: Cell Death due to Detachment 256
        Extrinsic and Intrinsic Pathways in Anoikis 257
        Integrin-Mediated Signalling Pathway 259
        PI3K/Akt Pathway 259
        MAPK/ERK Pathway 260
        Focal Adhesion Kinase (FAK) Signalling 260
        SRC Kinase Pathway 260
        Anoikis and Cancer 260
        Recent Advancements 261
    Ferroptosis: Iron-Dependent Cell Death Mechanism 262
        Iron Metabolism and Iron Regulatory Proteins 264
        Lipid Metabolism and Lipid Peroxidation 265
        Glutathione and Glutathione Peroxidase 4 (GPX4) 265
        Nuclear Factor Erythroid 2-Related Factor 2 (NRF2) Pathway 265
        Ferroptosis Regulators 265
        Recent Advancements 266
    Parthanatos: Death by PARP-1 Overactivation 267
        PARP Activation 268
        Poly(ADP-ribose) (PAR) Polymer Synthesis 268
        AIF Translocation and Nuclear Events 268
        Involvement of Other Proteins 268
        Mitochondrial Dysfunction 268
        Interactions with Apoptotic Pathways 269
        Recent Advancements 269
    NETosis: The Unique Cell Death Pathway 269
        Mechanism of NETosis 270
    ADCC: The Immune Response Cell Death 270
NON-PROGRAMMED CELL DEATH 272
    Necrosis and its Implications for Cell Death 272
    Coagulative Necrosis 272
        Preservation of Tissue Structure 273
        Denaturation of Proteins 273
        Nuclear Changes 273
    Liquefactive Necrosis 273
        Liquid Formation 274
        Inflammatory Response 274
    Caseous Necrosis 274
        Cheese-Like Appearance 275
        Loss of Tissue Structure 275
        Distinct Granulomas 275

Tissue Discoloration	276
MECHANISMS OF NECROSIS	277
Recent Advancements	279
Exploring mPTP-Mediated Necrosis	279
CELL DEATH: KEY FINDINGS AND FUTURE PROSPECTS	280
CONCLUSION	282
CONSENT FOR PUBLICATION	282
CONFLICT OF INTEREST	282
ACKNOWLEDGEMENTS	282
REFERENCES	282

CHAPTER 14 STEM CELLS: BREAKTHROUGHS IN MEDICINE AND THERAPEUTICS   289
Vanshikaa Karthikeyan, Janani Balaji, SriSri SatishKartik, Dannie Macrin and *K.N. Aruljothi*

INTRODUCTION	289
STEM CELL FUNCTIONS	290
TYPES OF STEM CELLS	290
Stem Cells Based on Differentiation Potential	290
Totipotent/Omnipotent Stem Cells	291
Pluripotent Stem Cells	291
Multipotent Stem Cells	291
Oligopotent Stem Cells	291
Unipotent Stem Cells	292
Stem Cells Based on Origin	292
Embryonic Stem Cells	292
Adult Stem Cells	293
Tissue-Resident Stem Cells	293
Induced-Pluripotent Stem Cells	294
Perinatal Stem Cells	294
STEM CELL NICHE	294
Types of Niche	295
Simple Niches	295
Complex Niches	295
Storage Niches	295
EVOLUTION OF STEM CELLS	296
SIMILARITIES AND DIFFERENCES BETWEEN PLANT AND ANIMAL STEM CELLS	296
Similarities	296
Differences	296
INTRODUCTION TO STEM CELL THERAPEUTICS	297
Types of Stem Cells used in Therapeutics	297
Embryonic Stem Cells	298
Adult Stem Cells	298
STEPS IN STEM CELL THERAPY	298
Determination of Stem Cell Source	298
Specification of Cell Dosage	298
Administrative Methods	299
Stem Cell Manipulation for Effective Treatment	299
MAJOR BREAKTHROUGH IN STEM CELL THERAPY	299
MAJOR TYPES OF STEM CELL THERAPY	301
Somatic Cell Nuclear Transfer	301
Implementation of Stem Cell Therapy in Cancer Treatment	302

Risks Associated with Stem Cell Therapy	304
RESEARCH PROSPECTS OF STEM CELL THERAPY	305
PLANT STEM CELLS	306
Plant Tissue Culture	306
Steps in Plant Tissue Culture	307
Pre-propagation	307
In vitro Cultivation	307
Culturing of Explants	307
Micropropagation	307
Hardening	307
Challenges in Plant Tissue Culture	307
INTRODUCTION TO EPIGENETICS IN STEM CELLS	308
KEY EPIGENETIC MECHANISMS IN STEM CELLS	309
DNA Methylation	309
Histone Modification	310
Chromatin Remodeling	310
EPIGENETIC MECHANISMS IN MESENCHYMAL STEM CELLS (HUNTINGTON'S DISEASE)	311
EPIGENETIC CELL REPROGRAMMING IN IPSCS	311
Ectopic Expression of Transcription Factors	311
CRISPR–Cas9-Based Genome Editing for Reprogramming	312
EPIGENETIC THERAPY TARGETING BONE MARROW STEM CELLS	313
RECENT TRENDS	313
Stem Cells in Neurodegenerative Diseases	313
Therapeutic Potential of Dental Stem Cells	314
Cytokines Regulate the Fates of Hematopoietic Stem Cells	314
Bioengineered Scaffolds to Deliver Stem Cells for Wound Healing	315
ASSESSING THE RISKS AND FUTURE POTENTIAL IN STEM CELL REPROGRAMMING	315
CONCLUSION	316
CONSENT FOR PUBLICATION	316
CONFLICT OF INTEREST	317
ACKNOWLEDGEMENTS	317
REFERENCES	317
CHAPTER 15 CANCER STEM CELLS: CATALYSTS OF CANCER PROGRESSION	321
Vedika Kartha, Saloni Semwal, Lakshmi Sai Varshini Yedavalli, Disha Kamath, S. Pooja, Dannie Macrin, Satish Ramalingam and K.N. Aruljothi	
INTRODUCTION TO CANCER STEM CELLS	322
History and Evolution of Cancer Stem Cells	322
Current Concerns and Challenges	322
CHARACTERISTICS OF CANCER STEM CELLS	323
Interaction between Immune and Cancer Stem Cells	324
Intrinsic Features of Cancer Stem Cells	325
BIOLOGICAL PROPERTIES OF CANCER STEM CELLS	325
Marker Expression	325
CD44	325
CD133	326
ALDH (Aldehyde Dehydrogenase)	326
Microenvironment	326
Extracellular Matrix	327

  Immune Cell 327
  Signaling Molecules 327
TYPES OF CELLS SIGNALLING PATHWAYS RESPONSIBLE FOR CANCER STEM CELLS 328
 Notch Pathway 328
 Hedgehog Pathway 328
CANCER HETEROGENEITY 329
THE ROLE OF CANCER STEM CELLS IN TUMOR DEVELOPMENT AND METASTASIS 330
 Mechanisms through which Cancer Stem Cells Mediate Tumor Progression and Metastasis 330
 Cancer Stem Cells and Metastasis 330
 Interaction with the Tumor Microenvironment (TME) 331
 Cancer Stem Cells and Angiogenesis in Tumor Progression 332
 Cancer Stem Cells in Metastatic Niche Formation 332
 Therapeutic Implications of Targeting Cancer Stem Cells in Metastasis 333
METHODS OF CANCER STEM CELL IDENTIFICATION AND ISOLATION 333
 Isolation using Cell Surface Markers 334
  Limitations 334
 Side Population (SP) Assay 335
  Limitations 335
 Aldehyde Dehydrogenase (ALDH) Assay 335
  Limitations 335
 Spheroid Formation Assay 336
  Limitations 336
 Combination Approaches 336
MECHANISM OF DRUG RESISTANCE AND METASTASIS 337
 ATP-binding Cassette Transporter (ABC Transporter) 337
 Apoptosis Avoidance 337
 Numb Protein 337
STRATEGIES TO TARGET CANCER STEM CELLS 337
 Target Cancer Stem Cells Markers 337
 Target miRNA/LncRNA to Cancer Stem Cells 338
 Cancer Stem Cells and Resistance to Therapies 338
 Superior DNA Repair 338
 Quiescence 338
CANCER STEM CELLS IN DIFFERENT CANCERS 339
 Cancer Stem Cells in Breast Cancer 339
  New Approaches to Treatment 340
 Cancer Stem Cells in Skin Cancer 341
  Importance of Cancer Stem Cells in Skin Cancer 341
  Methods for Identifying and Investigating Cancer Stem Cells 342
  Novel Strategies for the Management of Cancer Stem Cells 342
  Continuing and Preventive Care 342
  Cancer Stem Cells in Novel Therapies 343
 Cancer Stem Cells Play an Important Role in Colorectal Cancer 344
  Methods for Identification and Study of Cancer Stem Cells 344
  Novel Therapies Targeting Cancer Stem Cells 344
  Current Preventive Treatment 345
 Cancer Stem Cells in Cervical Cancer 345
  Introduction and Importance 345
  Characteristics of Cancer Stem Cells in Cervical Cancer 346

Mechanisms of Resistance to Therapy	346
Current Therapeutic Methods and Strategies	347
Targeted Mechanisms against Cancer Stem Cells	347
Targeting the Molecular Signaling Pathways	347
Targeting Cancer Stem Cell Markers	348
Targeting the Cancer Stem Cell Niche And The Quiescent State	348
Manipulation of miRNA Expression	348
Induction of Cancer Stem Cell Apoptosis	348
Induction of Cancer Stem Cell Differentiation	348
Cancer Stem Cells Targeting Therapy	348
Therapeutic Implications of Cancer Stem Cells in Cancer Therapy	348
CONCLUSION	349
ACKNOWLEDGEMENT	349
REFERENCES	349
SUBJECT INDEX	354

FOREWORD

It is with great pleasure that I write this foreword for Dr. Kandasamy Nagarajan ArulJothi's remarkable book, Cell Biology: Basics to Breakthroughs. Dr. ArulJothi is a distinguished researcher and academician whose contributions to the fields of genetics, molecular biology, and precision medicine have significantly advanced our understanding of complex cellular processes. With over 70 scientific publications in renowned journals, his work spans diverse areas, including cancer biology, dyslipidemia, exosomal biomarkers, and stem cell therapy, making him a key figure in the field of biomedical research. His active participation in international collaborations, editorial board memberships, and ground-breaking research reflects his unwavering commitment to scientific excellence.

What makes this book particularly valuable is its dual focus on education and research. It serves as an essential resource not only for undergraduate and graduate students seeking to build a strong foundation in cell biology but also for researchers and educators looking to stay abreast of the latest scientific developments. The clear organization, coupled with detailed explanations, illustrative diagrams, and real-world applications, makes complex cellular processes accessible and engaging.

A unique highlight of Cell Biology: Basics to Breakthroughs is the scientific contributions from world-renowned experts in the field, who bring diverse insights and expertise. The book features Dr. Krishnan Anand from the University of the Free State, South Africa, whose work in cancer biology and molecular pathology adds depth to the chapters on cancer stem cells and therapeutic interventions. Dr. Satish Ramalingam from SRM University, a distinguished expert in molecular diagnostics and translational medicine, provides valuable perspectives on cell signaling and emerging therapeutic targets. Additionally, Dr. Prakash Gangadharan from the School of Medicine, Kyungpook National University, Republic of Korea, offers his expertise in cellular mechanisms of disease and regenerative medicine, enriching the book with clinically relevant insights. Their collaborative editorial contributions enhance the scientific rigor and global relevance of this volume, making it a truly comprehensive resource for cell biology enthusiasts and professionals alike.

Dr. ArulJothi's ability to integrate basic principles with translational insights makes Cell Biology: Basics to Breakthroughs a timely and impactful contribution to the field. By highlighting the clinical and therapeutic relevance of cell biology, this book has the potential to inspire future researchers and contribute to the advancement of biomedical science. I am confident that this book will serve as an invaluable reference for students, scientists, and educators alike, offering profound insights into the intricate world of cells and their role in health, disease, and biotechnology.

Prof. Surajit Pathak, Ph.D.
Medical Biotechnology Department
Faculty of Allied Health Sciences
Chettinad Hospital & Research Institute
Chettinad Academy of Research and Education
Chennai, India

PREFACE

A remarkable trip to the fundamental building blocks of life, cell biology reveals the complicated and crucial mechanisms that regulate the survival, development, and evolution of organisms. This area of study is an incredible adventure into the fundamental building blocks of life. The purpose of this book, titled Cell Biology: Basics to Breakthroughs, is to provide readers with a clear and thorough overview of both the fundamentals of cell biology and the most recent advancements in the field. Students, researchers, and anybody else who has an interest in the inner workings of life are the target audience for this kind of instruction since it is designed to gradually expand knowledge, beginning with fundamental principles and progressing to cutting-edge study areas.

In the first few chapters of this book, the authors discus the origins of life as well as the structural and functional differences that exist between prokaryotic molecules and eukaryotic cells. By reading these chapters, you will gain detailed knowledge that lays the foundation for comprehending the fundamental ideas that underlie cellular biology generally.

After that, we will dig into the primary organelles that are responsible for defining the architecture and function of the cell. In each chapter, an essential component is discussed, including the nucleus, the endoplasmic reticulum, the Golgi apparatus, ribosomes, lysosomes, vacuoles, and plasma membranes. As the reader progresses through these chapters, they will get a comprehensive understanding of how the complex machinery of the cell works in harmony to maintain life forms. After that, we move on to semi-autonomous organelles like mitochondria and chloroplasts, focusing on the distinct functions that these organelles play in the generation of energy and the activities of the metabolic system. The topic of discussion continues with the cytoskeletal components and molecular transport, which are responsible for the conformation, structure, and mobility of the cells within the cell.

It is essential to modern biology to have a solid understanding of cellular communication, and cell signalling and transduction are the processes that explain the intricate ways in which cells detect and react to their surroundings. Consequently, this naturally leads to the regulation of the cell cycle and the processes that regulate cell division, which is then followed by a discussion of cell death, which is an essential step for preserving homeostasis. The clinical and therapeutic aspects of the subject matter are brought into emphasis in the final chapters of the book. It is the study of cancer biology that investigates the uncontrolled development of cells and the harmful effects of this expansion. The final chapters provide an in-depth examination of the intriguing world of stem cells and cancer stem cells, discussing the therapeutic potential of these cells as well as their increasing significance in contemporary medicine.

By the time you reach the conclusion of this book, you will not only gain a profound understanding of cell biology, but you will also have insights into the most recent achievements that are affecting research and therapy. Our intention is that this book will serve as both a resource and a source of inspiration for future investigations in the field of biology.

K. N. Aruljothi
Department of Genetic Engineering
College of Engineering and Technology
SRM Institute of Science and Technology
Kattankulathur, Chengalpattu - 603203
Tamil Nadu, India

ACKNOWLEDGEMENTS

This book represents a significant journey of learning, reflection, and progress, made possible through the invaluable support, guidance, and encouragement of numerous individuals. I wish to convey my sincere appreciation to God, my family, and my esteemed research mentors and colleagues.

I want to express my sincere gratitude to my research group, special kudos to Mr. K Kumaran, Ms. Kruthika, Ms. Sanjana D, Ms. Raksa, Ms. Srisri SK, Mr. Danyal Reyaz, Ms. Sayantani C, and Mr. Sasitharan T for their steadfast support and confidence in me. Their love, patience, and understanding during this journey have truly been invaluable to me. I appreciate your support in pursuing my dreams and the understanding you've shown in allowing me the time and space to do so.

I sincerely appreciate the invaluable editorial insights and contributions provided by my beloved associates, Dr. Satish Ramalingam, Dr. Krishnan Anand, Dr. Prakash Gangadaran, Dr. Kirubakaran Rangasamy, and Dr. Dannie Macrin. I truly appreciate their insightful suggestions, which have significantly enhanced my ideas and improved the book beyond what I could have accomplished alone.

I extend my deepest gratitude to my mentor, Prof. Devi Arikketh, whose mentorship, insightful guidance, and unwavering support were instrumental in shaping my research and personal growth.

My heartfelt thanks to my mentor, Prof. Sharon Prince, Professor and Head of the Cell Biology Division at the University of Cape Town, whose guidance and inspiration have profoundly shaped my passion and expertise in cell biology.

Finally, I would like to express my heartfelt gratitude to the Department of Genetic Engineering and SRM Institute of Science and Technology for their unwavering support and invaluable resources, which have been instrumental in the completion of this book.

List of Contributors

Abhinav Roy	Department of Genetic Engineering, SRM Institute of Science and Technology, Kattankulathur, Chengalpattu, Tamil Nadu, India
B. Surya	Department of Genetic Engineering, SRM Institute of Science and Technology, Kattankulathur, Chengalpattu, Tamil Nadu, India
Danyal Reyaz	Department of Genetic Engineering, SRM Institute of Science and Technology, Kattankulathur, Chengalpattu, Tamil Nadu, India
Dannie Macrin	Department of Computational Biology, Institute of Bioinformatics, Saveetha School of Engineering, SIMATS Saveetha Institute of Medical and Technical Sciences, Chennai, India
Disha Kamath	Department of Genetic Engineering, SRM Institute of Science and Technology, Kattankulathur, Chengalpattu, Tamil Nadu, India
Faizaan Khan	Department of Genetic Engineering, SRM Institute of Science and Technology, Kattankulathur, Chengalpattu, Tamil Nadu, India
Gargii Chatterjee	Department of Genetic Engineering, SRM Institute of Science and Technology, Kattankulathur, Chengalpattu, Tamil Nadu, India
Harin N. Ganesh	Department of Genetic Engineering, SRM Institute of Science and Technology, Kattankulathur, Chengalpattu, Tamil Nadu, India
Janani Balaji	Department of Genetic Engineering, SRM Institute of Science and Technology, Kattankulathur, Chengalpattu, Tamil Nadu, India
Krishnan Anand	Department of Chemical Pathology, School of Pathology, Faculty of Health Sciences, University of the Free State, Bloemfontein, South Africa
K.N. Aruljothi	Department of Genetic Engineering, SRM Institute of Science and Technology, Kattankulathur, Chengalpattu, Tamil Nadu, India
Kruthika Prakash	Department of Genetic Engineering, SRM Institute of Science and Technology, Kattankulathur, Chengalpattu, Tamil Nadu, India
K. Kumaran	Department of Genetic Engineering, SRM Institute of Science and Technology, Kattankulathur, Chengalpattu, Tamil Nadu, India
Keerthana Ramesh	Department of Genetic Engineering, SRM Institute of Science and Technology, Kattankulathur, Chengalpattu, Tamil Nadu, India
Keerthivasu Ramasamy	Department of Genetic Engineering, SRM Institute of Science and Technology, Kattankulathur, Chengalpattu, Tamil Nadu, India
Lakshmi Sai Varshini Yedavalli	Department of Genetic Engineering, SRM Institute of Science and Technology, Kattankulathur, Chengalpattu, Tamil Nadu, India
Meghana A. Shakthi	Department of Genetic Engineering, SRM Institute of Science and Technology, Kattankulathur, Chengalpattu, Tamil Nadu, India
Mashira Rahman	Department of Genetic Engineering, SRM Institute of Science and Technology, Kattankulathur, Chengalpattu, Tamil Nadu, India

Prakash Gangadaran	Department of Nuclear Medicine, School of Medicine, Kyungpook National University, Daegu, Republic of Korea Cardiovascular Research Institute, Kyungpook National University, Daegu, Republic of Korea BK21 FOUR KNU Convergence Educational Program of Biomedical Sciences for Creative Future Talents, School of Medicine, Kyungpook National University, Daegu, Republic of Korea
R. Kirubakaran	Department of Biotechnology, Vinayaka Mission's Kirupananda Variyar Engineering College, Vinayaka Mission's Research Foundation-DU, Salem, Tamil Nadu, India
Ranjini Sengupta	Department of Genetic Engineering, SRM Institute of Science and Technology, Kattankulathur, Chengalpattu, Tamil Nadu, India
Raksa Arun	Department of Genetic Engineering, SRM Institute of Science and Technology, Kattankulathur, Chengalpattu, Tamil Nadu, India
Rohan Vyas	Department of Genetic Engineering, SRM Institute of Science and Technology, Kattankulathur, Chengalpattu, Tamil Nadu, India
Ramya Lakshmi Rajendran	Department of Nuclear Medicine, School of Medicine, Kyungpook National University, Daegu, Republic of Korea Cardiovascular Research Institute, Kyungpook National University, Daegu, Republic of Korea BK21 FOUR KNU Convergence Educational Program of Biomedical Sciences for Creative Future Talents, School of Medicine, Kyungpook National University, Daegu, Republic of Korea
Sayantanis Chattopadhyay	Department of Genetic Engineering, SRM Institute of Science and Technology, Kattankulathur, Chengalpattu, Tamil Nadu, India
Soumyadeep Maiti	Department of Genetic Engineering, SRM Institute of Science and Technology, Kattankulathur, Chengalpattu, Tamil Nadu, India
Sanjana Dhayalan	Department of Genetic Engineering, SRM Institute of Science and Technology, Kattankulathur, Chengalpattu, Tamil Nadu, India
Shanmuga Priya	Department of Genetic Engineering, SRM Institute of Science and Technology, Kattankulathur, Chengalpattu, Tamil Nadu, India
S. Sahasra	Department of Genetic Engineering, SRM Institute of Science and Technology, Kattankulathur, Chengalpattu, Tamil Nadu, India
Shalini Roy	Department of Genetic Engineering, SRM Institute of Science and Technology, Kattankulathur, Chengalpattu, Tamil Nadu, India
Sparshika Mishra	Department of Genetic Engineering, SRM Institute of Science and Technology, Kattankulathur, Chengalpattu, Tamil Nadu, India
S. Gnanavel	Department of Biomedical Engineering, SRM Institute of Science and Technology, Kattankulathur, Chengalpattu, Tamil Nadu, India
Srisri Satishkartik	Department of Genetic Engineering, SRM Institute of Science and Technology, Kattankulathur, Chengalpattu, Tamil Nadu, India

S. Aswini	Department of Genetic Engineering, SRM Institute of Science and Technology, Kattankulathur, Chengalpattu, Tamil Nadu, India
Shambhavi Jha	Department of Genetic Engineering, SRM Institute of Science and Technology, Kattankulathur, Chengalpattu, Tamil Nadu, India
S. Manvi	Department of Genetic Engineering, SRM Institute of Science and Technology, Kattankulathur, Chengalpattu, Tamil Nadu, India
Saloni Semwal	Department of Genetic Engineering, SRM Institute of Science and Technology, Kattankulathur, Chengalpattu, Tamil Nadu, India
S. Pooja	Department of Genetic Engineering, SRM Institute of Science and Technology, Kattankulathur, Chengalpattu, Tamil Nadu, India
Satish Ramalingam	Department of Genetic Engineering, SRM Institute of Science and Technology, Kattankulathur, Chengalpattu, Tamil Nadu, India
T. Sasitharan	Department of Genetic Engineering, SRM Institute of Science and Technology, Kattankulathur, Chengalpattu, Tamil Nadu, India
Vasanth T.L. Kanth	Department of Genetic Engineering, SRM Institute of Science and Technology, Kattankulathur, Chengalpattu, Tamil Nadu, India
Vanshikaa Karthikeyan	Department of Genetic Engineering, SRM Institute of Science and Technology, Kattankulathur, Chengalpattu, Tamil Nadu, India
Vedika Kartha	Department of Genetic Engineering, SRM Institute of Science and Technology, Kattankulathur, Chengalpattu, Tamil Nadu, India

CHAPTER 1

Foundations of Life: Cells and Origin

Meghana A. Shakthi[1], Harin N. Ganesh[1], R. Kirubakaran[2] and K.N. Aruljothi[1,*]

[1] *Department of Genetic Engineering, SRM Institute of Science and Technology, Kattankulathur, Chengalpattu, Tamil Nadu, India*

[2] *Department of Biotechnology, Vinayaka Mission's Kirupananda Variyar Engineering College, Vinayaka Mission's Research Foundation-DU, Salem, Tamil Nadu, India*

Abstract: Various theories explain life's origin, including the Oparin-Haldane hypothesis, which suggests that life originated from simple organic molecules in Earth's early reducing atmosphere. This was also supported by the Miller-Urey experiment. The cell is the most fundamental unit of an organism. The cell theory states that all living organisms are composed of cells; an organism's basic unit is a cell, and cells arise from pre-existing cells. A basic cell consists of a nucleus, cytoplasm, cell membrane, and cell organelles. The cell organelles are suspended in the cytoplasm. Prokaryotic cells have an undefined region composed of genetic material called the nucleoid and are devoid of a membrane, unlike eukaryotes. Organelles present in all Eukaryotic cells are the Endoplasmic Reticulum, Ribosomes, Golgi Apparatus, Mitochondria, Plastids, and Vacuoles. Each organelle is specialized to function in a certain way, thereby regulating the cell's metabolism. There is a distinct difference between animal and plant cells. Some constituents are specialized for the plant cell, such as the Cell wall, Vacuoles, and the Plastids. Some are specialized for animal cells, such as Centrioles, lysosomes, Cilia, and Flagella. Cells are effectively detected, viewed, and characterized by numerous tools. The microscope plays an integral role in the world of Cell Biology. Since the invention of the standard microscope, there have been many variations to it, enhancing our ability to view microscopic structures.

Keywords: Abiogenesis, Cell organelles, Eukaryotes, Living being, Prokaryotes.

INTRODUCTION

Theory of Spontaneous Generation

An ancient theory proposed around 350 BC was widely believed until the seventeenth century. Greek Philosopher Aristotle articulated that living things could arise from non-living components. He did so by raising striking questions,

---

[*] **Corresponding author K.N. Aruljothi:** Department of Genetic Engineering, SRM Institute of Science and Technology, Kattankulathur, Chengalpattu, Tamil Nadu, India; E-mail: aruljotn@srmist.edu.in

K.N. Aruljothi, Prakash Gangadaran, Krishnan Anand, Satish Ramalingam, K. Kumaran & Kruthika Prakash (Ed.)
All rights reserved-© 2026 Bentham Science Publishers

such as how new fish are introduced in a freshly formed pond or frogs' unexpected appearance on the Nile River's banks [1]. He proposed that an organism can arise from non-living material if the material has pneuma or 'vital heat'.

For a long time, it was believed that living organisms could spontaneously arise out of their niche. It wasn't until the late 17th century that Francesco Redi, an Italian Scientist, refuted the Theory of Spontaneous Generation by providing solid proof that maggots did not appear spontaneously on a slab of rotten meat stored for days [2]. He disproved this by storing one slab in an airtight container, another in the open, and the other covered with gauze. He had observed that only that maggot arose on the uncovered meat. He later concluded maggots did not arise from the beef in the airtight contained. However, he noticed maggots on the gauze and concluded that the maggots were offspring of common flies.

Furthermore, Louis Pasteur (1859) continued to disprove the theory by boiling meat broth in a swan-necked flask. He aimed to prove that the downward curve of the flask prevents the particles from reaching the broth, hence hindering growth. When the flask was overturned, the particles could reach the broth much better, and it clouded immediately. The 'Law of Biogenesis', meaning life arises from previously existing life, had a good conclusive result, but not many favored Pasteur's findings [3].

It was not until John Tyndall's experiment in 1876 that Pasteur's conclusions were supported. Tyndall repeated Pasteur's experiments only to observe that some boiled growth media had remained sterile while others did not, despite boiling for a long time. He discovered that endospores (heat-resistant species that can develop into bacteria under favorable environmental conditions) could grow on a nutrient medium, posing a hindrance to arriving at an accurate conclusion [4]. Tyndall thus devised a sterile medium where even endospores could not survive, which required a series of mechanisms and steps to ensure sterility.

Abiogenesis for the Origin of Life

The notion that living organisms arose from non-living matter on Earth more than 3 billion years ago forms the backbone concept of abiogenesis. Russian biochemist Aleksandr Oparin and British scientist J.B.S. Haldane, in the 1920s, proposed individual ideas that formed a theory, and the basic outline of it is that non-living matter, with the help of an external source of energy, for example, UV radiation, can form organic molecules [5].

Oparin believed Coacervates were the precursors to the basic cell, and they may have been one of the very first entities to exhibit life-like properties, such as growth and reproduction. Coacervates are droplet-like structures formed during liquid-liquid phase separation [6]. Oparin, backed by many other prominent scientists, firmly believes that Coacervates are the precursors of cells.

Haldane believed that simple inorganic materials formed into more complex molecules in the presence of external energy sources, such as UV radiation, over time. The combination of both notions is what laid an essential foundation for abiogenesis.

The Miller-Urey experiment performed in 1953 by American scientists Harold.C.Urey and Stanley Miller provided solid proof for the Oparin-Haldane hypothesis [7]. The scientists sealed Hydrogen, Methane, Ammonia, and water in a glass flask and generated an electrical discharge, stimulating lightning. After a week of discharging electrical sparks, it was observed that the water had turned reddish and turbid, and yellow-brown deposits were on the electrodes inside the apparatus. This indicated the synthesis of amino acids – the building blocks of protein. The groundbreaking experiment and Oparin-Haldane theory paved the way for research focused on Astrobiology- the study of life in the universe.

Cell and its Contents

Following the invention of the microscope in the 16th century, Robert Hooke, in the year 1665, observed small compartments or honeycomb-like structures and named them cells, as the cellulose walls in the cork he observed reminded him of rooms or monasteries usually occupied by monks (cellula) (Fig. **1**). The Latin word 'cellula' means storeroom or chamber. All of these were published in his work Micrographia [8]. These small compartments would become all living organisms' basic, functional, and structural units.

In 1674, Anton Van Leeuwenhoek seemingly observed protists and, years later, bacteria, which he termed the coin 'animalcules,' meaning microscopic animals [9]. This was the first instance a living cell had been discovered. French Chemist Francoise Raspail laid an essential foundation for the Cell Theory – that cells arise from pre-existing cells, which he hypothesized from witnessing binary fission, wherein a single cell divides into two daughter cells [10].

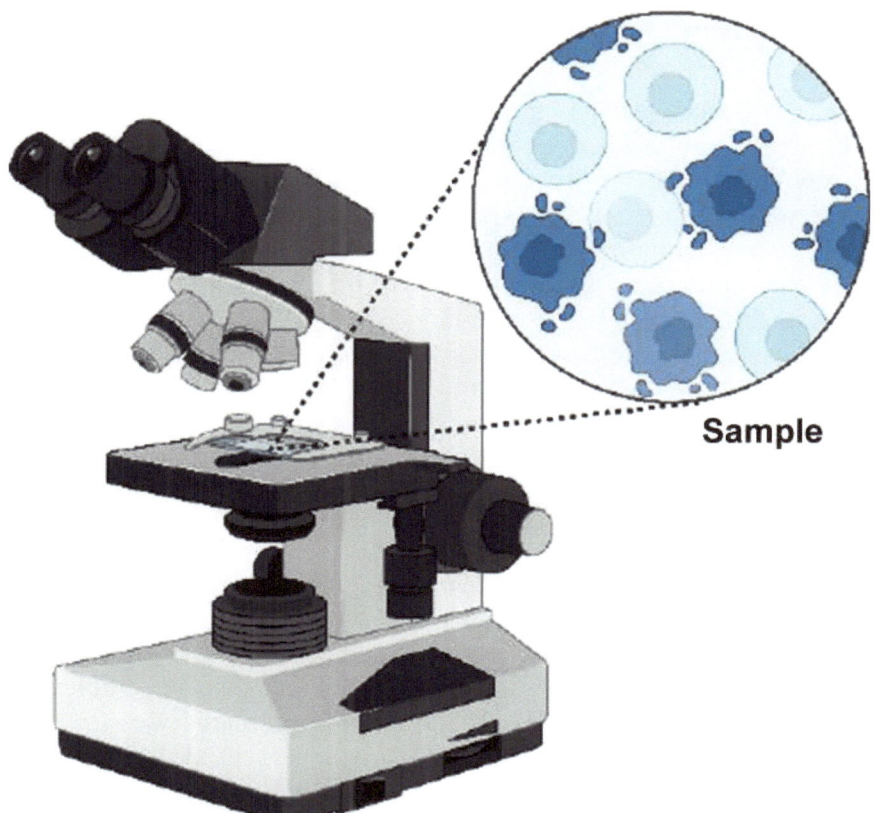

Fig. (1). The image illustrates a compound microscope examining a biological sample, with a magnified view showing various cell types observed under the lens.

The years 1836 and 1855 highlight the upcoming of the Cell Theory.

Matthias Jakob Schleiden and Theodor Schwann hypothesized three postulates

- The cell is a unit of structure in an organism
- A cell can exist as a separate entity as well as the building block of an organism
- Cells arise from spontaneous generation similar to that of crystallization.

Rudolf Virchow, in the year 1855, disproved the last postulate that cells arise from pre-existing cells, which was enunciated in his statement "Omnis Cellula e Cellula," translating to "Cells arise from cells". Cells are, hence, specialized structures that function on various levels. They absorb nutrients, derive energy, and use it for metabolic processes, which help in bodily functions. Cells can also replicate through the cell cycle.

A basic cell is classified on many bases. Two of those are:-

- Based on the presence or absence of a well-defined nucleus and nuclear membrane, cells are classified into two types: Prokaryotes and Eukaryotes. Prokaryotes ('Pro' – Primitive, 'Karyote' – Nucleus) and Eukaryotes ('Eu' – True, 'Karyote' – Nucleus). Prokaryotic cells do not possess a nucleus, nuclear membrane, or membrane-bound organelles. Instead, their genetic material is in a region of the cytoplasm called the nucleoid, which is not enclosed by a membrane and lacks a defined structure.
- Based on their localization and presence, they are classified into Plant and animal cells. Both exhibit unique characteristics that suit the requirements of the organism in which they are present.

A cell is the smallest unit in a living organism that can live independently and is irregular in structure. However, some cells have definite structures solely based on their functions. Almost every cell comprises three vital parts: the cell membrane, the nucleus, and the cytoplasm [11]. All cells have an outer membrane that protects them from foreign bodies. It is selectively permeable, allowing certain substances to pass through. It does so by the process of diffusion. The Nucleus is the control center of the cell. This is because the nucleus contains genetic information (chromosomes – Linear DNA), which codes for the body to perform vital functions; more specifically, it is essential for the organism's metabolism. The cytoplasm is fluid, makes up the cell's contents, and suspends the cell organelles. The Cytoplasm contains the cell's genetic material since a definite nuclear membrane is absent in Prokaryotes [12].

Cell Organelles

Cell Organelles are cellular constituents integrated into the cytoplasm, specific for functions such as intracellular signaling, defense against harmful foreign pathogens, *etc.* Cell organelles are categorized into membraneless, Single, and Double.

- Endoplasmic Reticulum – It is a single membrane-bound organelle composed of structures resembling tubules and sacs called cisternae. There are two types- Rough Endoplasmic Reticulum (RER) and Smooth Endoplasmic Reticulum (SER). The main difference between the two is the presence of ribosomes. RER has ribosomes on its surface, giving it a rough appearance, and SER lacks ribosomes on its surface, making it smooth. The Endoplasmic Reticulum is involved in protein, lipids, lipoproteins, and organic molecule synthesis and helps in efficient transport [13].

- Mitochondria – Mitochondria are rod-shaped and double membrane-bound. It is famously known as the 'powerhouse of the cell' as its primary function is to produce ATP- the cell's energy currency. ATP is a crucial molecule every cell utilizes for energy-dependent processes, including metabolism [14]. ATP production and consumption can be observed in metabolic pathways such as glycolysis and the Krebs cycle.
- Golgi Apparatus – Also known as the packaging unit of the cell, the Golgi Apparatus is an essential organelle of the cell, organized into a flattened sac-like structure called cisternae that helps mainly in packaging and modifying proteins and lipids [15].
- Vacuoles are membrane-bound organelles found in plant and animal cells, primarily used for storage, maintaining cell buoyancy, and providing structural support [16]. They are exceptionally prominent in plant cells, occupying around 75% of the cell's volume and typically in a central position. In contrast, vacuoles in animal cells are much smaller, fewer in number, and situated peripherally. Vacuoles also play a role in intracellular digestion and waste management.
- Ribosomes- Ribosomes are the smallest cell organelles in a cell, having diameters that range from 20nm to 30nm. They are one of the very few organelles that do not have a membrane. A ribosome has two subunits- one large and the other small- that are different in prokaryotic and eukaryotic cells. Their primary function is to synthesize proteins.

Organelle Biogenesis and the Endosymbiotic Theory

Organelle Biogenesis is the process of making new organelles (Fig. 2). As far as the history of our cellular organelles goes, they have arisen from endosymbiosis. Studies show that organelles like mitochondria and chloroplasts are morphologically similar to bacteria like Amoeba [17]. A cell organelle is a subcellular structure suspended in the cytoplasm, bound by the cell membrane. It is found predominantly in Eukaryotes. An organelle means 'little organ'. Much like the organs in our human body that carry out specialized functions, each cell organelle has a distinct function to carry out in the cell, which later on facilitates life processes on a larger scale in the organism's body.

Cell organelles, except ribosomes, are membrane-bound. These membranes comprise double phospholipid layers that insulate the compartmentalized organelles. Organelles are distinct functional structures within the cell, performing specific tasks essential for maintaining cellular life. Inclusions are stored materials within the cytoplasm, such as secretory products, pigment granules, and stored nutrients. While organelles actively contribute to cellular processes, inclusions primarily serve as storage sites or byproducts of cellular metabolism.

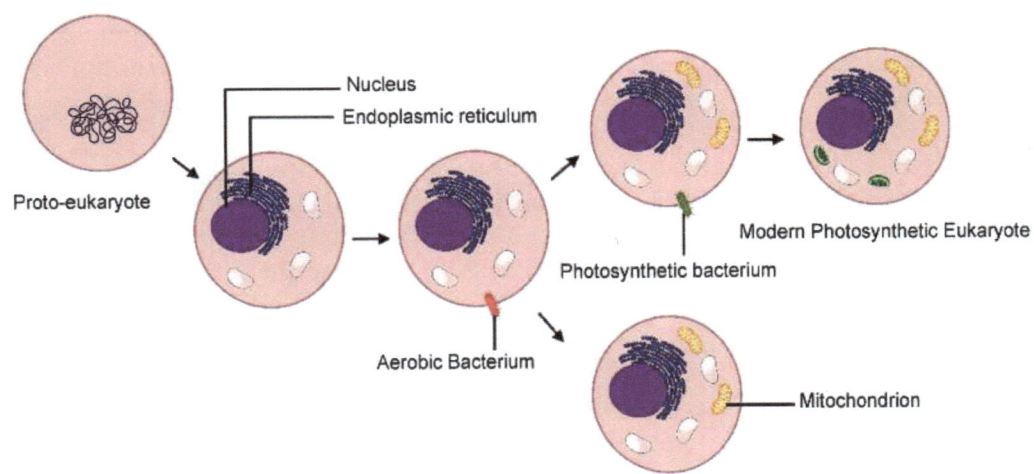

Fig. (2). Schematic representation of endosymbiotic theory, where a proto-eukaryote first engulfed an aerobic bacterium, forming mitochondria, and later engulfed a photosynthetic bacterium, leading to chloroplasts in photosynthetic eukaryotes. This process gave rise to modern heterotrophic and photosynthetic eukaryotic cells.

The evidence available for endosymbiotic events can occasionally become ambiguous and vague. There are two possibilities: Either some endosymbiotic events are being misinterpreted, or there is a problem with how some gene trees with numerous leaves are interpreted. Evidence that is not dependent on gene trees is needed. The best proof we currently have for the common ancestry of mitochondria and chloroplasts is protein import [12]. It is most likely the most convincing evidence for categorizing the number and type of secondary endosymbiotic interactions the red plastid experienced during evolution. Endosymbiotic theory can provide us with lots of information and can be highly beneficial in the future.

Chemical Composition of Cell

The chemical components of the cell include both molecular and ionic compounds. Major constituents include water and inorganic and organic molecules. These components are distributed throughout the cell in different parts.

- ***Water*** - Water is the primary molecule found in cells, making up at least 70% of the cell mass [18]. Water is a polar molecule and is crucial in understanding essential mechanisms. Water molecules can interact with positively or negatively charged ions and other polar molecules and form hydrogen bonds. These interactions make ions and polar compounds hydrophilic. Contrarily,

nonpolar molecules are not soluble in an aqueous environment (hydrophobic), as they cannot interact with water. As a result, nonpolar molecules usually associate closely with one another to reduce their contact with water.
- **Inorganic Ions** - While 99% of the cell is made of water, organic components, and water, the rest consists of inorganic ions such as Sodium, Potassium, Calcium, Magnesium, Chloride, and such. The inorganic ions play a significant role in cell signaling, pH determination, Fluid balance, Electrical activity, such as muscle contraction and activation of neurons, and Osmotic Pressure [19].
- Macromolecules - The organic molecules include nucleotides, lipids, proteins, and carbohydrates. A cell's dry weight comprises 80–90% of these macromolecules [18]. The functions of proteins are diverse. They act as enzymes, provide structural support, and maintain cell shape and integrity. Lipids are vital for energy storage; membrane formation facilitates cell recognition and signaling. Nucleic Acids store and help in the expression of genetic information.

Cytoplasmic Composition of the Cell

The cytoplasm contains the following vital components.

- Cytosolic Enzymes – Crucial in metabolic pathways such as glycolysis. Hexokinase, phosphofructokinase, and pyruvate kinase are cytosolic enzymes.
- Cytoskeletal Proteins – Actin, tubulin, and intermediate filaments are essential structural components that form a dynamic network that provides support and shape and helps facilitate movement.
- Ribosomes- Free Ribosomes are found floating in the cytoplasm. They are mainly involved in protein synthesis used within the cell.
- Dissolved Ions- Dissolved ions like potassium (K^+), sodium (Na^+), chloride (Cl^-), bicarbonate (HCO_3^-), magnesium (Mg^{2+}), and calcium (Ca^{2+}) are highly involved in osmoregulation, maintaining membrane potential and cell signaling.

Structural and Functional Aspects of the Cell

Cells are of different sizes and shapes, which are proportional to their function. The function of a cell is determined by its role in the organism and the forces that act upon it.

Cell Shape and How it Impacts Function

About 200 types of cells have varying shapes, influencing their structure and facilitating function. Some of them are listed as follows:

- **Spherical Cells:** Most unicellular organisms, including some bacteria, are approximately spherical to achieve maximum surface area for nutrient transfer.
- **Elongated Cells:** Neurons have long extensions (axons and dendrites) by which they can pass their signals effectively over long distances.
- **Flattened Cells:** Epithelial cells, such as skin cells, are wide and flat and form protective barriers.
- **Biconcave Cells:** RBCs are biconcave in shape with increased surface area to transport oxygen and to enable flexibility to pass through capillaries

Cell Size Variation

Cell size is relatively disparate between various organisms and cell types. Most cells are 5-15 μm in size, but some specialized cell types are considerably smaller or larger.

- They can be as small as 0.1–0.5 μm within a bacterium.
- The red blood cells (RBCs) are 5–8 μm in diameter.
- The largest single cell is the ostrich egg, with a diameter of up to 8 cm.

Cell size does not have anything to do with organism size. Elephant and mouse cells are virtually the same size, but there are more cells in larger organisms than in smaller organisms, not larger cell sizes. Specific specialized cells, like neurons, have cells that are a few meters long but tiny cell bodies.

Cell structure has a direct relationship with a cell's function. The above organelles indicate how a cell's structure can influence its function. The specialized structures of organelles help in intracellular and intercellular activities such as transportation, storage, and energy production. The cell membrane, an integral part of the cell, plays a crucial role in maintaining internal balance and has selective permeability, controlling the entry and exit of substances to and from the cell.

The cell wall, exclusively found in plants, fungal, prokaryotic, and algal cells, is rich in cellulose and composed of lignin, pectin, and hemicellulose. These biomaterials are known for their rigid nature. The Cell Wall maintains a cell's rigid and protects it from external damage [11]. An animal cell needs movement instead of the other types mentioned above. Hence, a cell wall would only hinder the animal cell. An erythrocyte or RBC (Red Blood Cell) portrays a classic structure-function relationship. The biconcave shape of the RBC allows it to have what is known as elastic deformity, facilitated by its flexible membrane and high surface-to-volume ratio. The elastic deformity is necessary for microcirculation through capillaries and transportation [20].

Cells have a variety of functions that contribute to an organism's metabolism. Primary functions of the cell include:

Cell Growth and Development: Cells grow and develop through various events. These events are collectively put together into what is known as the cell cycle. Each phase of the cell cycle helps increase the cell in different ways. For example, Interphase helps in preparing for the following phases and replicated chromosomes; the S phase is when chromosomes are duplicated, G1 and G2 are gap phases, the M phase is when segregated chromosomes are sorted into two separate nuclei, and cytokinesis is the last phase when the cell splits in two, giving rise to two daughter cells [11].

Cell Division: Cell division can occur in two notable ways. Cell division occurs by binary fission in some unicellular organisms, such as amoeba and prokaryotes. Binary fission is these organisms' primary reproduction mode [21]. It is similar to mitosis but less complex. Mitosis is observed in most eukaryotes and multicellular organisms. Mitosis helps in regeneration, growth, and also healing [22].

Cell Motility: Some organisms have cells specialized for immobile, like plant cells. Some cells perform their primary function with the help of their motility, i.e., the ability to move. Cell motility is driven by a protein called actin. This ability of cells also contributes to metastasis, enabling studies to properly understand the behavioral aspects of cancer cells [23].

Energy Production in Cells: The mitochondria are vital organelles in the cell tasked with producing energy. It produces ATP (Adenosine Triphosphate), the source of energy, which is obtained through a series of processes – glycolysis, Citric Acid Cycle, and Oxidative Phosphorylation [24]. Cells consume the ATP produced by performing various functions, such as mechanical work, muscle contraction, and passing electrical signals in neural cells.

Cellular Transport: The lipid bilayer in cells plays a significant role in transportation as it is a selectively permeable membrane. It facilitates diffusion and active, selective transport against the concentration gradient for many essential components such as ions (Na^+, K^+, Ca^{2+}), glucose, and amino acids with the help of transport proteins. An important feature is aquaporins' presence, which mainly facilitates water transport across the membrane. This also enables the cell to maintain homeostasis [25].

Integration of Cells into Tissues

The cells in multicellular animals are arranged into groups termed tissues, like the connective, nerve, and muscular tissues seen in vertebrates. A tissue sample is observed to have a network of proteins and other vital molecules that help support cells, known as the Extracellular Matrix. This matrix provides strength to supportive tissues [26]. Cells can link to one another directly or through the matrix, which is one way to connect them. Cell junctions, which connect cells in the flexible, mobile tissues of animals, become involved. These junctions either transmit pressures from one cell's cytoskeleton to the next or from a cell's cytoskeleton to the extracellular matrix.

However, mechanics alone cannot fully explain how tissues are organized. Blood vessels, nerves, and other elements must be created from various specialized cell types for an animal tissue to function. All of the tissue's constituent parts must be appropriately arranged and coordinated, and many necessitate ongoing upkeep and replenishment. As cells age, they must be replaced with fresh ones that are the proper type, location, and quantity.

Extracellular Matrix and Connective Tissues

Animals and plants both underwent independent multicellular organization evolution, and the building blocks of their tissues are distinct. Animals must be robust and nimble because they feed on other living things and are frequently the prey of different creatures. Animals need tissues that aid them in their mobility; they must quickly change their shape to be flexible. In contrast, plants are sedentary. Despite plant cells being weaker and easily damageable, their extracellular matrix ensures stiffness in plants.

The cell wall, which encloses, safeguards, and regulates the form of each plant's cells, is the supporting matrix in plants. The extracellular matrix is created, secreted, and controlled by plant cells. For example, a leaf's cell wall can be thin and flexible or thick and stiff like that of a piece of wood. Nonetheless, all tissues have the same fundamental concept – They comprise numerous tiny boxes adhered to each other, each containing a cell inside [27].

Animal tissues have a wider variety. They contain cells and extracellular matrix, just like plant tissues, but they are arranged in various ways. The extracellular matrix is usually abundant in nature, but in tissues such as bone or tendons. The extracellular matrix is needed in sparse quantities as the ECM does not carry mechanical strain but rather the cytoskeletons of the cells (Fig. **3**) [28].

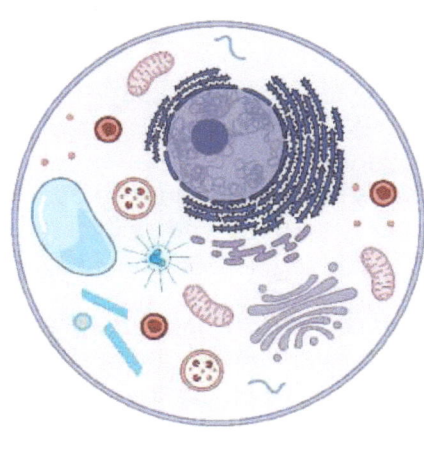

Fig. (3). The image depicts a plant cell (left) with a rigid cell wall, large central vacuole, and chloroplasts for photosynthesis, while the animal cell (right) lacks a cell wall and chloroplasts but contains centrioles and lysosomes for cellular functions.

Plant and Animal Cells

Plant Cell

The first cell viewed under the microscope back in 1663 was a cell of a cork, a plant cell. The plant cell's characteristic thick cell wall and numerous plant cells making a hive-like structure led to one of the most significant discoveries in the history of biology. The plant cell is more rigid and more potent than an animal cell. The main component of a plant cell is cellulose, a crucial biomaterial now widely used in the paper industry.

A key feature distinguishing plant cells from animal cells is the vacuole, which occupies the center of the plant cell. Other distinguishing features are plastids, especially chloroplasts that give the plant its green pigment, enabling photosynthesis, as well as a rigid cell wall that provides the cell with structural integrity.

A vacuole is a sac filled with cell sap and protected by a membrane. It is present in all plants and additionally present in fungi. Vacuoles can take up between 30% and 90% of the volume of a cell [29]. Using water to create hydrostatic pressure, store nutrients, and non-nutrient compounds, and degrade complex molecules all contribute to the plant's stiffness.

A tonoplast, a membrane barrier, constrains each vacuole. This membrane is unique because it can quickly stretch to produce an organelle that can take up as much as 95% of the volume of the cell after enclosing a tiny amount of fluid for a short period, during which water is taken in. Yet, none of these results in the tonoplast's integrity as an active membrane being compromised. All of the other cell organelles are forced against the sturdy cellulose cell wall during this process without suffering any damage [30].

Chloroplast - One of the three plastids is larger than mitochondria by size. Both organelles have similar functions, but there is a distinct difference between them. The chloroplast is the site of photosynthesis, the process by which plants gain energy. The energy chloroplasts gain from the sunlight transforms into chemical energy, helping plants respond to adverse environmental conditions such as a sudden change in pH, extreme heat and dehydration, and harmful pathogens [31].

- It helps form flat sac-like discs called thylakoids. These discs contain the pigment known as chlorophyll, which is vital to photosynthesis. It helps produce energy in the form of ATP and NADPH.
- The thylakoids are arranged in stacks. A stack is called a granum. Multiple grana are called grana.
- The stromal lamellae arise from the thylakoid membrane, interconnecting grana.
- The stroma is the space that contains all the above-mentioned components, similar to that of the cytoplasm of a cell [11].

Cell Wall - Cell walls are a specific type of extracellular matrix integral to every plant cell. It performs various functions such as intercellular networking, Osmotic regulation, defense responses against foreign organisms, and interactions with microorganisms. The cell wall comprises a complex extracellular matrix composed of cellulose microfibrils, lignin, and hemicellulose [32]. The cell wall is known to have three layers – the primary cell wall, the secondary cell wall, and the middle lamella. In contrast, the cell wall comprises various components that can be broadly divided into three categories. Its complex composition gives the cell tensile strength [12].

Evolution of Photosynthesis

Plants, specifically the chloroplasts in plant cells, are responsible for absorbing sunlight and using it to produce glucose and oxygen. Oxygen and Glucose are necessary for all living organisms.

Extensive studies suggest that organisms that did not need oxygen existed billions of years ago. They were called anoxygenic. Archaebacteria were such organisms that did not need oxygen to survive. This was only shortly after the origin of life. Not long after, cyanobacterial-like organisms developed into chloroplasts, necessitating the process of photosynthesis. Although there is very little information centered on photosynthesis, some hypotheses provide insight into the topic [33].

The Granick Hypothesis proposed by Sam Granick suggests that photosynthesis was primarily a simple pathway that has evolved into a more complex one with new enzymes being added over time in a stepwise manner. These enzymes were likely added to the equation through horizontal gene transfer or genetic mutations. Comparative Genomics supports this notion strongly with the help of the characterization of photosynthetic genes [34]. Photosynthesis has profoundly impacted our planet even while evolving nonlinearly.

Animal Cell

Animal cells are diversified as opposed to plant cells. The former has a varied number of functions in an animal. Mobility is a distinct factor, and the animal cell is designed to have a flexible membrane devoid of rigid cell walls. The animal cell has specialized features in each organ of the human body. For example, the neurons (nerve cells) have axons and dendrites that help transmit signals, cardiomyocytes (muscle cells in the heart) have more mitochondria to produce more energy, and muscle cells, specifically skeletal muscle cells, are often long and slender for better reach. There are many types of cells in the human body, such as the ones mentioned.

Some components in the animal cell differentiate it from the plant cell, such as centrioles, lysosomes, cilia, and flagella.

Centrioles – Centrioles are membrane-less organelles. They are cylindrical and are made up of 9 sets of triplet microtubules. These microtubules are responsible for the structure of the centriole and the formation of other organelles like cilia and flagella. Centrioles are the organelles crucial for cell division. It produces spindle fibers during mitosis, which help divide the chromosomes [35].

Lysosome – Lysosome is a membrane-bound organelle. It is tasked with the primary responsibility of digestion in all animal cells. It contains enzymes, commonly known as lysozymes, that specialize in breaking down biological substances such as proteins, lipids, *etc.* Lysosomes can digest both materials taken up by the cell and the non-functional materials in the cell. They digest the extracellular materials with the help of endocytosis, which is otherwise elaborated as the process in which a cell uptakes material through its membrane, where the foreign substance is engulfed. Lysosomes are also known as the 'suicidal bags' of the cell. They break down the cell's parts if found damaged to regulate the cell's functioning. This is known as Autophagy [36].

Cilia – Cilia are tiny hair-like organelles that are often found together. These 'bundles' are otherwise commonly known as tufts. Unlike other organelles mentioned before, Cilia are not found inside the cell but outside. They are found in both unicellular and multicellular organisms and are commonly known to function in the motility of an organism. In a unicellular organism such to *Paramecium*, the cilia control the direction and the speed. Cilia in multicellular organisms do not help in the organism's movement; instead, they help move the fluid or materials that pass the cilia. A prime example is cilia in the lungs that remove dust particles, harmful microorganisms, and other pathogenic substances.

Flagella – Flagella are like Cilia, but not found in tufts. Flagella are also hair-like appendages that are more whip-like. It is also involved in the motility of a cell. The flagella are commonly found in bacteria. In an animal cell, the sperm cell is the only cell that has a flagellum. Sperm motility is essential as the flagella aid the sperm in fusing with the ovum, which in turn carries out fertilization [37].

Cells Under the Microscope

The early cell biologists observed tissues and cells at first, opening them up or splitting them up to see what was within. What they observed looked to them to be an impenetrable mystery; it was a collection of tiny, hardly perceptible items whose relationship to the characteristics of living stuff. However, this visual analysis was the starting point for comprehending cells and is still vital to cell biology. It wasn't until the invention of the microscope in the seventeenth century that cells became visible. All that was known about cells for hundreds of years after that was learned using this equipment: Hans and Zacharias Janssen are credited for the invention of the microscope [38], a device that enables the human eye to observe organisms, cells, and non-living entities that cannot be viewed under the naked eye.

The invention of the Electron Microscope in the 20th century by Ernst Ruska and Max Knoll opened a new possibility. The interior of cells was now clearly visible. The prototype had a magnification of 400. The electron Microscope replaces light with beams of electrons, and its wavelength is 100,000 times shorter than that of visible light. Hence, electron microscopes have a resolution 1000x better than light microscopes.

There are two main types of Electron Microscopes:

Transmission Electron Microscope: Used to view thin samples, the Transmission Electron Microscope works on the basic principle of electrons scattering when they hit a material. On scattering, the electrons create a contrast, hence capturing an image. TEM consists of an electron gun to generate a beam of electrons, electromagnetic lenses to focus the beam, and a vacuum system to ensure a clear picture, as electrons can deflect when coming into contact with air molecules [39].

Scanning Electron Microscope: The Scanning Electron Microscope produces an image with the help of electrons generated by the sample. Most SEMs have a resolution of about 10nm, enabling the user to view the surfaces of samples easily [40]. A photomultiplier detects and amplifies the electrons generated upon interacting with the sample's atoms. The amplified signals are used to visualize a 3D image of the surface.

CONCLUSION

Life, in all its complexity, begins with the cell. Understanding how cells function gives us a deeper appreciation of the intricate processes that sustain life. But how did life itself begin? While scientists have proposed many ideas, the exact answer remains a mystery. Research continues to explore how simple molecules might have come together to form the first living systems. With discoveries and advancing technology, we are gradually uncovering the fascinating story of life's origins, one piece at a time.

ACKNOWLEDGEMENTS

We want to acknowledge Ms. Kruthika P, Mr. Kumaran K, Ms. Raksa Arun, Ms. Srisri SatishKartik, Ms. Sanjana Dhayalan, and Mr. Sasitharan T for their contribution to proofreading and editing works.

REFERENCE

[1]   Dunn PM. Aristotle (384–322 bc): philosopher and scientist of ancient Greece. Arch Dis Child Fetal Neonatal Ed 2006; 91(1): F75-7.
      [http://dx.doi.org/10.1136/adc.2005.074534] [PMID: 16371395]

[2] Parke EC. Flies from meat and wasps from trees: Reevaluating Francesco Redi's spontaneous generation experiments. Stud Hist Philos Sci Part Stud Hist Philos Biol Biomed Sci 2014; 45(1): 34-42.
[http://dx.doi.org/10.1016/j.shpsc.2013.12.005] [PMID: 24509515]

[3] Schwartz M. Louis Pasteur and molecular medicine: a centennial celebration. Mol Med 1995; 1(6): 593-5.
[http://dx.doi.org/10.1007/BF03401596] [PMID: 8529126]

[4] Weed LA. John Tyndall and His Contribution to the Theory of Spontaneous Generation. Ann Med Hist 1942; 4(1): 55-62.
[PMID: 33943710]

[5] Lazcano A. Historical development of origins research. Cold Spring Harb Perspect Bio 2010.
[http://dx.doi.org/10.1101/cshperspect.a002089]

[6] Matsuo M, Kurihara K. Proliferating coacervate droplets as the missing link between chemistry and biology in the origins of life. Nat Commun 2021; 12(1): 5487.
[http://dx.doi.org/10.1038/s41467-021-25530-6] [PMID: 34561428]

[7] Xie X, Backman D, Lebedev AT, *et al.* Primordial soup was edible: abiotically produced Miller-Urey mixture supports bacterial growth. Sci Rep 2015; 5(1): 14338.
[http://dx.doi.org/10.1038/srep14338] [PMID: 26412575]

[8] Lawson I. Crafting the microworld: how Robert Hooke constructed knowledge about small things. Notes Rec 2016; 70: 23-44
[http://dx.doi.org/10.1098/rsnr.2015.0057]

[9] Kutschera U. Antonie van Leeuwenhoek (1632–1723): Master of Fleas and Father of Microbiology. Microorganisms 2023; 11(8): 1994.
[http://dx.doi.org/10.3390/microorganisms11081994] [PMID: 37630554]

[10] Schultz M. Rudolf Virchow. Emerg Infect Dis 2008; 14(9): 1480-1.
[http://dx.doi.org/10.3201/eid1409.086672]

[11] Alberts B, Johnson A, Lewis J, Raff M, Roberts K, Walter P. Molecular Biology of the Cell. 4th edition. New York: Garland Science; 2002. Available from: https://www.ncbi.nlm.nih.gov/books/NBK21054/

[12] Cooper G. The Cell: A Molecular Approach. Sunderland, MA: Sinauer Associates 2000.

[13] Crawford JM, Bioulac-Sage P, Hytiroglou P. Structure, Function, and Responses to Injury. In: Burt ID, Ferrell LD, Hübscher SG (Eds.)MacSween's Pathology of the Liver. Elsevier 2018; pp. 1-87.

[14] Dunn J, Grider MH. Physiology, adenosine triphosphate. StatPearls 2023.

[15] Mohan AG, Calenic B, Ghiurau NA, Duncea-Borca RM, Constantinescu AE, Constantinescu I. The Golgi Apparatus: A Voyage through Time, Structure, Function and Implication in Neurodegenerative Disorders. Cells 2023; 12(15): 1972.
[http://dx.doi.org/10.3390/cells12151972] [PMID: 37566051]

[16] Wada Y. Vacuoles in mammals. Bioarchitecture 2013; 3(1): 13-9.
[http://dx.doi.org/10.4161/bioa.24126] [PMID: 23572040]

[17] Zimorski V, Ku C, Martin WF, Gould SB. Endosymbiotic theory for organelle origins. Curr Opin Microbiol 2014; 22: 38-48.
[http://dx.doi.org/10.1016/j.mib.2014.09.008] [PMID: 25306530]

[18] Geoffrey M. The Cell: A Molecular Approach. 2nd ediion. Sunderland (MA); 2000.

[19] Jomova K, Makova M, Alomar SY, *et al.* Essential metals in health and disease. Chem Biol Interact 2022; 367: 110173.
[http://dx.doi.org/10.1016/j.cbi.2022.110173] [PMID: 36152810]

[20] Diez-Silva M, Dao M, Han J, Lim CT, Suresh S. Shape and Biomechanical Characteristics of Human Red Blood Cells in Health and Disease. MRS Bull 2010; 35(5): 382-8.
[http://dx.doi.org/10.1557/mrs2010.571] [PMID: 21151848]

[21] Chien AC, Hill NS, Levin PA. Cell size control in bacteria. Curr Biol 2012; 22(9): R340-9.
[http://dx.doi.org/10.1016/j.cub.2012.02.032] [PMID: 22575476]

[22] McIntosh JR. Mitosis. Cold Spring Harb Perspect Biol 2016; 8(9): a023218.
[http://dx.doi.org/10.1101/cshperspect.a023218] [PMID: 27587616]

[23] Stuelten CH, Parent CA, Montell DJ. Cell motility in cancer invasion and metastasis: insights from simple model organisms. Nat Rev Cancer 2018; 18(5): 296-312.
[http://dx.doi.org/10.1038/nrc.2018.15] [PMID: 29546880]

[24] Dunn J, Grider MH. Physiology, Adenosine Triphosphate. [Updated 2023 Feb 13]. In: StatPearls [Internet]. Treasure Island (FL): StatPearls Publishing; 2025 Jan-. Available from: https://www.ncbi.nlm.nih.gov/books/NBK553175/

[25] Cho I, Jackson MR, Swift J. Roles of Cross-Membrane Transport and Signaling in the Maintenance of Cellular Homeostasis. Cell Mol Bioeng 2016; 9(2): 234-46.
[http://dx.doi.org/10.1007/s12195-016-0439-6] [PMID: 27335609]

[26] Frantz C, Stewart KM, Weaver VM. The extracellular matrix at a glance. J Cell Sci 2010; 123(24): 4195-200.
[http://dx.doi.org/10.1242/jcs.023820] [PMID: 21123617]

[27] Shah DU, Reynolds TPS, Ramage MH. The strength of plants: theory and experimental methods to measure the mechanical properties of stems. J Exp Bot 2017; 68(16): 4497-516.
[http://dx.doi.org/10.1093/jxb/erx245] [PMID: 28981787]

[28] Stavolone L, Lionetti V. Extracellular Matrix in Plants and Animals: Hooks and Locks for Viruses. Front Microbiol 2017; 8: 1760.
[http://dx.doi.org/10.3389/fmicb.2017.01760] [PMID: 28955324]

[29] Tan X, Li K, Wang Z, Zhu K, Tan X, Cao J. A Review of Plant Vacuoles: Formation, Located Proteins, and Functions. Plants 2019; 8(9): 327.
[http://dx.doi.org/10.3390/plants8090327] [PMID: 31491897]

[30] Jin Y, Weisman LS. The vacuole/lysosome is required for cell-cycle progression. Elife 2015; 4: e08160.
[http://dx.doi.org/10.7554/eLife.08160] [PMID: 26322385] [PMCID: PMC4586482]

[31] Jensen PE, Leister D. Chloroplast evolution, structure and functions. F1000Prime Rep 2014; 6: 40.
[http://dx.doi.org/10.12703/P6-40] [PMID: 24991417]

[32] Hatfield RD, Rancour DM, Marita JM. Grass Cell Walls: A Story of Cross-Linking. Front Plant Sci 2017; 7: 2056.
[http://dx.doi.org/10.3389/fpls.2016.02056] [PMID: 28149301]

[33] Blankenship RE. Early evolution of photosynthesis. Plant Physiol 2010; 154(2): 434-8.
[http://dx.doi.org/10.1104/pp.110.161687] [PMID: 20921158]

[34] Ward LM, Shih PM. Granick revisited: Synthesizing evolutionary and ecological evidence for the late origin of bacteriochlorophyll *via* ghost lineages and horizontal gene transfer. PLoS One 2021; 16(1): e0239248.
[http://dx.doi.org/10.1371/journal.pone.0239248] [PMID: 33507911]

[35] Winey M, O'Toole E. Centriole structure. Philos Trans R Soc Lond B Biol Sci 2014; 369(1650): 20130457.
[http://dx.doi.org/10.1098/rstb.2013.0457] [PMID: 25047611]

[36] Glick D, Barth S, Macleod KF. Autophagy: cellular and molecular mechanisms. J Pathol 2010; 221(1): 3-12.

[http://dx.doi.org/10.1002/path.2697] [PMID: 20225336]

[37] Inaba K. Sperm flagella: comparative and phylogenetic perspectives of protein components. Mol Hum Reprod 2011; 17(8): 524-38.
[http://dx.doi.org/10.1093/molehr/gar034] [PMID: 21586547]

[38] Wollman AJM, Nudd R, Hedlund EG, Leake MC. From *Animaculum* to single molecules: 300 years of the light microscope. Open Biol 2015; 5(4): 150019.
[http://dx.doi.org/10.1098/rsob.150019] [PMID: 25924631]

[39] Malatesta M. Transmission electron microscopy as a powerful tool to investigate the interaction of nanoparticles with subcellular structures. Int J Mol Sci 2021; 22(23): 12789.
[http://dx.doi.org/10.3390/ijms222312789] [PMID: 34884592]

[40] Fischer ER, Hansen BT, Nair V, Hoyt FH, Dorward DW. Scanning electron microscopy. Curr Protoc Microbiol 2012; Chapter 2: Unit 2B.2.
[PMID: 22549162]

CHAPTER 2

Prokaryotes vs. Eukaryotes: Comparative Structural and Functional Insights

Kruthika Prakash[1], Sayantani Chattopadhyay[1], T. Sasitharan[1], Soumyadeep Maiti[1], Sanjana Dhayalan[1], K. Kumaran[1], R. Kirubakaran[2,*] and K.N. Aruljothi[1]

[1] *Department of Genetic Engineering, SRM Institute of Science and Technology, Kattankulathur, Chengalpattu, Tamil Nadu, India*

[2] *Department of Biotechnology, Vinayaka Mission's Kirupananda Variyar Engineering College, Vinayaka Mission's Research Foundation-DU, Salem, Tamil Nadu, India*

Abstract: Organisms are fundamentally divided into prokaryotes and eukaryotes, except for viruses. Further classifications are depicted in various classifications, among which the most recent one is Whittaker's classification. Prokaryotes, the primitive organisms, gave rise to eukaryotes, and this transformation led to simple single-celled organisms evolving into colonies and multicellular cells. Prokaryotic cells, which include bacteria and archaea, exhibit diverse metabolic processes such as autotrophy and heterotrophy. In contrast, eukaryotic cells are characterized by the presence of a well-defined nucleus and membrane-bound organelles, unlike prokaryotic cells. In contrast, eukaryotes developed around 2 billion years ago and possess complex cellular structures, including a well-defined nucleus and various organelles, such as mitochondria and chloroplasts. The transition from unicellular to multicellular life is a significant evolutionary milestone that involves various adaptations and mechanisms, including cell-to-cell communication, adhesion, and coordinated growth. Genetic conservation and epigenetic mechanisms play pivotal roles in the development of multicellular structures, as demonstrated in organisms like fungi and metazoans. Eukaryotic cells, such as those from yeast and mammalian sources, are pivotal in biotechnological applications, including the production of recombinant proteins and gene therapy. Their ability to properly fold and process proteins allows for the creation of functional biopharmaceuticals and vaccines that simulate pathogen structures to invoke robust immune responses. Notable eukaryotic microorganisms like algae and fungi are also increasingly recognized for their potential in sustainable biofuel production. Since genes serve as the backbone for almost all cells, they can be manipulated to be more user-friendly.

---

[*] **Corresponding author R. Kirubakaran:** Department of Biotechnology, Vinayaka Mission's Kirupananda Variyar Engineering College, Vinayaka Mission's Research Foundation-DU, Salem, Tamil Nadu, India; Tel: 9095288654; E-mail: rangasamykirubakaran@gmail.com

K.N. Aruljothi, Prakash Gangadaran, Krishnan Anand, Satish Ramalingam, K. Kumaran & Kruthika Prakash (Ed.)
All rights reserved-© 2026 Bentham Science Publishers

Keywords: Biofuel production, Characteristics, Eukaryotes, Evolution, Evolutionary biology, Gene therapy, Genes, Multicellularity, Prokaryotes, Recombinant proteins, Whittaker's classification.

INTRODUCTION TO PROKARYOTES

With a simpler cell structure than eukaryotes, prokaryotes are unicellular creatures devoid of membrane-bound organelles or a nucleus; their genetic material is arranged in a single circular DNA molecule that floats freely inside the cell. Common examples of prokaryotes, bacteria, and archaea lack compartmentalization, which helps them to adapt and flourish in many habitats, as shown in Fig. **1**. Among the first living entities on Earth, prokaryotes have greatly affected metabolic reactions, forming the ecosystems of the planet [1]. Originally discovered in the late 17th century with the development of the microscope, Dutch scientist Antonie van Leeuwenhoek saw "animalcules"—microbes—in 1676, hence establishing the basis for microbiology [2]. While Carl Woese's genetic categorization in the 1970s separated Archaea from Bacteria, redefining our knowledge of prokaryotes, advances in staining and microscopy techniques helped scientists like Louis Pasteur and Robert Koch identify bacteria as agents of illness.

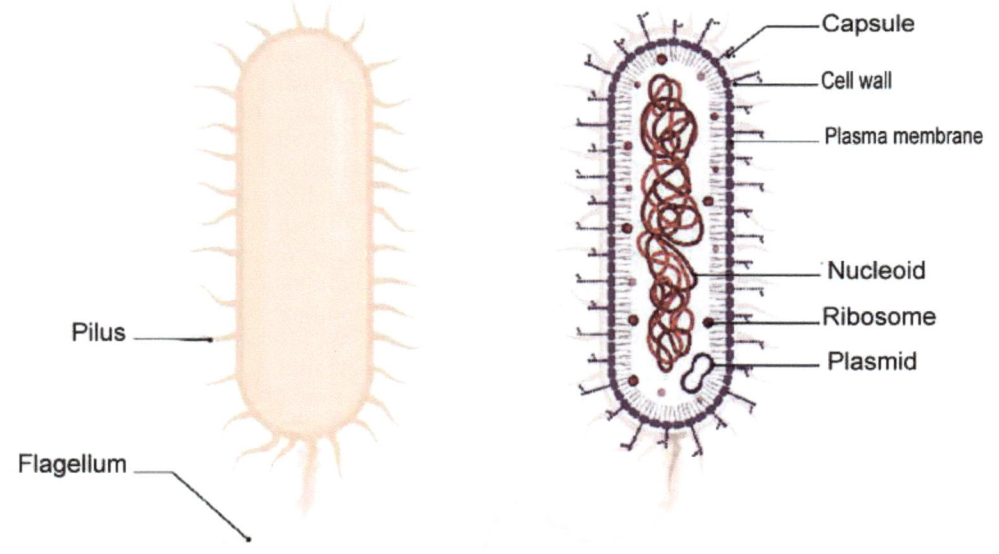

Fig. (1). Bacterium, a prokaryote

Crucially for plant development, prokaryotes drive biogeochemical cycles that help break down organic matter and recycle nutrients and nitrogen fixation, supporting ecosystems. Beyond their function in the surroundings, prokaryotes are essential for human life since they help digestion and enable waste breakdown and fermenting [3]. With their research opening the path for industrial uses, bioremediation, and genetic engineering, they also synthesize vitamins, antibiotics, and other chemicals, thereby stressing the need for prokaryotes to maintain the environment and improve health and quality of life.

Prokaryotic Diversity

Prokaryotes exhibit different forms, structures, metabolic pathways, and ecological responsibilities among their tremendous and varied diversity. Their adaptability lets them live practically anywhere on Earth, from human bodies to deep waters. Their taxonomy, form, metabolic strategies, and ecological niches are examined here.

Classification of Prokaryotes

Separated by genetic, biochemical, and physiological traits, bacteria and Archaea represent the two main domains of prokaryotes. Among the most well-known prokaryotes—from helpful gut microorganisms to disease-causing pathogens—are bacteria [4]. Living in almost every environment on Earth, they are pretty varied and have cell walls typically made of peptidoglycan, a unique polymer providing structural support. Previously believed to be bacteria, Carl Woese identified archaea as a discrete domain based on their distinct genetic and metabolic properties. Like bacteria, Archaea can survive in hostile conditions because their cell membranes consist of ether-linked lipids rather than ester-linked lipids. Usually, extremophiles survive in hot springs, salt flats, and deep-sea vents. Archaea can also be found in more common environments like the human intestine [5].

Morphological Diversity

The range of forms, sizes, and configurations that prokaryotes exhibit enhances their adaptability and usefulness in many contexts. Among the most often occurring forms are cocci (spherical-shaped bacteria such as *Streptococcus*), bacilli (rod-shaped cells like *Escherichia coli*), spirilla (spiral-shaped cells like *Spirillum*), and vibrios (comma-shaped cells like *Vibrio cholerae*). Usually small, ranging from 0.2 to 5 micrometers, prokaryotic cells can vary in size depending on their environment and genes. Division patterns affect their cellular structures as well; they produce linear chains (as seen in streptococci), grape-like clusters (staphylococci), pairs (diplococci), and long filaments that increase surface area for nutrient absorption [6].

Metabolic Diversity

With their incredible metabolic diversity, prokaryotes can adapt ecologically and use various energy and carbon sources. Often using carbon dioxide as a carbon source, autotrophs create their food; examples include certain archaea, which use chemosynthesis in hostile conditions, and cyanobacteria, which photosynthesize. While many bacteria in soil and water are heterotrophic, feeding on organic materials, human gut bacteria derive their nutrients from broken-down food; homotrophs obtain carbon by consuming organic molecules [7]. Chemotrophs use inorganic substances such as hydrogen sulfide or ammonia, and chemoorganotrophs depend on organics; they get energy by oxidizing inorganic or organic molecules. Cyanobacteria, oxygenic phototrophs that release oxygen during photosynthesis, demonstrate that phototrophs employ light energy to generate food. Other phototrophic prokaryotes, such as purple sulfur bacteria, engage in anoxygenic photosynthesis, producing no oxygen [8].

Ecological Diversity

In different ecological niches, prokaryotes interact with other creatures differently and adjust to a broad spectrum of environmental variables. Usually unable to survive in most other life forms, extremophiles are prokaryotes found in severe environments. These comprise thermophiles, which are found in high-temperature environments like hot springs; halophiles, which are suited to high-salt environments like salt flats; acidophiles, which can survive in acidic environments like mine drainage areas; and methanogens, a type of archaea that generates methane and usually resides in anaerobic environments like swamps and the guts of some animals. Usually offering mutual benefits, symbionts are prokaryotes forming symbiotic interactions with other species. While in the human microbiome, prokaryotes promote digestion, synthesize vitamins, and help control immunological processes, nitrogen-fixing bacteria in the roots of legumes transform atmospheric nitrogen into a usable form, supporting plant growth [9].

Pathogens are prokaryotes that strike humans, animals, and plants, causing diseases. Two prominent examples are *Vibrio cholerae*, the bacterium causing cholera, and *Mycobacterium tuberculosis*, the bacterium causing tuberculosis. By employing pathogenic prokaryotes, medicine has evolved significantly, and antibiotics and vaccinations have resulted [10].

The simplest type of cellular life, prokaryotic cells, lack a nucleus and other membrane-bound organelles. Their two main domains are Bacteria and Archaea.

Though simple, prokaryotic cells show a fantastic diversity and adaptability that help them live and flourish in many contexts, including severe ecosystems and the human body (Fig. 1).

Cell Envelope

Surrounded by a multi-layered structure, the cell envelope keeps the cell form and offers the necessary protection for the prokaryotic cell. It usually has three layers. Comprising carbohydrates and/or polypeptides, the glycocalyx layer forms on the outside. This layer can form as a firmly bound capsule or as a somewhat detached slime layer. The glycocalyx helps cells avoid the host immune system, shields against desiccation, and promotes cell adhesion, among other things. Underneath the glycocalyx is the cell wall, a stiff construction that gives structural support and protects the cell from osmotic lysis. While archaeal cell walls vary and may be built from materials including pseudo-peptidoglycan, S-layer proteins, or polysaccharides, bacterial cell walls are essentially made of peptidoglycan [11]. Comprising a phospholipid bilayer enclosing the cell's cytoplasm, the innermost layer is the plasma membrane. Along with cell signaling and energy generation, it controls the movement of molecules into and out of the cell [12].

Cytoplasm and External Structures

The cytoplasm is a gel-like material filling the cell and housing several cellular components. Among these elements are ribosomes, which oversee protein synthesis. Comprising two subunits, 30S and 50S, prokaryotic ribosomes are smaller than their eukaryotic counterparts, 70S vs. 80S. Glycogen, polyphosphate, sulfur, and gas vesicles are among the molecules stored in inclusion bodies—granules within the cytoplasm. Usually, one circular chromosome, the nucleoid, is a separate area within the cytoplasm where the genetic material of the cell dwells. Unlike a eukaryotic nucleus, the nucleoid lacks a membrane [13].

Additionally, exterior structures supporting movement, adhesion, and other purposes are prokaryotic cells. Comprising a basal body, hook, and filament, flagella are long, filamentous appendages used for motility. Whereas archaeal flagella can rotate flexibly or structurally, flagella in bacteria rotate like propellers. Short, hair-like pili appendages help bacteria conjugate—passing genetic material between cells—and cling to surfaces. Though often shorter and more numerous, fimbriae resemble pili and primarily help cells adhesively to surfaces and other cells. Some cells may also feature well-organized, thick layers of polysaccharides or polypeptides encircling the cell wall. These capsules offer extra defense against desiccation, phagocytosis, and antibiotics [14].

Prokaryotic Genetics

Prokaryotic genetics is a fascinating discipline exploring the structure, function, and inheritance of genetic material in prokaryotic organisms. Unlike eukaryotic cells, prokaryotes usually have a single, circular chromosome with extra extrachromosomal DNA components called plasmids. Across various habitats, this genetic configuration helps prokaryotes survive and adapt.

Genome Structure

Prokaryotes have a single, circular chromosome that is their basis for genome organization. Supercoiled to fit within the cell, this chromosome carries all the necessary genes for the survival and reproduction of the organism. Apart from the main chromosome, prokaryotes sometimes include plasmids—small circular DNA molecules that reproduce apart from the chromosome [15]. Under some environmental situations, plasmids are helpful since they can carry genes that give desirable features such as antibiotic resistance, toxin generation, or the ability to use particular nutrients. Moreover, horizontal gene transfer allows plasmids to be passed between cells, promoting genetic variation and adaptation in prokaryotic populations [16].

Gene Transfer Mechanisms

Multiple mechanisms for horizontal gene transfer have evolved in prokaryotes to enable them to obtain fresh genetic material from other species, hence facilitating their fast adaptability to changing surroundings. Usually mediated by a conjugation pilus, conjugation includes direct cell-to-cell contact between a donor and a recipient cell. Conjugation increases genetic variety by utilizing plasmid-bearing genes for features like antibiotic resistance or virulence factors transmitted from the donor to the recipient cell. Under transformation, prokaryotic cells can absorb naked DNA from their surroundings—a process dependent on particular proteins on the cell surface to enable DNA absorption. Either lysed cells or the free release of this DNA into the surroundings lets prokaryotes assimilate fresh genes. Transduction is another type of gene transfer mediated by bacteriophages and viruses that infect bacteria. During the lytic cycle, phage particles sometimes bundle bits of bacterial DNA rather than viral DNA. These phages can pass the bacterial DNA when infecting fresh bacterial cells, promoting genetic variety [17].

Mutation and Adaptation

Survival and bacterial evolution depend much on mutation and adaptation. In prokaryotes, mutations—changes in the DNA sequence—are the leading causes of genetic diversity. These modifications can arise naturally or be brought about by outside influences, including UV light or chemical mutagens. Although most mutations could be damaging, others could be helpful and give the organism a selection advantage. High mutation rates, fast generation times, and horizontal gene transfer let prokaryotes respond rapidly to environmental changes. Prokaryotes' success and capacity to colonize and flourish in many environments depend critically on their fast adaptation [18].

Prokaryotes and Human Health

Though minuscule, prokaryotes have a significant effect on human health and well-being. Some bacteria, for example, *Rhizobium* and *Azotobacter*, may fix nitrogen—that is, transform atmospheric nitrogen into a form plants can use. Support for food security and sustainable farming methods depends on this process, which is also necessary for plant development and agricultural output. In bioremediation, a technique used to treat environmental pollutants, prokaryotes, too, are vital. Helping to rehabilitate contaminated habitats, some bacteria can break down sewage, harmful chemicals, and oil spills [19]. Prokaryotes help to produce several fermented goods in the food industry. While yeast (a eukaryote with comparable microbial qualities) ferments dough for bread and beer, bacteria are utilized to convert milk into yogurt and cheese [20]. Apart from food processing, these techniques generate distinctive flavors and textures that are the mainstay of many cuisines.

Introduction to Eukaryotes

Eukaryotes are organisms that have a well-defined nucleus and other membrane-bound cell organelles [21]. Prokaryotes emerged around 3.5 billion years ago. Following that, Eukaryotes appeared around 2 billion years ago. Their name comes from the Greek "eu," meaning well or sound, and "karyon," meaning kernel.

The eukaryotes are a diverse lineage, consisting of microorganisms like Amoeba to the largest animal that has ever lived, the Blue Whale. Multicellularity in some form has evolved independently at least 25 times within the eukaryotes [22]. Complex multicellular organisms, not counting the aggregation of amoebae to form slime molds, have developed within only Plants, Fungi, and Animals.

Characteristics of Eukaryotes

One of the most essential features of a eukaryotic cell is membrane-bound cell organelles like mitochondria and chloroplasts (reference in Fig. **2**). Among them, chloroplast is found in plant cells, eukaryotic algae, and some photosynthetic eukaryotic entities, such as Euglena. The acquisition of these membrane-bound organelles is a critical step in the evolution of eukaryotes [23]. It is believed that these organelles were acquired by a process known as endosymbiosis [24]. It is believed that the primitive eukaryotic cells transformed themselves by a process in which the primitive eukaryotes and these prokaryotes engulfed a prokaryote, integrated within the cell, and exhibited a symbiotic relationship [25]. This theory is supported by the fact that both mitochondria and chloroplasts have their own DNA, separate from the eukaryotic DNA [26].

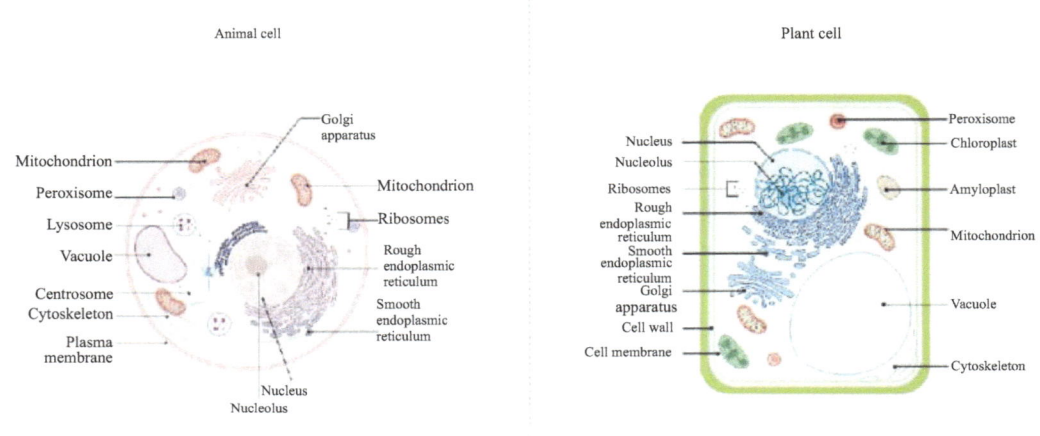

Fig. (2). Eukaryotic cells.

In a nutshell, some of the basic features of both prokaryotes and eukaryotes are listed below. Table **1** illustrates the anatomical differences between the structures of both prokaryotes and eukaryotes.

WHITTAKER CLASSIFICATION - A BREAKTHROUGH IN THE STUDY OF EUKARYOTE ANATOMY

The Whittaker classification, introduced by Robert H. Whittaker in 1969, is a commonly employed framework for classifying living species into five kingdoms, considering their cellular structure, method of obtaining nutrients, and other essential traits. The Whittaker classification consists of five kingdoms: Monera, Protista, Fungi, Plantae, and Animalia, as in Table **2**. This classification categorizes eukaryotes into Protista, Fungi, Plantae, and Animalia kingdoms. Eukaryotes have cells with a true nucleus and membrane-bound organelles [28].

Table 1. Basic characteristics of prokaryotes and eukaryotes [27].

Features	Prokaryotic cell	Eukaryotic cell
Multicellularity	Unicellular	Both unicellular and Multicellular
Size	0.5-100 µm	10-150 µm
Cell Wall	Present made of peptidoglycan	Present in only plants(cellulose) and Fungi(chitin)
Membrane-bound organelles	Absent	Present
Reproduction	Asexually	Sexually
Chromosome	Single Chromosome	Several Chromosome
Nucleus	Nucleoid (membrane-less)	Well-defined nucleus

Table 2. RH Whittaker classification.

Characters	Monera	Protista	Fungi	Plantae	Animalia
Cell type	Prokaryotic	Eukaryotic	Eukaryotic	Eukaryotic	Eukaryotic
Cell wall	Non-cellular	Present in some	Non-cellulose-based cell wall	Cell wall (presence of cellulose)	Absent
Nuclear membrane	Absent	Present	Present	Present	Present
Body organization	Cellular	Cellular	Multicellular	Tissues and organs (vascular system, sex organs)	Tissues/organs/organ system
Mode of nutrition	Autotrophic and heterotrophic	Autotrophic (photosynthesis) and heterotrophic	Heterotrophic	Autotrophic	Heterotrophic

DEVELOPMENT OF MULTICELLULAR ORGANISMS

It was generally believed that single cells were the first to evolve, but the oldest fossils contained multicellular cyanobacterial filaments 3.5 billion years ago. The transition from unicellular forms to multicellular forms is unusual among the significant evolutions that have occurred numerous times. Many multicellular forms are still being discovered (Fig. **2**). The condition of growth of multicellularity has arisen from the organic kingdoms, and the evidence is the development of sponges from choanoflagellates [29].

This is also true for microorganisms. The cellular slime molds, myxomycetes, and myxobacteria are the only forms of multicellularity. This transformation from Unicellularity to Multicellularity can be viewed from the repetition of various experiments, but the result of the repetition created some challenges for biologists in identifying the similarities. Nevertheless, how multicellularity is defined, and its multiple origins, raise numerous unanswered questions [30].

The evidence proves that adopting multicellularity misleads the basic biophysical process of generating multicellular phenotypes. In addition, the standard multilevel selection theory of multicellularity concludes that a unicellular "bottleneck" is required to reduce the intercellular conflict; it is to be noted that asexual life cycles, too, have a dormancy (for example, cyanobacterial heterocyst) [31].

Multilevel selection theory (MLS) characterizes "individuality" during the evolutionary transition from an ancestor of unicellular organisms to a descendant of multicellular organisms. This theory identifies two stages: the alignment-of-fitness phase (MLS1), which minimizes cell-cell conflict, and the export-of-fitness phase (MLS2), where cells remain interdependent, requiring a shift from individual cells to a unified multicellular entity [32].

The development of multicellular organisms occurs through cell division, which serves multiple roles, including growth, repair, and reproduction. Reproduction is not exclusive to eukaryotic, multicellular organisms; it occurs in both prokaryotes and eukaryotes through two primary forms: **asexual reproduction** (such as binary fission in prokaryotes and mitotic division in some eukaryotes) and **sexual reproduction** (involving gamete formation and fertilization in eukaryotes). In multicellular eukaryotes, gametes are produced through **meiosis**, a specialized form of cell division that ensures genetic diversity. Genetic, ecological, and environmental factors influence the development of multicellular organisms, including humans, plants, and fungi [33]. As a result, this leads to the formation of new materials possessing the characteristics of "dynamical patterning modules," physical forces with effects called up by the materials' gene products and other molecules [34].

The development of multicellular organisms is ultimately through the division of the cells. This development is explained only through cellular growth, thereby excluding reproductive growth. These cells, which are present in multicellular organisms, get clubbed to form tissue and organs in those organisms. Cell division in multicellularity is also meant to repair and replace injured cells [35].

In the case of the development of multicellularity, cell division shall lead to reproduction, but it cannot occur at the same time. The prime example is humans;

the gametes are produced by the process called meiosis (cell division), and they fuse to form a zygote, hence the development of offspring, which is nothing but the process of reproduction. The cases for developing multicellular organisms include humans, plants, fungi, etc. This development process includes ecological factors as well as genetic and environmental factors.

If we take fungi as an example of the development of multicellular organisms, mycelial and pseudomycelial growth, fruiting bodies, etc., are the structures of multicellular organisms. Ustilaginomycetes are the fungi that usually show unicellular as sporidia. Also, it has some structures due to its development as a multicellular organism under environmental conditions, such as nutrient starvation, nitrogen starvation, acidic media culture, or fatty acids with carbon as the source [36]. So, this development can be seen only by conserving some essential genes and molecular mechanisms involved, such as Histone acetyltransferases and Histone deacetylases, which are largely observed with epigenetic patterns. Other important genes include autophagy-related genes, stress-related genes, virulence genes, mating-type genes, and signal transduction genes.

Biotechnological Applications of Eukaryotes

Eukaryotic cells, particularly yeast and mammalian cells, are widely used to produce recombinant proteins engineered through genetic modification by inserting a foreign gene into a host cell. These proteins have essential applications in medicine, diagnostics, and research. Eukaryotic cells are preferred because they can accurately fold and modify proteins, ensuring that the resulting biopharmaceuticals are functional and correctly processed [37]. Gene therapy involves the introduction of a functional gene into a patient's cells to address genetic abnormalities. Eukaryotic viruses, such as lentiviruses or adenoviruses, are commonly used as viral vectors to transport therapeutic genes into specific cells [38]. Additionally, eukaryotic expression systems are widely used in vaccine production, where they aid in generating antigens that closely mimic the pathogen's structure, triggering an effective immune response [39]. Beyond medical applications, eukaryotic microorganisms, including algae and fungi, are employed in biofuel production due to their ability to convert sunlight and carbon dioxide into biofuels. Furthermore, tissue engineering, a rapidly advancing field, relies on eukaryotic cells to create artificial organs and tissues for drug testing and transplantation [40].

CONCLUSION

Prokaryotes and eukaryotes together form the foundation of life, each with unique characteristics that highlight their evolutionary significance and adaptability.

Among Earth's earliest life forms, prokaryotes thrive in diverse environments despite lacking membrane-bound nuclei and organelles. They play essential roles in nitrogen fixation, bioremediation, and food production, sustaining ecosystem balance. However, their misuse in industry and agriculture has contributed to antibiotic resistance, underscoring the need for responsible management.

Eukaryotes exhibit greater complexity from unicellular amoebae to multicellular plants, fungi, and animals due to their membrane-bound organelles and true nuclei, enabling multicellularity and specialized functions. Insights into eukaryotic biology, particularly endosymbiosis and multicellularity, have deepened our understanding of evolution and ecological roles. Additionally, eukaryotes are central to biotechnological advancements, including tissue engineering, gene therapy, vaccine production, and biofuel development. Studying prokaryotes and eukaryotes provides valuable insights into life's complexity, ecological sustainability, and innovations that enhance human health and environmental well-being.

REFERENCES

[1] David L. Kirchman. Processes in Microbial Ecology. OUP Oxford 2012.

[2] Porter JR. Antony van Leeuwenhoek: tercentenary of his discovery of bacteria. Bacteriol Rev 1976; 40(2): 260-9.
[http://dx.doi.org/10.1128/br.40.2.260-269.1976] [PMID: 786250]

[3] Bodaker I, Sharon I, Suzuki MT, et al. Comparative community genomics in the Dead Sea: an increasingly extreme environment. ISME J 2010; 4(3): 399-407.
[http://dx.doi.org/10.1038/ismej.2009.141] [PMID: 20033072]

[4] Otieno DO. Biology of Prokaryotic Probiotics. In: Liong, MT. (Ed) Probiotics. Microbiology Monographs, vol 21. Springer, Berlin, Heidelberg 2011: pp. 1-28.
[https://doi.org/10.1007/978-3-642-20838-6_1]

[5] Jain S, Caforio A, Driessen AJ. Biosynthesis of archaeal membrane ether lipids. Front Microbiol 2014; 5: 641.
[http://dx.doi.org/10.3389/fmicb.2014.00641] [PMID: 25505460]

[6] Uluçay O. Mikroorganizmalar ve Bakteri Çeşitliliği.Matematik ve Fen Bilimleri Üzerine Araştırmalar. Özgür Yayınları 2023.
[http://dx.doi.org/10.58830/ozgur.pub81.c476]

[7] Anemaet IG, Bekker M, Hellingwerf KJ. Algal photosynthesis as the primary driver for a sustainable development in energy, feed, and food production. Mar Biotechnol (NY) 2010; 12(6): 619-29.
[http://dx.doi.org/10.1007/s10126-010-9311-1] [PMID: 20640935]

[8] Richa Salwan VS, Ed. Physiological and Biotechnological Aspects of Extremophiles. 1st ed., 2020.

[9] Colwell RR. Colwell. Infectious disease and environment: cholera as a paradigm for waterborne disease. Int Microbiol 2004; 7(4): 285-9.

[10] Długonski J. Microbial Biotechnology in the Laboratory and Practice: Theory, Exercises, and Specialist Laboratories; Jagiellonian University Press 2022: p. 553..

[11] Banfalvi G. Permeability of biological membranes. Permeability of Biological Membranes. Springer International Publishing 2016: p. 1–263.
[http://dx.doi.org/10.1007/978-3-319-28098-1]

[12] Mayer F. Cytoskeletons in prokaryotes. Cell Biol Int 2003; 27(5): 429-38.
[http://dx.doi.org/10.1016/S1065-6995(03)00035-0] [PMID: 12758091]

[13] Joseph W. Lengeler, Gerhart Drews, Hans G. Schlegel; Biology of the Prokaryotes. Wiley; December 1998.

[14] Liò P. Investigating the relationship between genome structure, composition, and ecology in prokaryotes. Mol Biol Evol 2002; 19(6): 789-800.
[http://dx.doi.org/10.1093/oxfordjournals.molbev.a004136] [PMID: 12032235]

[15] Eberhard WG. Evolution in bacterial plasmids and levels of selection. Q Rev Biol 1990; 65(1): 3-22.
[http://dx.doi.org/10.1086/416582] [PMID: 2186429]

[16] Thomas CM, Nielsen KM. Mechanisms of, and barriers to, horizontal gene transfer between bacteria. Nat Rev Microbiol 2005; 3(9): 711-21.
[http://dx.doi.org/10.1038/nrmicro1234] [PMID: 16138099]

[17] Reddy GC, Goyal RK, Puranik S, Waghmar V, Vikram KV, Sruthy KS. Biofertilizers Toward Sustainable Agricultural Development.Plant Microbe Symbiosis. Springer International Publishing 2020; pp. 115-28.
[http://dx.doi.org/10.1007/978-3-030-36248-5_7]

[18] Tofalo R, Fusco V, Böhnlein C, et al. The life and times of yeasts in traditional food fermentations. Critical Reviews in Food Science and Nutrition 2019; 60(18): 3103-32.
[http://dx.doi.org/10.1080/10408398.2019.1677553]

[19] Baum B, Spang A. On the origin of the nucleus: a hypothesis. Microbiol Mol Biol Rev 2023; 87(4): e00186-21.
[http://dx.doi.org/10.1128/mmbr.00186-21] [PMID: 38018971]

[20] Bozdag GO, Zamani-Dahaj SA, Day TC, et al. De novo evolution of macroscopic multicellularity. Nature 2023; 617(7962): 747-54.
[http://dx.doi.org/10.1038/s41586-023-06052-1] [PMID: 37165189]

[21] Brunk CF, Marshall CR. Opinion: The Key Steps in the Origin of Life to the Formation of the Eukaryotic Cell. Life (Basel) 2024; 14(2): 226.
[http://dx.doi.org/10.3390/life14020226] [PMID: 38398735]

[22] Lyman GH, Lyman CH, Kuderer NM. The Nature, Origin, and Evolution of Life: Part IV Cellular Differentiation and the Emergence of Multicellular Life. Cancer Invest 2024; 42(4): 275-7.
[http://dx.doi.org/10.1080/07357907.2024.2302201] [PMID: 38175037]

[23] Song Q, Zhao F, Hou L, Miao M. Cellular interactions and evolutionary origins of endosymbiotic relationships with ciliates. ISME J 2024; 18(1): wrae117.
[http://dx.doi.org/10.1093/ismejo/wrae117] [PMID: 38916437]

[24] Butenko A, Lukeš J, Speijer D, Wideman JG. Mitochondrial genomes revisited: why do different lineages retain different genes? BMC Biol 2024; 22(1): 15.
[http://dx.doi.org/10.1186/s12915-024-01824-1] [PMID: 38273274]

[25] Zhukov A, Popov V. Eukaryotic Cell Membranes: Structure, Composition, Research Methods and Computational Modelling. Int J Mol Sci 2023; 24(13): 11226.
[http://dx.doi.org/10.3390/ijms241311226] [PMID: 37446404]

[26] Hagen JB. Five Kingdoms, More or Less: Robert Whittaker and the Broad Classification of Organisms. Bioscience 2012; 62(1): 67-74.
[http://dx.doi.org/10.1525/bio.2012.62.1.11]

[27] Maldonado M. Choanoflagellates, choanocytes, and animal multicellularity. Invertebr Biol 2004; 123(1): 1-22.
[http://dx.doi.org/10.1111/j.1744-7410.2004.tb00138.x]

[28] Panstruga R, Antonin W, Lichius A. Looking outside the box: a comparative cross-kingdom view on

the cell biology of the three major lineages of eukaryotic multicellular life. Cell Mol Life Sci 2023; 80(8): 198.
[http://dx.doi.org/10.1007/s00018-023-04843-3] [PMID: 37418047]

[29] Haselkorn R. Cyanobacteria. Curr Biol 2009; 19(7).

[30] Marín C. Three types of units of selection. Evolution (N Y) 2024; 78(3): 579-86.

[31] Romero H, Aguilar PS, Graña M, Langleib M, Gudiño V, Podbilewicz B. Membrane fusion and fission during eukaryogenesis. Curr Opin Cell Biol 2024; 86: 102321.
[http://dx.doi.org/10.1016/j.ceb.2023.102321] [PMID: 38219525]

[32] Benítez M, Hernández-Hernández V, Newman SA, Niklas KJ. Dynamical Patterning Modules, Biogeneric Materials, and the Evolution of Multicellular Plants. Front Plant Sci 2018; 9: 871.
[http://dx.doi.org/10.3389/fpls.2018.00871] [PMID: 30061903]

[33] Niklas KJ, Newman SA. The many roads to and from multicellularity. J Exp Bot 2020; 71(11): 3247-53.
[http://dx.doi.org/10.1093/jxb/erz547] [PMID: 31819969]

[34] Martínez-Soto D, Ortiz-Castellanos L, Robledo-Briones M, León-Ramírez CG. Molecular Mechanisms Involved in the Multicellular Growth of Ustilaginomycetes. Microorganisms 2020; 8(7): 1072.
[http://dx.doi.org/10.3390/microorganisms8071072] [PMID: 32708448]

[35] Bachhav B, de Rossi J, Llanos CD, Segatori L. Cell factory engineering: Challenges and opportunities for synthetic biology applications. Biotechnol Bioeng 2023; 120(9): 2441-59.
[http://dx.doi.org/10.1002/bit.28365] [PMID: 36859509]

[36] Biju TS, Priya VV, Francis AP. Role of three-dimensional cell culture in therapeutics and diagnostics: an updated review. Drug Deliv Transl Res 2023; 13(9): 2239-53.
[http://dx.doi.org/10.1007/s13346-023-01327-6] [PMID: 36971997]

[37] Kohn DB, Chen YY, Spencer MJ. Successes and challenges in clinical gene therapy. Gene Ther 2023; 30(10-11): 738-46.
[http://dx.doi.org/10.1038/s41434-023-00390-5] [PMID: 37935854]

[38] Cai R, Gimenez-Camino N, Xiao M, Bi S, DiVito KA. Technological advances in three-dimensional skin tissue engineering. Rev Adv Mater Sci 2023; 62(1): 20220289.
[http://dx.doi.org/10.1515/rams-2022-0289]

[39] Khudainazarova NS, Granovskiy DL, Kondakova OA, et al. Prokaryote- and Eukaryote-Based Expression Systems: Advances in Post-Pandemic Viral Antigen Production for Vaccines. Int J Mol Sci. 2024; 25(22): 11979.
[http://dx.doi.org/10.3390/ijms252211979] [PMID: 39596049] [PMCID: PMC11594041]

[40] Lobo-Moreira AB, Xavier-Santos S, Damacena-Silva L, Caramori SS. Trends on Microalgae-Fungi Consortia Research: An Alternative for Biofuel Production? Front. Microbiol 2022; 13: 903737.
[http://dx.doi.org/10.3389/fmicb.2022.903737]

CHAPTER 3

Nucleus: the Control Centre of the Cell

K. Kumaran[1], Harin N. Ganesh[1] and K.N. Aruljothi[1,*]

[1] *Department of Genetic Engineering, SRM Institute of Science and Technology, Kattankulathur, Chengalpattu, Tamil Nadu, India*

Abstract: The nucleus is a crucial organelle in eukaryotic cells, serving as the primary center for genetic material and essential cellular processes like replication, transcription, and gene expression. The nucleus, surrounded by a double-membrane nuclear envelope, includes critical elements like chromatin, the nucleolus, and nuclear pore complexes, each integral to functions such as gene expression, RNA processing, and ribosome assembly. This document offers a comprehensive examination of nuclear structure and its functional dynamics, highlighting the importance of Cajal bodies in RNA metabolism and the influence of nuclear organization on gene regulation. Furthermore, it underscores the consequences of nuclear irregularities in several illnesses, such as laminopathies and chromatin remodeling disorders, which may have significant health ramifications. As research advances, novel treatment methods aimed at nuclear functions, including gene editing technologies and techniques for treating laminopathies, are becoming increasingly prominent. These observations highlight the nucleus's essential function in preserving cellular integrity and operation and its significance in health and disease management.

Keywords: Cajal bodies, CRISPR-Cas9, DNA replication, Eukaryotic cells, Gene expression, Genetic material, Hutchison-Gilford Progeria Syndrome, laminopathies, nuclear matrix, nuclear pore complexes, RNA processing, Spinal muscular atrophy, Treacher Collins Syndrome.

INTRODUCTION

The nucleus is a membrane-bound organelle that acts as the command center of eukaryotic cells, ensuring genetic information integrity and controlling cellular activity. It is usually the most visible organelle in the cell, occupying a large percentage of the cellular volume. Gene expression, DNA replication, and cell division occur exclusively because of the eukaryotic cell's nucleus. In the prokaryotes, a nucleoid is present instead of a nucleus. The genetic material known as DNA (deoxyribonucleic acid) will be present as chromatin, a mixture of

---

[*] **Corresponding author K.N. Aruljothi:** Department of Genetic Engineering, SRM Institute of Science and Technology, Kattankulathur, Chengalpattu, Tamil Nadu, India; E-mail: aruljotn@srmist.edu.in

K.N. Aruljothi, Prakash Gangadaran, Krishnan Anand, Satish Ramalingam, K. Kumaran & Kruthika Prakash (Ed.)
All rights reserved-© 2026 Bentham Science Publishers

proteins and DNA that is arranged into chromosomes during cell division. The DNA that codes for everything from protein synthesis to cell growth and reproduction is found in the nucleus in eukaryotes / nucleoid in prokaryotes and includes all the instructions for the cell to function. Despite the more straightforward organization of genetic material, prokaryotes effectively regulate gene expression and adapt rapidly to environmental changes. Their streamlined genome organization and mechanisms, like operons (clusters of genes under a single promoter), enable efficient and coordinated gene expression. This efficiency contributes to their high adaptability and survival across diverse environments. At the same time, eukaryotic cells have a highly organized nucleus with complex DNA packaging that facilitates intricate regulation of gene expression and cellular specialization. In contrast, prokaryotic cells lacking a defined nucleus possess a more straightforward yet efficient system of DNA organization that supports rapid growth and adaptability. The presence of a membrane-bound nucleus is the characteristic that sets eukaryotic cells apart from prokaryotic cells, which have genetic material distributed throughout the cytoplasm and no clearly defined nucleus [1].

History

Over the decades, the discovery and comprehension of the nucleus have undergone substantial changes, playing a pivotal role in advancing cell biology. Robert Brown, a Scottish botanist, coined the term "nucleus" for the first time in a biological context in 1831. Brown saw a tiny, thick, spherical structure inside the cells of orchids and other plants while investigating plant cells under a microscope. He called this structure the "nucleus," which is Latin for "kernel" or "core." Though Brown did not wholly comprehend its purpose then, he recognized it as a recurring and essential aspect of the cell. In the years that followed, many scientists verified Brown's findings and started to make assumptions regarding the function of the nucleus. Matthias Schleiden and Theodor Schwann contributed to the cell theory in the middle of the 19th century. This theory postulated that all living things comprise cells, with the nucleus considered a crucial component [2].

Expanding on this idea, Rudolf Virchow proposed in 1855 that all cells originate from pre-existing cells, highlighting the significance of the nucleus in cell division and inheritance. Improvements in staining methods and microscopy during the late 19th century gave scientists a better understanding of the nucleus. Walther Flemming's 1880s discovery of chromosomes inside the nucleus demonstrated their function in mitosis or cell division. Later, August Weismann put out the theory that the nucleus contained genetic material, which impacted the field of genetics. Thomas Hunt Morgan and his associates' work showed that genes found on chromosomes inside the nucleus were the units of heredity, significantly advancing our knowledge of the function of the nucleus in the early 20th century. This discovery confirmed the nucleus's function in genetic information storage and inheritance [3]. The nucleus's function as the cell's command center was further demonstrated in 1953 with the discovery of the DNA structure by James Watson and Francis Crick. It was discovered that DNA, which is found in the nucleus of eukaryotic cells, is the molecule in charge of storing and transferring genetic information. This discovery made modern molecular biology possible by establishing the nucleus as the primary locus of gene control, replication, and expression [4].

STRUCTURE AND FUNCTION

Structure of Nucleus in Eukaryotes

The nucleus comprises several parts that form a highly ordered structure supporting its role as the cell's command center. The nuclear envelope is a two-layered barrier that surrounds the cell nucleus and keeps it separate from the cytoplasm. This envelope consists of an inner and outer membrane, both lipid bilayers. The outer membrane is continuous with the endoplasmic reticulum, facilitating the transport of materials between the nucleus and cytoplasm. Nuclear pores are embedded within the nuclear envelope, which regulate the passage of molecules such as nucleic acids and proteins into and out of the nucleus. A gel-like substance fills the nuclear space inside the nucleoplasm, providing structural support and a medium for suspending various nuclear components. Among these components, chromatin, a complex of DNA and proteins, plays a crucial role in packaging genetic material and regulating gene expression. Another vital structure within the nucleoplasm is the nucleolus, a dense, spherical body primarily responsible for producing and assembling ribosomes (Fig. 1) [1].

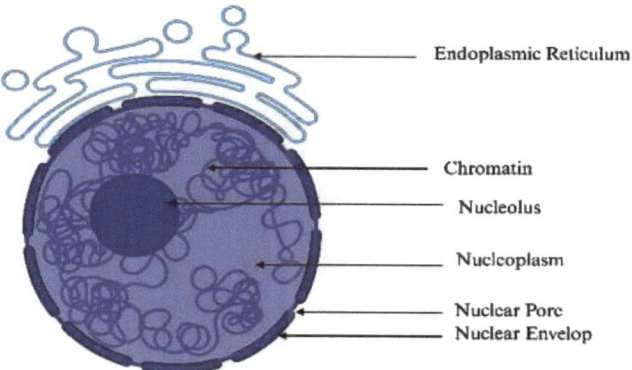

Fig. (1). Structure of the nucleus in eukaryotes.

- *Nuclear Envelope*: A nucleus is protected by a double membrane, the nuclear envelope, which keeps the cytoplasm outside. The two membranes are separated by a distance of 20 to 40 nm, and each membrane is a phospholipid bilayer linked to proteins. The continuous inner and outer membranes at specific locations form the nuclear pore complexes, which provide controlled transport between the cytoplasm and nucleus. The endoplasmic reticulum is attached to the outer membrane (ER). Proteins on the inner membrane, which faces the nucleoplasm, aid in keeping the nucleus in its proper form [5].
- *Nucleoplasm*: The nucleoplasm, a gel-like material found inside the nuclear envelope, acts as a medium for a variety of nuclear functions by offering a fluid environment that makes it easier for ions, molecules, and enzymes necessary for activities like transcription, DNA replication, and RNA processing to pass through. Many nuclear bodies are unique sub-structures with specialized roles in controlling RNA processing, gene expression, and other essential cellular processes in the nucleoplasm. The Cajal bodies, Polycomb bodies, and Gemini of coiled bodies (GEMs) are some of these nuclear bodies [6]. The spherical subnuclear structures called Cajal bodies, or coiled bodies, have a diameter ranging from 0.3 to 1.0 micrometers. They are especially prevalent in rapidly proliferating cells, such as those seen in tumors and embryonic tissues, and in cells with high metabolic activity, such as neurons [7].
- *Nuclear Pore Complexes*: Large protein structures called nuclear pore complexes (NPCs) are enmeshed in the nuclear envelope. They make it easier for some molecules, such as proteins and RNA, to move from the nucleus to the cytoplasm. Nucleoporins, a class of proteins that make up these complexes, establish a channel that controls molecular transit [8].
- *Chromatin*: DNA and histone proteins comprise the highly ordered structure known as chromatin, which is found inside the nucleus. It has two types: heterochromatin, which is more condensed and usually not transcriptionally active, and euchromatin, which is less condensed and actively involved in transcription. This organization makes gene expression, DNA replication, and DNA repair easier. Retains genetic data in DNA, which is accessible for use in transcription and replication processes [9].
- *Nucleolus*: One of the most noticeable substructures inside the nucleus is the nucleolus, which is mainly responsible for ribosomal subunit assembly and rRNA production. It is a dense region devoid of a membrane where rRNA transcription takes place concurrently with ribosomal protein assembly, which aids in the production of ribosomes. Accountable for the biosynthesis of ribosomes [10].
- *Nuclear Matrix*: In addition to providing structural support, the network of fibers and proteins known as the nuclear matrix may also be involved in DNA replication and gene expression.

- **Nuclear Lamina:** Between the peripheral chromatin and inner nuclear membrane, composed of nuclear lamina, which contains lamins, that interacts with numerous proteins. The interaction between the protein lamins and chromatin-associated proteins compounds the nuclear lamina as a highly complex structure. These play a crucial role in chromatin organization, nuclear migration, and other nuclear developmental roles [11, 12].

Function of Nucleus in Eukaryotes

Cajal bodies are more effective in performing nuclear functions, especially those about RNA metabolism, because of the high concentration of coilin proteins they contain. Small nuclear ribonucleoprotein (snRNP) synthesis, maturation, and modification are one of the primary roles of Cajal bodies. These snRNPs play a crucial role in the spliceosome process, which creates mature messenger RNA (mRNA) by connecting exons and deleting introns from pre-messenger RNA (pre-mRNA). Cajal bodies are essential for controlling gene expression because they aid in creating and modifying snRNPs [5 - 7].

Other RNA-protein complexes, such as small nucleolar RNPs (snoRNPs), which are necessary for the processing and chemical modification of ribosomal RNA (rRNA), are also assembled and modified by Cajal bodies in addition to snRNP biogenesis. Cajal bodies are essential parts of the cell's regulatory machinery because of their widespread participation in several facets of RNA metabolism. This highlights the significance of Cajal bodies in guaranteeing precise and efficient RNA processing and gene expression [13]. Therefore, Cajal bodies play a significant role in the cell's ability to control and regulate RNA-based activities through their connection with coilin proteins and their processing of snRNP and snoRNP. Their prevalence in cells with high metabolic or proliferative activity indicates how crucial it is to meet the demands of these settings, where sustaining cellular integrity and function depends on accurate and fast RNA processing and gene expression [12].

Structure of Genetic Material in Prokaryotes

Prokaryotes don't have proper nuclei; examples of these organisms are bacteria and archaea. Instead, they contain a more basic structure known as a nucleoid, the cytoplasmic area with an uneven shape, and house the cell's genetic material. The prokaryotic genome is contained in a single, circular, double-stranded DNA molecule called the nucleoid, which is not surrounded by a nuclear membrane like the nucleus in eukaryotic cells. Unlike in eukaryotes, this DNA is not linked to histone proteins, yet some proteins aid in the organization and compacting of the DNA. Prokaryotes also lack a nucleolus since ribosome synthesis occurs in the cytoplasm [14]. To fit within the restricted area of the cell, the DNA within the nucleoid is compressed by supercoiling and linked to specific proteins [15].

Function of Genetic Material in Prokaryotes

The nucleoid plays a vital function in gene expression in addition to storing and copying DNA. The nucleoid's DNA is used as a template for transcription, which converts it into messenger RNA (mRNA). The cytoplasm then uses this mRNA for protein translation. These proteins provide a wide range of biological tasks for the cell, including sustaining structural integrity, controlling cellular activities, and catalyzing biochemical events. The nucleoid also has a regulatory function because it enables the cell to regulate which genes are expressed at any given time. Prokaryotic cells can react quickly to changes in their environment, such as the presence of stressors or the availability of nutrients, by modifying the structure and supercoiling of DNA. Therefore, the nucleoid effectively combines genetic storage, expression, and control to enable the survival and adaptability of the cell, even though it lacks the compartmentalization of a eukaryotic nucleus [16].

DISORDERS ASSOCIATED WITH NUCLEUS

Abnormalities in the nucleus are linked to several illnesses and disorders, especially those that affect its structural elements, genetic makeup, and regulatory systems. Mutations in nuclear DNA, flaws in nuclear proteins, or problems with nuclear activities, including DNA repair, gene expression, and RNA processing, can all lead to these nuclear-related illnesses [1].

- A subset of illnesses called laminopathies is caused by genetic abnormalities in the genes that code for **nuclear lamins**, which give the nuclear envelope structural support. For instance, children with Hutchinson-Gilford Progeria Syndrome (HGPS), which is brought on by mutations in the LMNA gene that result in an aberrant form of lamin A known as progerin, age more quickly than usual because of malformed nuclear envelopes and malfunctioning cells. Similarly, mutations in nuclear lamins or related proteins cause Emery-Dreifuss Muscular Dystrophy (EDMD), which is characterized by muscle atrophy, joint contractures, and cardiac abnormalities. Dilated cardiomyopathy is another disorder associated with LMNA mutations that affect the heart's capacity to pump blood effectively, causing it to expand and weaken.
- Neurological diseases can also result from mutations affecting nuclear structures involved in RNA splicing, known as **Cajal bodies**. For example, motor neurons in the spinal cord degenerate in Spinal Muscular Atrophy (SMA), a hereditary disorder resulting in atrophy and muscle weakening. Mutations in the SMN1 gene, which is essential for the assembly of snRNPs in Cajal bodies and

interferes with RNA splicing, are frequently linked to this condition [17, 18].
- Another group of nuclear-related illnesses, known as **chromosome instability syndromes,** is caused by imbalances in chromatin organization and DNA repair, which makes a person more vulnerable to cancer and DNA damage. For instance, mutations in the BLM gene might result in genomic instability, increased sister chromatid exchanges, and poor DNA repair, which can cause Bloom Syndrome. Similarly, Fanconi anemia is brought on by mutations in genes linked to DNA repair pathways. This disorder impairs the body's capacity to repair DNA cross-links, which leads to physical defects, bone marrow failure, and an increased risk of cancer.
- Numerous diseases can also result from defects in nuclear **envelope proteins**. Charcot-Marie-Tooth Mutations in the *LMNB1* gene, which codes for lamin B1, cause type 2B1, a neurological condition characterized by gradual muscular weakening and atrophy. Greenberg Mutations in the LAMIN B receptor (LBR) gene cause skeletal dysplasia, often called Hydrops-Ectopic Calcification-Moth-Eaten (HEM) skeletal dysplasia. This condition causes skeletal deformities as well as early fetal mortality from nuclear membrane instability.
- Finally, diseases like Treacher Collins Syndrome and Dyskeratosis Congenita are associated with **nucleolus anomalies,** the location of ribosomal RNA production, and ribosome assembly. *TCOF1* gene mutations impact ribosomal RNA synthesis and induce craniofacial deformities in Treacher-Collins syndrome. Mutations in genes such as *DKC1* cause dyskeratosis congenita, which impairs telomere preservation and ribosome synthesis, resulting in deformities of the skin, bone marrow failure, and an elevated risk of cancer [18].
- Mutations in genes that control chromatin shape and modification impact gene expression, resulting in **chromatin remodeling diseases**. For instance, mutations in the MECP2 gene, which codes for a protein that binds to methylated DNA and controls chromatin shape, can result in Rett Syndrome, a neurodevelopmental condition. These mutations cause developmental regression, intellectual incapacity, and motor abnormalities, mainly in females. They also affect the control of genes crucial for brain development. A further illustration is Coffin-Siris Syndrome, a rare hereditary illness associated with abnormalities in genes like *SMARCB1* or *ARID1B* that encode parts of the SWI/SNF chromatin remodeling complex. Developmental delays, intellectual incapacity, characteristic facial traits, and deformities of the fingers and toes are the hallmarks of this condition.
- Gene control, nuclear integrity, and nuclear-cytoplasmic communication can all be hampered by **abnormalities in the nuclear membrane**. The progeroid disease known as Néstor-Guillermo Progeria Syndrome (NGPS) is brought on by mutations in the *BANF1* gene, which produces the barrier-to-autointegration factor (BAF) protein. BAF plays a critical role in chromatin organization and

nuclear envelope stability. Mutations in *BANF1* cause nuclear abnormalities and signs of premature aging, such as low height and joint contractures. Still, they do not cause the cardiovascular problems associated with Hutchinson-Gilford Progeria Syndrome. Another instance is the Pelger-Huët Anomaly, a hereditary condition marked by atypical neutrophil nuclear morphologies resulting from mutations in the LBR gene. Severe instances can exhibit bone abnormalities, developmental delays, and other problems, even though they are often asymptomatic.

- Stress reactions, cell cycle control, and ribosome synthesis can all be impacted by **nucleolar dysfunctions**. One unusual liver illness, for example, is North American Indian Childhood Cirrhosis (NAIC), which is associated with mutations in the *CIRH1A* gene (UTP4), which is involved in ribosome synthesis and nucleolar function. Children affected experience severe cirrhosis and liver fibrosis, which eventually results in liver failure. Similarly, abnormalities in genes encoding ribosomal proteins, such as RPS19 and RPL11, cause Diamond-Blackfan Anaemia (DBA), a rare hereditary bone marrow failure condition. These mutations affect ribosome synthesis in the nucleolus, which leads to anemia, impaired erythropoiesis, and a higher risk of cancer.

RECENT RESEARCH IN THE NUCLEUS

Recent developments in chromatin remodeling and nuclear dynamics fundamentally alter our knowledge of how the nucleus controls biological processes. A significant field of study is chromatin architecture, which refers to the three-dimensional (3D) arrangement of chromatin in the nucleus and its essential function in controlling the expression of genes. This study investigates chromatin's spatial organization's effects on biological functions, including transcription, DNA replication, and repair [2, 10, 19].

Advanced computer models have been used in recent research to mimic the three-dimensional structure of chromatin, offering extensive insights into the organization of chromatin in animals such as fruit flies or Drosophila melanogaster. These models enable researchers to forecast the behavior of chromatin domains, such as Lamina-Associated Domains (LADs), about the nuclear envelope. They are based on experimental methods like Hi-C, which maps chromatin connections throughout the genome. The 3D architecture of the genome is stabilized during interphase by chromatin areas known as LADs, which connect to the nuclear lamina. This structure resembles a mesh underneath the nuclear membrane. An interesting discovery is that LADs often bind and release from the nuclear envelope, generating a dynamic chromatin environment that changes nuclear organization and gene expression during the cell cycle [20].

There are several ramifications for this dynamic interaction between LADs and the nuclear envelope. Nuclear architecture appears to actively regulate gene expression since chromatin attachment to the nuclear periphery is often correlated with gene repression and dissociation, which can result in gene activation. Furthermore, the dynamic behavior of chromatin may also affect other biological processes, such as the timing of replication, chromatin remodeling, and DNA repair [21, 22].

Recent research has also focused on topoisomerase II activity, essential for chromatin disentanglement. Promoting transitory connections that control DNA supercoiling during activities like transcription and replication aids in maintaining the spatial organization of chromatin. Numerous human disorders have been related to genomic instability and chromatin misfolding, which can result from disruptions in these topological relationships.

Moreover, studies employing CRISPR-based genome editing and live-cell imaging improve our ability to view chromatin dynamics in real-time. Through interactions with nuclear structures, chromatin loops, and Topologically Associating Domains (TADs) influence gene expression under various developmental stages and environmental situations, as this research reveals [23, 24].

Therapeutics

Treatments that target the nucleus are becoming more popular as scientists investigate chromatin organization and nuclear dynamics' roles in disease processes. These treatments target aging-related illnesses, malignancies, and genetic abnormalities by interfering with the nucleus's vital functions, including DNA repair, chromatin remodeling, and gene expression control [25, 26].

Gene Editing: Platforms for editing genes, such as CRISPR-Cas9, allow for the exact repair of genetic mutations that cause illnesses by targeting specific DNA sequences inside the nucleus. By correcting faulty genes inside the patient's nucleus, CRISPR has been extensively investigated to treat hereditary diseases such as sickle cell anemia and muscular dystrophy. CRISPR-based "base editing" and "prime editing" are two recent developments that enable even more precise adjustments to nuclear DNA without causing double-strand breaks and lower the likelihood of unintentional genetic alterations [23, 24, 27, 28]

Gene Therapy for Laminopathies: Hutchinson-Gilford Progeria Syndrome (HGPS) and other laminopathies are caused by mutations in nuclear lamins, proteins that keep the nuclear envelope structurally intact. Antisense oligonucleotides (ASOs) are one therapeutic method targeting these illnesses by lowering progerin levels, a mutant lamin A protein seen in HGPS. These treatments seek to cure or moderate the condition's premature aging signs by lowering progerin accumulation in the nucleus [29, 30].

Chromatin Remodelling: Mutations in nuclear processes, such as chromatin remodeling, frequently lead to dysregulated gene expression, which is the cause of cancer. The goal of medications that target the regulatory systems of the nucleus is to restore standard chromatin structure and function. Examples of these medications are bromodomain inhibitors and histone deacetylase inhibitors (HDAC inhibitors). These medications show promise in treating malignancies such as leukemia, lymphoma, and solid tumors because they alter the nuclear architecture to either repress oncogenes or reactivate tumor-suppressor genes [9].

RNA-Based Therapeutics: Another developing field of treatment is RNA processing in the nucleus, particularly for splicing error-related illnesses such as Spinal Muscular Atrophy (SMA). Spliceosome-targeting medications and RNA splicing modulators aim to fix incorrect splicing processes inside the nucleus. Among the most well-known treatments is nusinersen (Spinraza), an antisense oligonucleotide that, in people with SMA, modifies RNA splicing to boost the synthesis of the functional SMN protein [31, 32].

Nuclear Pores for Drug Delivery: Molecules enter and exit the nucleus through the nuclear pore complex (NPC). Drug delivery innovations are working to find more effective ways to deliver medicinal compounds straight to the nucleus. This approach is beneficial for treating cancer because it allows for more precise delivery of medications that target nuclear enzymes or DNA through nuclear import/export mechanisms.

CONCLUSION

This chapter represents the structure of the nucleus/nucleoids and its function predominantly in cell division. The recent trends in research and the future perspective provide new light on nuclear plasticity and how cells dynamically rearrange their chromatin in response to different physiological requirements. We are learning new rules governing nuclear architecture as scientists work to improve these models. These insights will aid in our comprehension of the fundamental processes driving chromatin organization and how it affects cell function. These discoveries have broad ramifications for comprehending disorders, including cancer, laminopathies, and chromosomal instability syndromes associated with nuclear anomalies.

ACKNOWLEDGEMENTS

We want to acknowledge Ms. Kruthika P, Ms. Raksa Arun, Ms. Srisri SatishKartik, Ms. Sanjana Dhayalan, and Mr. Sasitharan T for their contribution to proofreading and editing works.

REFERENCES

[1] Minchin S, Lodge J. Understanding biochemistry: structure and function of nucleic acids. Essays Biochem 2019; 63(4): 433-56.
[http://dx.doi.org/10.1042/EBC20180038] [PMID: 31652314]

[2] Field MC, Rout MP. Pore timing: the evolutionary origins of the nucleus and nuclear pore complex. F1000 Res 2019; 8: 369.
[http://dx.doi.org/10.12688/f1000research.16402.1] [PMID: 31001417]

[3] Baum B, Spang A. On the origin of the nucleus: a hypothesis. Microbiol Mol Biol Rev 2023; 87(4): e00186-21.
[http://dx.doi.org/10.1128/mmbr.00186-21] [PMID: 38018971]

[4] Devos DP, Gräf R, Field MC. Evolution of the nucleus. Curr Opin Cell Biol 2014; 28(100): 8-15.
[http://dx.doi.org/10.1016/j.ceb.2014.01.004] [PMID: 24508984]

[5] Lee GE, Byun J, Lee CJ, Cho YY. Molecular Mechanisms for the Regulation of Nuclear Membrane Integrity. Int J Mol Sci 2023; 24(20): 15497.
[http://dx.doi.org/10.3390/ijms242015497] [PMID: 37895175]

[6] Taliansky ME, Love AJ, Kołowerzo-Lubnau A, Smoliński DJ. Cajal bodies: Evolutionarily conserved nuclear biomolecular condensates with properties unique to plants. Plant Cell 2023; 35(9): 3214-35.
[http://dx.doi.org/10.1093/plcell/koad140] [PMID: 37202374]

[7] Hertzog M, Erdel F. The Material Properties of the Cell Nucleus: A Matter of Scale. Cells 2023; 12(15): 1958.
[http://dx.doi.org/10.3390/cells12151958] [PMID: 37566037]

[8] Li Y, Zhu J, Zhai F, Kong L, Li H, Jin X. Advances in the understanding of nuclear pore complexes in human diseases. J Cancer Res Clin Oncol 2024; 150(7): 374.
[http://dx.doi.org/10.1007/s00432-024-05881-5] [PMID: 39080077]

[9] Wang X. Editorial: Chromatin architecture in gene regulation and disease. Front Genet 2023; 14: 1254865.
[http://dx.doi.org/10.3389/fgene.2023.1254865] [PMID: 37662842]

[10] Dubois ML, Boisvert FM. The Nucleolus: Structure and Function.The Functional Nucleus. Cham: Springer International Publishing 2016; pp. 29-49.
[http://dx.doi.org/10.1007/978-3-319-38882-3_2]

[11] Bihani A, Avvaru AK, Mishra RK. Biochemical Deconstruction and Reconstruction of Nuclear Matrix Reveals the Layers of Nuclear Organization. Mol Cell Proteomics 2023; 22(12): 100671.
[http://dx.doi.org/10.1016/j.mcpro.2023.100671] [PMID: 37863319]

[12] Soujanya M, Bihani A, Hajirnis N, Pathak RU, Mishra RK. Nuclear architecture and the structural basis of mitotic memory. Chromosome Res 2023; 31(1): 8.
[http://dx.doi.org/10.1007/s10577-023-09714-y] [PMID: 36725757]

[13] Rodrigues A, MacQuarrie KL, Freeman E, et al. Nucleoli and the nucleoli–centromere association are dynamic during normal development and in cancer. Mol Biol Cell 2023; 34(4): br5.
[http://dx.doi.org/10.1091/mbc.E22-06-0237] [PMID: 36753381]

[14] Love AJ, Yu C, Petukhova NV, Kalinina NO, Chen J, Taliansky ME. Cajal bodies and their role in plant stress and disease responses. RNA Biol 2017; 14(6): 779-90.

[http://dx.doi.org/10.1080/15476286.2016.1243650] [PMID: 27726481]

[15] Murat D, Byrne M, Komeili A. Cell biology of prokaryotic organelles. Cold Spring Harb Perspect Biol 2010; 2(10): a000422-2.
[http://dx.doi.org/10.1101/cshperspect.a000422] [PMID: 20739411]

[16] Fefilova AS, Antifeeva IA, Gavrilova AA, Turoverov KK, Kuznetsova IM, Fonin AV. Reorganization of cell compartmentalization induced by stress. Biomolecules 2022 Oct 8; 12(10): 1441.
[http://dx.doi.org/10.3390/biom12101441] [PMID: 36291650] [PMCID: PMC9599104]

[17] Malik I, Kelley CP, Wang ET, Todd PK. Molecular mechanisms underlying nucleotide repeat expansion disorders. Nat Rev Mol Cell Biol 2021; 22(9): 589-607.
[http://dx.doi.org/10.1038/s41580-021-00382-6] [PMID: 34140671]

[18] M. Ellerby Lisa. Repeat expansion disorders: mechanisms and therapeutics. Neurotherapeutics 2019; 6(4): 924-7. ISSN 1878-7479. Available from: https://www.sciencedirect.com/science/article/pii/S1878747923009595.
[http://dx.doi.org/10.1007/s13311-019-00823-3]

[19] Lammerding J. Mechanics of the Nucleus.Comprehensive Physiology. Wiley 2011; pp. 783-807.
[http://dx.doi.org/10.1002/cphy.c100038]

[20] Zhang P, Sun Y, Ma L. ZEB1: At the crossroads of epithelial-mesenchymal transition, metastasis and therapy resistance. Cell Cycle 2015; 14(4): 481-7.
[http://dx.doi.org/10.1080/15384101.2015.1006048] [PMID: 25607528]

[21] El Bezawy R, Tinelli S, Tortoreto M, et al. miR-205 enhances radiation sensitivity of prostate cancer cells by impairing DNA damage repair through PKCε and ZEB1 inhibition. J Exp Clin Cancer Res 2019; 38(1): 51.https://jeccr.biomedcentral.com/articles/10.1186/s13046-019-1060-z
[http://dx.doi.org/10.1186/s13046-019-1060-z] [PMID: 30606223]

[22] Yuan M, Eberhart CG, Kai M. RNA binding protein RBM14 promotes radio-resistance in glioblastoma by regulating DNA repair and cell differentiation. Oncotarget 2014; 5(9): 2820-6.
[http://dx.doi.org/10.18632/oncotarget.1924] [PMID: 24811242]

[23] Lim JM, Kim HH. Basic Principles and Clinical Applications of CRISPR-Based Genome Editing. Yonsei Med J 2022; 63(2): 105-13.
[http://dx.doi.org/10.3349/ymj.2022.63.2.105] [PMID: 35083895]

[24] Chehelgerdi M, Chehelgerdi M, Khorramian-Ghahfarokhi M, et al. Comprehensive review of CRISPR-based gene editing: mechanisms, challenges, and applications in cancer therapy. Mol Cancer 2024; 23(1): 9.
[http://dx.doi.org/10.1186/s12943-023-01925-5] [PMID: 38195537]

[25] Corey DR, Damha MJ, Manoharan M. Challenges and Opportunities for Nucleic Acid Therapeutics. Nucleic Acid Ther 2022; 32(1): 8-13.
[http://dx.doi.org/10.1089/nat.2021.0085] [PMID: 34931905]

[26] Kulkarni JA, Witzigmann D, Thomson SB, et al. The current landscape of nucleic acid therapeutics. Nat Nanotechnol 2021; 16(6): 630-43.
[http://dx.doi.org/10.1038/s41565-021-00898-0] [PMID: 34059811]

[27] Matano M, Date S, Shimokawa M, et al. Modeling colorectal cancer using CRISPR-Cas9–mediated engineering of human intestinal organoids. Nat Med 2015; 21(3): 256-62.
[http://dx.doi.org/10.1038/nm.3802]

[28] Drost J, Van Boxtel R, Blokzijl F, et al. Use of CRISPR-modified human stem cell organoids to study the origin of mutational signatures in cancer. Science 2017; 358(6360): 234-8.
[http://dx.doi.org/10.1126/science.aao3130]

[29] Liu Z, Zhu L, Roberts R, Tong W. Toward Clinical Implementation of Next-Generation Sequencing-Based Genetic Testing in Rare Diseases: Where Are We? Trends Genet 2019; 35(11): 852-67.
[http://dx.doi.org/10.1016/j.tig.2019.08.006] [PMID: 31623871]

[30] Bedinghaus JM. Genetic testing. N Engl J Med 2002; 351-6.
[http://dx.doi.org/10.1056/NEJMoa012113]

[31] Zhang X, Xie K, Zhou H, et al. Role of non-coding RNAs and RNA modifiers in cancer therapy resistance. Mol Cancer 2020; 19(1): 1-26.
[http://dx.doi.org/10.1186/s12943-020-01171-z]

[32] Lukkani LK, Naorem LD, Muthaiyan M, Venkatesan A. Identification of potential key genes related to idiopathic male infertility using RNA-sequencing data: an *in-silico* approach. Human Fertility 2022; 26(5): 1149-63
[http://dx.doi.org/10.1080/14647273.2022.2144771]

CHAPTER 4

Endoplasmic Reticulum: The Cellular Factory

Shanmuga Priya[1], K. Kumaran[1], Sanjana Dhayalan[1], Krishnan Anand[2,*] and K.N. Aruljothi[1]

[1] *Department of Genetic Engineering, SRM Institute of Science and Technology, Kattankulathur, Chengalpattu, Tamil Nadu, India*

[2] *Department of Chemical Pathology, School of Pathology, Faculty of Health Sciences, University of the Free State, Bloemfontein, South Africa*

Abstract: The endoplasmic reticulum is an organelle that performs dynamic functions in many critical cellular processes. This chapter discusses the evolutionary aspects of the Smooth ER (SER), the tasks of RER, their involvement in cellular stress responses, and implications for human health and disease. Complexities of ER stress and their relationship to conditions such as ulcerative colitis and type 2 diabetes will be discussed. The chapter discusses the therapeutic potential of targeting ER stress pathways as therapy for UC. We also discuss the relationship between the ER and autophagy, a cellular degradation and recycling process. Discussion is taken upon the role of the mitochondria-associated ER membrane (MAM) in calcium signaling and its implications for various cellular processes. This chapter summarizes the roles of ER, trying to unveil the complexity of the ER functions and the importance of the ER in both human health and disease.

Keywords: Apoptosis, Autophagy, Endoplasmic Reticulum, ER Stress, Glycosylation, Homeostasis, Lipid Synthesis, Mitochondria, Mitochondria-Associated ER Membrane (MAM), Protein Folding, Quality Control, Smooth ER, Type 2 Diabetes, Ulcerative Colitis.

INTRODUCTION

The endoplasmic reticulum encompasses a membrane-bound organelle with the framework of a network of sacs and tubules. It is continuous with the nuclear envelope. The ER is vital in shipping the synthesized proteins to the Golgi apparatus for further processing. Apart from synthesizing proteins, the ER synthesizes the lipids that essentially serve as the membrane in all organisms. The endoplasmic reticulum can be classified into two types based on the compounds they synthesize: Smooth Endoplasmic Reticulum (SER) and Rough Endoplasmic

---

** **Corresponding author Krishnan Anand:** Department of Chemical Pathology, School of Pathology, Faculty of Health Sciences, University of the Free State, Bloemfontein, South Africa; E-mail: organicanand@gmail.com*

K.N. Aruljothi, Prakash Gangadaran, Krishnan Anand, Satish Ramalingam, K. Kumaran & Kruthika Prakash (Ed.)
All rights reserved-© 2026 Bentham Science Publishers

Reticulum (RER). The Rough ER has attached ribosomes that help in protein synthesis. Smooth ER lacks ribosomes, which are key drivers in protein synthesis. Therefore, the SER is involved in the production of lipids, and the RER plays a role in Protein synthesis. Apart from synthesizing proteins and lipids, the ER plays other vital roles in transporting proteins and lipids, storing Ca^{2+} ions, steroid production, stress regulation mechanisms (endoplasmic reticulum stress), and protein folding machinery [1]. Fig. (**1**), below, denotes the structural orientation of the endoplasmic reticulum. Rough ER has ribosomes on its surface, whereas smooth endoplasmic reticulum lacks ribosomes.

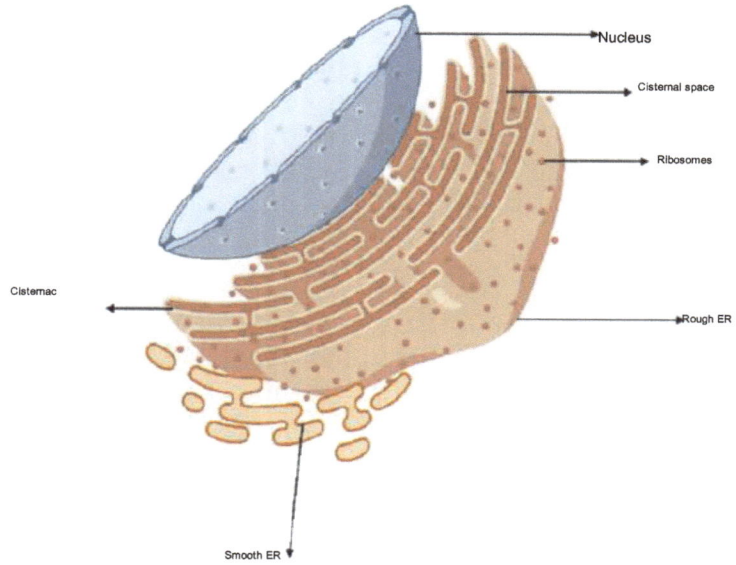

Fig. (1). Structural orientation of endoplasmic reticulum.

FUNCTIONS OF RER

The protein translocation in the RER occurs in two ways,

1. Where the protein, while being synthesized by the ribosomes, undergoes the translocation and translation simultaneously in ER, called the co-translational translocation.
2. The proteins synthesized by the free ribosomes are translocated into the ER using the SEC pathway.

The ER synthesizes three significant proteins: secretory proteins, integral membrane proteins, and Lumenal resident proteins. The secretory proteins are involved in folding proteins and delivering specific proteins. The integral membrane proteins are derived from the ER, and almost 30% of the entire human genome codes for the integral membrane proteins. The last category of proteins is called the luminal resident proteins of the ER, Golgi, Nuclear envelope, and Lysosomes. Therefore, RER plays a significant role in the biosynthesis of proteins. An example of the integral membrane protein is Single Carboxy-terminal TM spans (tail-anchored membrane proteins). Most organelles' protein requisite are satisfied by the protein synthesis at the RER. Therefore, molecules for protein synthesis must reach the ER for processing, followed by proper folding. The protein folding machinery folds the protein synthesized by the ribosomes at the RER, called the chaperones. Folding is essential for building a fully structural and functional unit of protein. Therefore, proteins reach a thermodynamically stable state. Also, proteins are not independent in their functioning. Their interactions can range from short and temporary within their functional group to stable oligomers in biological synthesis. The majority of proteins in the secretory pathway include oligomers. Therefore, the proteins must be folded at the late translational steps [2].

The Process of Protein Folding Associated with Molecular Chaperones

The chaperon-assisted protein folding is followed by the assembly of multi-subunit proteins, formation of the disulfide bond, and N-glycosylation (initial stages); in the case of the proteins of the plasma membrane, glycolipid anchors are added to the proteins. The Lumenal proteins in the ER play the primary role in assisting the folding of proteins and assembly of the polypeptide units that are newly translocated. The molecular chaperones, being proteins, act on a non-native protein to stabilize or bring about its native conformation. These molecular chaperones do not appear in the final three-dimensional confirmation of the proteins. The major ER resident chaperones are Hsp70 Bip, lectin chaperones like calnexin, and calreticulin (folding glycoproteins). Bip releases the abnormally folded proteins, which are then removed from the ER by the Endoplasmic reticulum-associated degradation process [3]. In the process of N-glycosylation, asparagine residues of proteins are glycosylated that are present within the ER. By this time, the translation is still in process (co-translational). In the first step, the oligosaccharides with 14 sugar residues are added to the acceptor molecules of protein, that is, the asparagine residue. The lipid carrier dolichol is the site for synthesizing the oligosaccharide anchored to the ER membrane. From here, the oligosaccharide unit is transferred to the asparagine residues by an enzyme associated with a membrane called the oligosaccharide transferase. Three glucose units are removed from the long oligosaccharide, and the protein remains in the endoplasmic reticulum. As discussed earlier, the protein from the ER has to reach the Golgi apparatus for further modification, packing, and shipping. Glycosylation is significant as it provides signals to promote protein folding, followed by adequately folded proteins being shipped to the Golgi for additional modifications. Its significance is that it prevents the formation of protein aggregates inside the ER.

ER as the Quality Control Center

Most of the proteins synthesized in the ER are rapidly degraded. This may be attributed to several issues, primarily improper folding of proteins or the proteins residing in the ER getting folded for an exceptionally long time while adequately folded. These misfolded proteins are identified and then taken to the cytosol from the ER, where the ubiquitin-proteasome system finally degrades them. From the previous discussion about chaperones and enzymes within the ER lumen and their functions associated with protein folding. They act as sensors of misfolded proteins. Find the two essential chaperones in the ER lumen and ER membrane.

Two glucose residues are removed following the glycoprotein that exits from the trans-Locon. This allows the binding of glycoprotein to calnexin and its assisted protein folding. The extent of protein folding is assessed as the glycoprotein interacts with the protein folding sensor. This is done by monitoring the hydrophobic regions that are exposed. For an incorrectly folded protein, a glucose residue is added back by the glycosyltransferase (protein folding sensor). If the protein is severely misfolded, the protein undergoes ER degradation by the EDEM-1, reorganizing the protein and removing mannose residues (Fig. **2**).

ER Stress

ER stress represents an impact of the cellular response to an overload of unfolded proteins in the endoplasmic reticulum. It is caused by increased synthesis of proteins, protein misfolding, or environmental stressors. When ER stress persists or worsens, it may evoke the Unfolded Protein Response (UPR), which seeks to restore homeostasis inside the cell but can lead to cell death if not resolved. The mRNA stage of gene expression plays a critical role in responding to ER stress. Covalent modification of initiation factors enables cells to regulate protein synthesis rates with dynamic rapidity in response to changes in intracellular conditions and external stimuli. Different cell types have different requirements concerning protein synthesis and secretion; thus, nutrients and energy sources must be provided differently to facilitate protein folding. High-proliferation cells, such as acinar cells, islet beta cells, plasma cells, and malignant cells, are a few that rely on efficient ER protein folding and quality control to allow only the right proteins into the secretory pathway. These various causes of intracellular changes and extracellular stimuli often lead to the accumulation of misfolded proteins. Other factors that can result include increased protein synthesis, impaired ubiquitination and proteasomal degradation, defective autophagy, energy imbalance, availability problems of nutrients, inflammatory challenges, and hypoxia. ER stress is versatile based on cell physiological and pathological conditions.

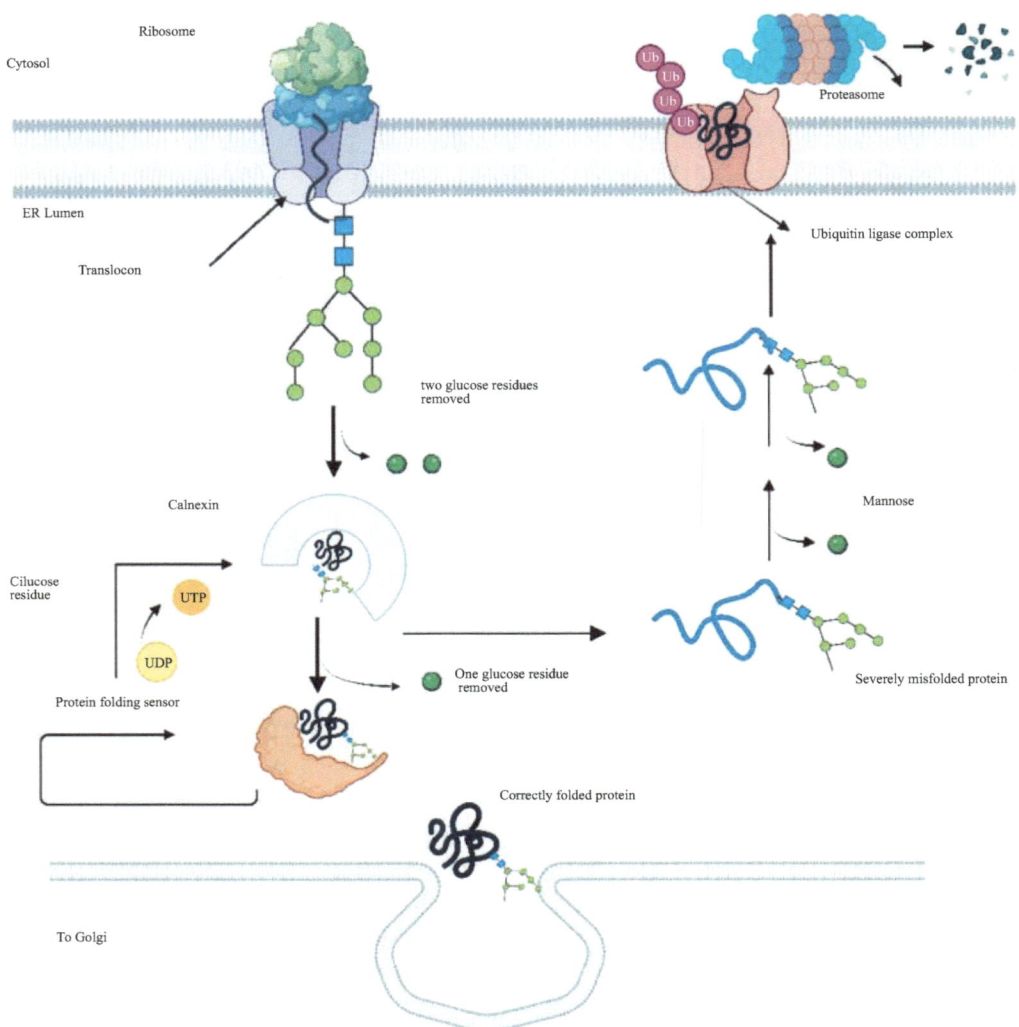

Fig. (2). The glycoprotein folding by calnexin.

The revocation of endoplasmic reticulum protein folding is triggered by the activation of exogenous or endogenous elements, and it invigorates a cell stress reaction called endoplasmic reticulum stress (Fig. **3**). ER stress restores endoplasmic reticulum homeostasis through incorporated flagging named the

Unfolded Protein Response in ER (UPRER). The pro-survival function of UPRER is transformed into a lethal signal that is sent to and carried out by mitochondria in the presence of severely toxic or prolonged ER stress. Both apoptotic and autophagic cell death depend on mitochondria. In this way, the endoplasmic reticulum is crucial in detecting and organizing pressure pathways to keep up with, by and large, physiological homeostasis. Cancer cells, on the other hand, exhibit dysregulation of this function, which prevents them from responding to apoptosis-inducing stimuli, such as therapeutic agents. The associations between endoplasmic reticulum pressure and mitochondrial apoptosis depict potential disease-helpful targets [4].

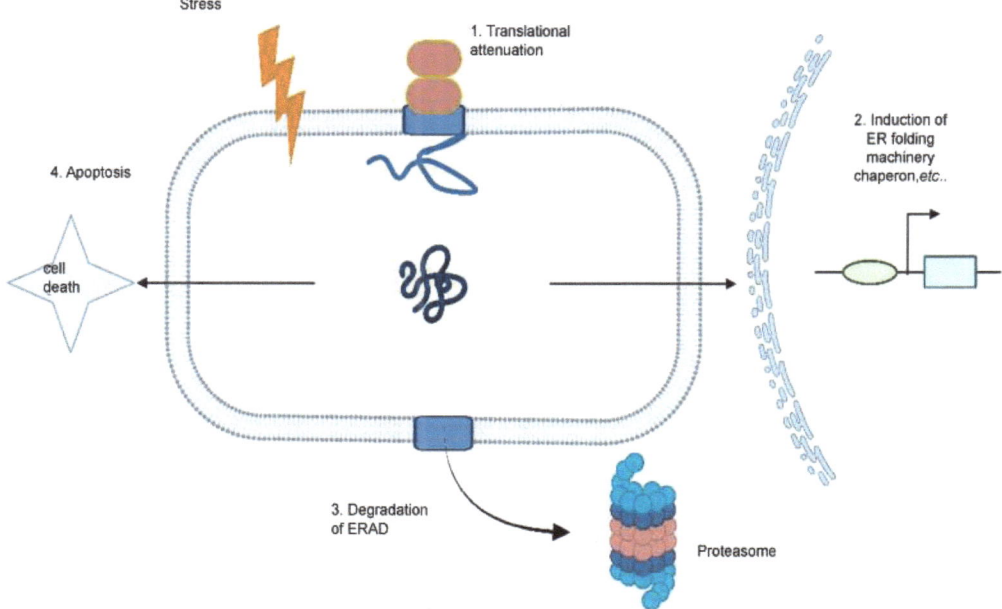

Fig. (3). The successive process in the endoplasmic reticulum in response to external stress. 1. Translational attenuation 2. Induction of ER folding machinery 3. ERAD (Degradation) 4. Apoptosis.

Diabetes Type II and ER Stress in Autophagy

Type 2 diabetes is a medical condition distinguished by persistent inflammation. The involvement of cytokines such as interleukin 1beta (IL1beta) in pancreatic inflammation, leading to diabetes and disturbance of glucose homeostasis, has been unequivocally established by multiple investigations. Recent findings suggest that islet β-cells can release cytokines, generating various inflammations. This process involves the induction of thioredoxin-interacting protein (TXNIP) by Endoplasmic Reticulum (ER) stress. This observation provides additional evidence that Endoplasmic Reticulum (ER) stress can potentially contribute to innate immunity by stimulating the NOD-like receptor (NLRP) 3/caspase1 inflammasome, as well as promoting the development of diabetes by the release of cytokines. A new finding has revealed that the autophagy process is also essential in the context of endoplasmic reticulum stress and inflammation. The intracellular catabolic system known as autophagy plays a crucial role in maintaining the standard structure of islets and intracellular insulin levels and facilitating a form of specialized cell death. We looked at the ER, and its autophagy-deficient cells had ER distention because ER stress plays a significant role in developing diabetes as a disease. Electron microscopy was used to examine the ultrastructural changes and showed cisternal distension of the endoplasmic reticulum (ER), even in normal-looking β cells of Atg7δβ-cell mice at lower magnifications. Inflammation is a secondary response that supports the aging process in mammals. A decrease in autophagic capacity, which aids in cellular survival against nutrient deprivation and the turnover of damaged organelles, is linked to aging. This peculiarity is connected with type 2 diabetes, tumorigenesis, neurodegenerative infections, cardiomyopathy, and numerous other immune system and inflammatory illnesses [5]. Identifying prolonged chronic inflammation can aggravate the tissues and is indicative of increasing diabetes type.

Therapeutics

ER Stress – Therapeutics and Curative Treatment for Acute Spinal Cord Injury

Targeting ER stress and halting cell death-related inflammation could be a key strategy for treating Spinal Cord Injuries (SCI), one of the most severe conditions that can cause long-standing disabilities. Several kinds of literature to date have established the relationship between cell death (apoptosis) and various diseases. However, the factors that intertwine cell death, ER, and diseases remain unclear. One of the most severe conditions, Spinal Cord Injuries (SCI), can cause long-standing handicaps. A recent study of chloroquine as a potential compound against severe illness caused by SCI has proven to have inhibitory effects against cell death-related inflammation, and it also reduces ER-related tension due to an increase in the severity of SCIs.

An analysis was also done to bring in ER stress-related factors, such as p62 and MRP (Microtubule-Related Protein). Chloroquine has brought in inhibitory effects against I-kBÎ± (post-translationally modified), which further led to the inhibition of the localization of p62. This restricted the action of ATF4 factors that play a significant role in autophagy-related illness. Therefore, chloroquine treatment remains one of the novel approaches to cell death and endoplasmic reticulum-related illness [6].

Therapeutic for Ulcerative Colitis

Ulcerative colitis, or UC, is characterized by persistent inflammation and affects the large intestine. Recent studies have shown that stress and autophagy within the Endoplasmic Reticulum (ER) play a significant role in developing ulcerative colitis. In this event, many misfolded proteins are accumulated in the ER; this might lead to a condition discussed earlier called ER stress, which affects the cell's activities. Activation of the Unfolded Protein Response (UPR) is how cells tend to react to ER stress; the purpose of this reaction is to assist in reestablishing the equilibrium of the cell. Persistent ER stress, on the other hand, may lead to the apoptosis of the cell. A cellular mechanism known as autophagy is responsible for breaking down damaged components. The elimination of the misfolded proteins can assist in the reduction of ER stress. Disrupted autophagy, on the other hand, can potentially be a factor in the development of Ulcerative Colitis (UC) by causing damage to the intestinal lining and causing the immune system to operate improperly. Regarding UC, the interaction between ER stress and autophagy is complicated because ER stress can stimulate autophagy. Cells can activate specific pathways that are known to promote autophagy when they are subjected to ER stress. These pathways include the ATF6-DAPK1 and REDD1 pathways. Autophagy and removing misfolded proteins are two ways in which autophagy can assist in alleviating ER stress and protect intestinal cells. Autophagy can also help preserve intestinal cells. UC can be caused by autophagy that is not regulated correctly. It is possible for inflammation and damage to the intestinal lining to occur due to either excessive or insufficient autophagy.

Several clinical perspectives may be built in the future, and we are inclined to consider this aspect as a treatment solution for UC that could be developed by targeting ER stress and autophagy. When researchers have a better knowledge of the complex interaction between these two processes, they can better design new medicines to help improve the lifestyle of people with ulcerative colitis [7]. The endoplasmic reticulum is involved in everything from the manufacturing of proteins to the synthesis of lipids. The Endoplasmic Reticulum (ER) plays a significant role as a cytoplasmic cell organelle. It also plays a vital role in deciding how the body reacts to stress, whether from within or outside. The ER is responsible for activating adaptive stress responses to maintain balance when confronted with obstacles such as nutritional overload, common in obesity. The strategies included in these responses include those regulating carbs, lipids, and proteins. On the other hand, in conditions such as obesity, these reactions fail to fall under fully regulated mechanisms, leading to many metabolic disorders.

To create treatments for metabolic disorders, we must thoroughly understand how the endoplasmic reticulum deals with stress. Even though natural substances have provided insights, it isn't easy to practically implement this information in medicine. The Endoplasmic Reticulum (ER) plays a complex function in various biological activities, which means that manipulating it can have unforeseen consequences. It is still possible for us to make progress despite these hindrances. By studying animal models, researchers are investigating potential methods to target specific ER activities, which have shown promising results for treating metabolic disorders. Discovering innovative approaches to alleviate stress in the endocrine system and promote metabolic health appears to have a bright future [8].

SMOOTH ENDOPLASMIC RETICULUM

As discussed earlier, the smooth endoplasmic reticulum, contiguous with the rough endoplasmic reticulum, differs from the RER in the absence of ribosomes. The Smooth Endoplasmic Reticulum is the site for lipid synthesis. In the first step, the Fatty acids are brought by Coenzyme-A to glyceraldehyde-3-phosphate. This results in the formation of the phospholipids (phosphatidic acids). The phosphatidic acid can be replaced or added to the sub-chains to form phosphatidyl choline, phosphatidyl inositol, phosphatidyl serine, and phosphatidyl ethanolamine. From earlier knowledge about the cell membrane and its constituents, we can see that these phospholipids constitute most of the cell membrane. As the formation of the phospholipid continues, the phospholipids are added to the same side of the membrane. However, the phospholipid bilayer should be distributed evenly on both the lumen and the cytosol. As the SERs are highly hydrophobic, they are impermeable to the polar heads of the phospholipids. Therefore, the enzyme flippases act by flipping the bilayer from cytosolic to luminal. This leads to an even distribution of the phospholipids; hence, the stability of SER is maintained. Phospholipids are primary components of cell membranes and serve essential functions. These mainly constitute glycerophospholipids and sphingolipids. These lipids' production occurs in several different organelles within the cell. Acyltransferases are one class of enzymes that are extremely important in phospholipid production. The addition of fatty acid chains to various lipid backbones is accomplished by considering fatty acyl-CoA as a substrate. The Endoplasmic Reticulum (ER), mitochondria, and peroxisomes are some of the cell organelles responsible for this activity.

Mitochondrion acts as the site for the production of phospholipids, and some examples of these phospholipids include cardiolipin (CL) and phosphatidylglycerol (DGPGro). Peroxisomes that are majorly involved in lipid metabolism are responsible for producing an essential kind of glycerophospholipid known as plasmalogens. Producing phospholipids is a multi-step process requiring an intricate network of enzymes and pathways. Some acyltransferases are directly involved in making new phospholipids, while others are responsible for altering existing acyl chains. Several acyltransferases are described here. Consequently, this guarantees cell membranes' variety and functionality [9].

Evolutionary Aspects of SER

The smooth endoplasmic reticulum was initially studied from thick sections of rat spinal and chick ciliary ganglia by saturating with heavy metals (0.5-2 µm thick). Further, studies under the microscope with voltage ranging from (10-5v) low to high (10-6v) voltage converged to the result that SER is a contiguous membrane that runs from the perikaryon to the axon terminal. Synaptic vesicles were prominently found at the tip of the tin tubules of SER [4]. A study on rat liver endoplasmic reticulum for 11 days is conducted 3 days before birth and 8 days after birth. In this period, the ER increased in size, and the increases were much higher for the smooth-surfaced parts following birth. The RNA-to-protein ratio decreased, whereas the lipoprotein/protein ratio increased with age. It remains relatively constant during the study period, with rates comparable to those recorded in the adult liver.

The data indicate that ER growth and development in rat liver cells occur rapidly during the neonatal period and are associated with protein and lipid composition changes. The unsaturated fat structure of absolute phosphatides goes through, notwithstanding, exceptional changes post-partum. During the time of quick ER advancement, the in vivo fusion of leucine-C14 and glycerol-C14 into the proteins and lipids of microsomal layers is higher in the harsh than in the smooth-surfaced microsomes for the central hours after the infusion of the name. After approximately 10 hours, the circumstance is turned around. These outcomes emphatically propose that a new layer is orchestrated in the rough ER and, in this way, moved to the smooth endoplasmic reticulum [4].

Calcium Storage in Smooth Endoplasmic Reticulum

This study aimed to study the morphology of the smooth endoplasmic reticulum (SER) within the axons of rodent dorsal root ganglia, thick (1 µm), as well as the thin sections of the dorsal roots were processed with lead and copper staining techniques to visualize the SER. Osmium-pyroantimonate staining was also used to locate Ca storage sites within the axons.

The results showed the existence of SER in thick and thin sections of dorsal root axons. The calcium stores were predominantly at the SER, while smaller amounts were in the mitochondria. If dorsal roots were preincubated in a calcium-containing solution known to depolarize the axons, then calcium stores were observed within the inner surface of the axolemma and within both SER and mitochondria. These findings establish a role for the SER in calcium storage and regulation in the dorsal root axons. The ability of the SER to store and release calcium might be crucial for the proper functioning of these neurons in transmitting sensory information to the central nervous system.

The Mitochondrial-Associated Endoplasmic Reticulum Membrane (MAM)

The endoplasmic reticulum has an extended Function in molding specific membrane-bound compartments that remain adjacent to the Outer Mitochondrial Membrane (OMM). This complex is known to have been held alongside OMM through specific proteins called tethers. These MAMs are associated with several physiological roles, such as transporting lipids, Ca^{2+}, and some aspects of mitochondria. They also serve essential functions in cell death (apoptosis), the formation and functioning of the inflammasome, and autophagy. The change in any associated component of MAMs will lead to an increased risk of damage to cellular homeostasis via unstable Ca^{2+} flux, producing reactive oxygen species in the mitochondrion. This may pave the way for advancing metabolic infections by activating MAM-related cytokines produced by the NLRP3 (Nod-like receptors). We will also discuss several diseases induced through this mechanism [5].

In the case of neurodegenerative disorders such as Parkinson's disease and Alzheimer's disease, the accumulation of metabolites intensifies the neurodegeneration. For example, when we consider Alzheimer's, it is caused by the accumulation of apolipoprotein (APP). Considering the MAMs and status of ER as discussed earlier, the proinflammatory cytokine signaling that leads to disease conditions here may lead to excessively induced ER stress that further increases the production of cytokines such as IL-1 β, which is involved in neural cell damage [6].

Understanding the Role of MAM in Cardiovascular Disease

The role that Mitochondrial-associated membranes play is essential in maintaining cardiovascular health. It is positioned between the endoplasmic reticulum and mitochondria, coordinating and communicating these two needed organelles, but its specific role in cardiovascular disease is somewhat complex and multifaceted. Another big challenge in the study of MAMs is that different cell types of the cardiovascular system, including cardiomyocytes, endothelial cells, and VSMCs, might have differential effects of MAMs. Cardiomyocytes generally depend on mitochondrial respiration for their energy, whereas most of the energy of endothelial cells and VSMCs is produced through glycolysis. The targeted therapeutic strategy would require a well-understood functionality of MAMs in each cell type. In Balancing MAM Formation and Function, MAMs can be functional or dysfunctional. In some ways, the proximity of the ER and mitochondria increases ATP but leads to an overload of calcium in the mitochondria and triggers apoptosis. Precise control over the formation and function of MAMs is needed to exclude deleterious effects. The Potential of MAMs as Therapeutic Targets Despite these challenges, MAMs have a high potential as therapeutic targets for cardiovascular diseases. Studies have identified the presence of MAM proteins in the pathogenesis of these disorders, thereby indicating the importance of MAM proteins. The elucidation of the molecular basis for MAM function and dysfunction may pave the way for more specific and effective treatments for these disorders. While MAMs indeed open excellent prospects, there are a lot of limitations and uncertainties that remain. The interactions of MAM proteins with the rest of the cellular constituents are so complex that it is unclear. Many of the MAM-associated membrane proteins are only partly characterized to date. Disruption of MAM function results in a range of pathological effects, including calcium dysregulation, autophagy defects, and apoptosis. Conclusion: MAMs are essential in maintaining cardiovascular health, but remain poorly characterized in terms of their functions and involvement in pathology. The challenges will be overcome, and a better understanding of MAM biology will be achieved to serve as the foundation upon which innovative approaches to therapy for cardiovascular diseases may be developed [6-9].

RECENT TRENDS IN ER RESEARCH

Several studies related to the endoplasmic reticulum and diseases related to its malfunction have been discussed in recent years. Endoplasmic reticulum-related metal complexes have been intensively studied to have shown up as an effective therapeutic and anticancer agent [10]. There are studies related to xenobiotics (drug-related compounds that are known to affect the biochemical equilibria of living organisms) that cause liver-related diseases as a result of prolonged ER stress that resulted in the dysfunction of the endoplasmic reticulum in hepatocytic cells [11]. Over the last few years, the molecular mechanism of toxicity has been extensively explored to discover signal molecules or genes that enhance oxidative stress, inflammation, and autophagy. In mammals, the most exposed organ is the lung because the primary entry route for nanoparticles in the human body is through the respiratory tract. Silica nanoparticles induce pulmonary inflammation in mice by damaging lysosomes as well as through the disruption of autophagy. Unlike silica nanoparticles, MSNs trigger liver inflammation by activating the NF-κB pathway, IL-1β, and TNFα, but protective autophagy against inflammation. New research now demonstrates that nanoparticles also lead to endoplasmic reticulum stress (ERS), another harmful mechanism of cytotoxicity. ERS is also associated with neurological systems, cardiovascular disease, and kidney disease. Therefore, attempts are made to elucidate the mechanism of nanoparticle toxicity by elaborating on ERS and analyzing the role of this process in nanoparticle toxicity to provide new insights for future nanoparticle treatments [12]. Recent studies have indicated that the transcription factor TOX is necessary to develop fully exhausted CD^{8+} T cells. This makes them do this by modulating the chromatin structure and regulating the RNA transcriptome. The ER stress, however, controls the changes in chromatin in the tumor and the RNA. Therefore, ER stress may have roles in controlling TOX-induced CD8+ T-cell exhaustion that are yet to be explored thoroughly. We have only begun to understand how ER stress regulates immunity and tolerance in healthy and sick individuals. Work conducted here will eventually make new immunotherapies possible [13]. Extracellular vesicles both induce and exacerbate ER stress. ER stress and EVs can interact with each other through several means, depending on cell type, extent of ER stress, and the context of the disease. These interactions can profoundly affect many disease processes, including tumor growth, cell proliferation, angiogenesis, immune responses, and even apoptosis. Many diseases are caused by protein accumulation due to ER stress, but how the two communicate in many ways is still unknown. The complicated means of disease initiation make it impossible to conclude that ER stress causes the release and movement of EVs or that EVs cause ER stress [14].

CONCLUSION

The endoplasmic reticulum is a transport path between the Nucleus and the Cytoplasm. ER, stress, and EVs are part of a cycle, and how many of the EVs work is still unknown. Blocking parts of this cyclic process allows future studies to treat some diseases. More research must be done on how EVs can be used in biotech to reduce ER stress. Additional convergence points in signaling pathways can be found, and medicines for treating autoimmune diseases and other UPR-mediated illnesses can be developed. In the future, the interaction of ER stress with EVs will potentially become significant in all realms of medical practice.

ACKNOWLEDGEMENTS

We want to acknowledge Ms. Kruthika P, Ms. Raksa Arun, Ms. Srisri SatishKartik, and Mr. Sasitharan T for their contribution in proofreading and editing the work.

REFERENCES

[1] Perkins HT, Allan V. Intertwined and Finely Balanced: Endoplasmic Reticulum Morphology, Dynamics, Function, and Diseases. Cells 2021; 10(9): 2341.
[http://dx.doi.org/10.3390/cells10092341] [PMID: 34571990]

[2] Pobre KFR, Poet GJ, Hendershot LM. The endoplasmic reticulum (ER) chaperone BiP is a master regulator of ER functions: Getting by with a little help from ERdj friends. J Biol Chem 2019; 294(6):

2098-108.
[http://dx.doi.org/10.1074/jbc.REV118.002804] [PMID: 30563838]

[3] Adams, B.M., Canniff, N.P., Guay, K.P., Hebert, D.N. (2021). The Role of Endoplasmic Reticulum Chaperones in Protein Folding and Quality Control. In: Agellon, L.B., Michalak, M. (eds) Cellular Biology of the Endoplasmic Reticulum. Progress in Molecular and Subcellular Biology, vol 59. Springer, Cham.
[http://dx.doi.org/10.1007/978-3-030-67696-4_3]

[4] Chen X, Cubillos-Ruiz JR. Endoplasmic reticulum stress signals in the tumour and its microenvironment. Nat Rev Cancer 2021; 21(2): 71-88.
[http://dx.doi.org/10.1038/s41568-020-00312-2] [PMID: 33214692]

[5] Degechisa ST, Dabi YT, Gizaw ST. The mitochondrial associated endoplasmic reticulum membranes: A platform for the pathogenesis of inflammation-mediated metabolic diseases. Immun Inflamm Dis 2022; 10(7): e647.
[http://dx.doi.org/10.1002/iid3.647] [PMID: 35759226]

[6] Wu F, Wei X, Wu Y, et al. Chloroquine Promotes the Recovery of Acute Spinal Cord Injury by Inhibiting Autophagy-Associated Inflammation and Endoplasmic Reticulum Stress. J Neurotrauma 2018; 35(12): 1329-44.
[http://dx.doi.org/10.1089/neu.2017.5414] [PMID: 29316847]

[7] Qiao D, Zhang Z, Zhang Y, et al. Regulation of Endoplasmic Reticulum Stress-Autophagy: A Potential Therapeutic Target for Ulcerative Colitis. Front Pharmacol 2021; 12: 697360.
[http://dx.doi.org/10.3389/fphar.2021.697360] [PMID: 34588980]

[8] Lemmer IL, Willemsen N, Hilal N, Bartelt A. A guide to understanding endoplasmic reticulum stress in metabolic disorders. Mol Metab 2021; 47: 101169.
[http://dx.doi.org/10.1016/j.molmet.2021.101169] [PMID: 33484951]

[9] Fagone P, Jackowski S. Membrane phospholipid synthesis and endoplasmic reticulum function. J Lipid Res 2009; 50(Suppl) (Suppl.): S311-6.
[http://dx.doi.org/10.1194/jlr.R800049-JLR200] [PMID: 18952570]

[10] Huang C, Li T, Liang J, Huang H, Zhang P, Banerjee S. Recent advances in endoplasmic reticulum targeting metal complexes. Coord Chem Rev 2020; 408: 213178.
[http://dx.doi.org/10.1016/j.ccr.2020.213178]

[11] Zhang Y, Qi Y, Huang S, et al. Role of ER Stress in Xenobioti-Induced Liver Diseases and Hepatotoxicity. Oxid Med Cell Longev 2022; 2022(1): 4640161.
[http://dx.doi.org/10.1155/2022/4640161] [PMID: 36388166]

[12] Li B, Zhang T, Tang M. Toxicity mechanism of nanomaterials: Focus on endoplasmic reticulum stress. Sci Total Environ 2022; 834: 155417.
[http://dx.doi.org/10.1016/j.scitotenv.2022.155417] [PMID: 35472346]

[13] Li A, Song NJ, Riesenberg BP, Li Z. The Emerging Roles of Endoplasmic Reticulum Stress in Balancing Immunity and Tolerance in Health and Diseases: Mechanisms and Opportunities. Front Immunol 2020; 10: 3154.
[http://dx.doi.org/10.3389/fimmu.2019.03154] [PMID: 32117210]

[14] Ye J, Liu X. Interactions between endoplasmic reticulum stress and extracellular vesicles in multiple diseases. Front Immunol 2022; 13: 955419.
[http://dx.doi.org/10.3389/fimmu.2022.955419] [PMID: 36032078]

CHAPTER 5

Golgi Apparatus: Shaping and Shipping Cellular Proteins

S. Sahasra[1], Sanjana Dhayalan[1], K. Kumaran[1] and K.N. Aruljothi[1,*]

[1] *Department of Genetic Engineering, SRM Institute of Science and Technology, Kattankulathur, Chengalpattu, Tamil Nadu, India*

Abstract: The Golgi body is an organelle in charge of altering proteins and lipids after translation and getting them ready for transport to spots within and outside the cell. This part of our study investigates its makeup, focusing on three key sections— cis side, middle section, and trans side, and how they contribute to sugar attachment to proteins, lipid processing, and protein arrangement. Furthermore, the chapter discusses how it plays a role in cell functions like secretion and creating lysosomes, and its impact on diseases like neurodegenerative conditions, infections, and cancer. It also covers progress in treatments and research related to the Golgi apparatus, including its involvement in the dynamics of COVID-19.

Keywords: Cancer, Cisternae, COVID-19, Golgi Apparatus, Lysosome Formation, Neurodegenerative Diseases, Phosphorylation, Post-Translational Modification, Protein Sorting, RAB GTPase, Sphingomyelin, TGN (Trans Golgi Network), A-Synuclein.

INTRODUCTION

The Golgi apparatus is a fundamental organelle and a key feature of eukaryotic cells that distinguishes them from prokaryotes. It consists of membrane-bound vesicles attached to stacked, flattened sacs known as cisternae. It helps in the post-translational modification and sorting of proteins and lipids. It packages these modified molecules into vesicles, which are then transported to specific locations inside and outside the cell through endocytic and exocytic pathways. As a result, it is vital for preserving cellular homeostasis and facilitating proper cell functions, making it crucial for intracellular transport and protein secretion. The Golgi apparatus can be traced back to the Last Eukaryotic Cell Ancestor (LECA), implying that the presence of the Golgi is nearly two billion years old, originating around the time of the divergence of major eukaryotic lineages [1]. The endosym-

---

[*] **Corresponding author K.N. Aruljothi:** Department of Genetic Engineering, SRM Institute of Science and Technology, Kattankulathur, Chengalpattu, Tamil Nadu, India; E-mail: aruljotn@srmist.edu.in

K.N. Aruljothi, Prakash Gangadaran, Krishnan Anand, Satish Ramalingam, K. Kumaran & Kruthika Prakash (Ed.)
All rights reserved-© 2026 Bentham Science Publishers

biotic theory has been turned down, and the trending view is that the Golgi and other organelles of the endomembrane system evolved through autogenous means, meaning that they originated from inside the cell and not from endosymbiosis. Functional diversity exists in the different eukaryotes, from the ring Golgi in the malaria parasite to the mobile Golgi in the plants. This merely underlines the functional plasticity of the Golgi and its adaptation to various ecological niches.

STRUCTURE OF GOLGI APPARATUS

It is worth noting that the early description of the Golgi apparatus was "apparato reticulare interno", meaning the internal reticular apparatus was based on its morphology. It is a membrane-bound organelle, made of a series of cisternae. These stacks often appear as a series of flattened pancakes or discs. The size of the Golgi stacks can differ, with each stack containing a variable number of cisternae. These stacks are encircled by a network of vesicles and tubules, which are essential for the transport of materials to and from the Golgi apparatus.

The Golgi apparatus has three compartments, called "cis", "medial," and "trans". The cis compartment is the cisternae near the Endoplasmic Reticulum (ER), the medial compartment is the central layers of cisternae, and the trans compartment is the cisternae farthest from the ER. There are two Golgi networks, which are the Cis Golgi network and the Trans Golgi network; these comprise the outermost cisternae located at the cis and trans faces of the Golgi, respectively. These networks are responsible for sorting the proteins and the lipids that are received at the cis face from the ER and released at the trans face by the Golgi (Fig. 1).

CIS Face

The cis face in the Golgi apparatus is closest to the ER. It is located to facilitate the transfer of the newly synthesized protein and lipids from the ER. Its primary function is to receive materials (protein and lipid) that need to be processed in the Golgi apparatus. It also initiates modification of the incoming molecules, like glycosylation, sulfation, and phosphorylation. It recognizes the specific signals on the incoming molecules, which decides its destination within the Golgi stacks, thus it is the "receiving a sorting hub" of the organelle. The Golgi apparatus consists of stacked cisternae linked by tubular structures, forming a network referred to as the Golgi ribbon. At the entrance of the Golgi, known as the cis-Golgi, Vesicular Tubular Clusters (VTCs) act as a transitional area between the endoplasmic reticulum and the stacked cisternae [2].

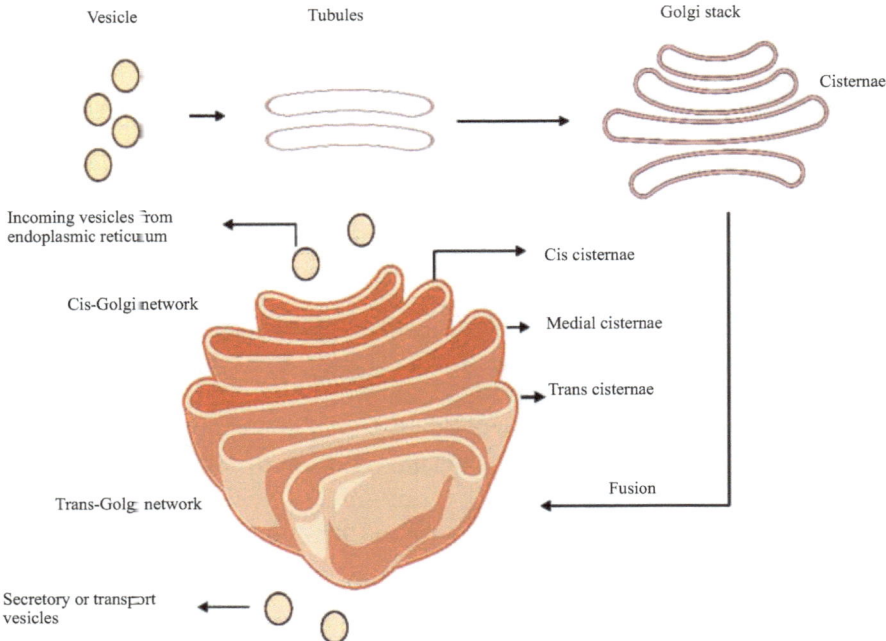

Fig. (1). The Golgi apparatus is composed of stacked cisternae interconnected by tubules, receiving vesicles from the endoplasmic reticulum and sorting proteins for their final destinations.

MEDIAL Face

The medial face is positioned between the cis face and the trans face within the Golgi stacks. The primary role of the medial face is to modify proteins and lipids. It is at this stage that many post-translational modifications occur, including glycosylation (addition of sugar molecules), sulfation, and phosphorylation, which are crucial for the functional maturation of proteins and the alteration of lipid molecules. The medial cisternae contain various enzymes that facilitate these modifications. It also helps in pH regulation within the cisternae, which is crucial for the functioning of the enzymes involved in various modifications. The medial face is responsible for further processing materials destined for secretory vesicles that will carry proteins and lipids to the cell membrane or for the formation of lysosomes, which contain enzymes for breaking down cellular waste.

TRANS Face

The trans face is positioned at the opposite end of the Golgi apparatus from the cis face and is called the "shipping" face, as it is responsible for directing modified molecules to their final destinations within or outside the cell. The primary role of the trans face is to sort out the proteins and lipids that have undergone modifications in the earlier faces of the Golgi apparatus and to package them. These molecules are sorted based on their specific destinations, which can include transport to the cell's surface, lysosomes, or other organelles. There will be slight modifications made to the proteins to ensure that molecules are properly processed and sorted. The trans side of the Golgi apparatus also contributes to the formation of lysosomes, which are intracellular structures that break down cellular waste products. Lysosomes are formed through a process in which enzymes and proteins are sorted and packaged in vesicles at the trans network. The clathrin-coated Trans-Golgi Network (TGN) is a cellular structure that forms clathrin-coated vesicles. These vesicles contain lysosomal enzymes tagged with Mannose-6-Phosphate (M6P). The enzymes are bound to Mannose-6-Phosphate Receptors (MPRs) and are transported to endosomes [2].

In many "lower" eukaryotic organisms, individual cells typically contain one or more separate Golgi stacks. In contrast, in vertebrates, the Golgi stacks are often interconnected to form a continuous structure known as the Golgi ribbon. This single continuous Golgi structure is typically located near the centrosome. Surrounding the Golgi stacks is a network of vesicles and tubules. These membranous structures transport materials between the faces or compartments, facilitating the transportation of molecules among the diverse cisternae and stacks, and ultimately, to their designated locations within or beyond the cell. This polarized configuration is of utmost importance for the sequential manipulation of molecules as they traverse the Golgi apparatus.

FUNCTION OF THE GOLGI APPARATUS

Protein Glycosylation

Glycosylation is the primary function of the Golgi apparatus. It is a fundamental post-translational and co-translational modification process where glycans (sugars) are enzymatically attached to proteins. This modification plays an important role in controlling the folding, stability, and transport of proteins through the secretory pathway. Additionally, the Golgi complex is surrounded by a variety of transport vesicles and tubules, which are involved in the movement of cargo between the ER, Golgi, and other cellular compartments. These vesicles and tubules are often coated with proteins such as clathrin, COPI, and COPII, which are important for their formation, targeting, and fusion with the appropriate membranes. Clathrin delivers new lysosomal enzymes from the Golgi to the lysosome [3]. COPI vesicles move cargo between the Golgi and the ER as well as within the Golgi itself, while COPII vesicles transport newly synthesized secretory proteins out of the ER [4].

Most of the proteins are synthesized in the ER and subjected to a quality control process. In this process, misfolded ones are degraded, whereas correctly folded proteins are tagged to be exported. Thereafter, the proteins are packaged into COPII-coated transport vesicles budded off from the ER's exit sites. Following their budding, the vesicles lose their COPII coat and undergo homotypic fusion to form larger transport carriers (Fig. 2). These carriers then travel to the Golgi apparatus, a structure consisting of stacks called saccules and interconnecting tubules. Upon arrival at the cis-Golgi network, sorting and modification of the proteins ensue. During the passage through the medial and trans compartments of the Golgi, proteins are continuously modified. Sorting for their final destinations and packaging into transport vesicles occur at the Trans Golgi Network (TGN). These vesicles bud off from the TGN and fuse with target membranes, such as lysosomes, the plasma membrane, or other cellular locations, ensuring that proteins make it to their final destinations [5].

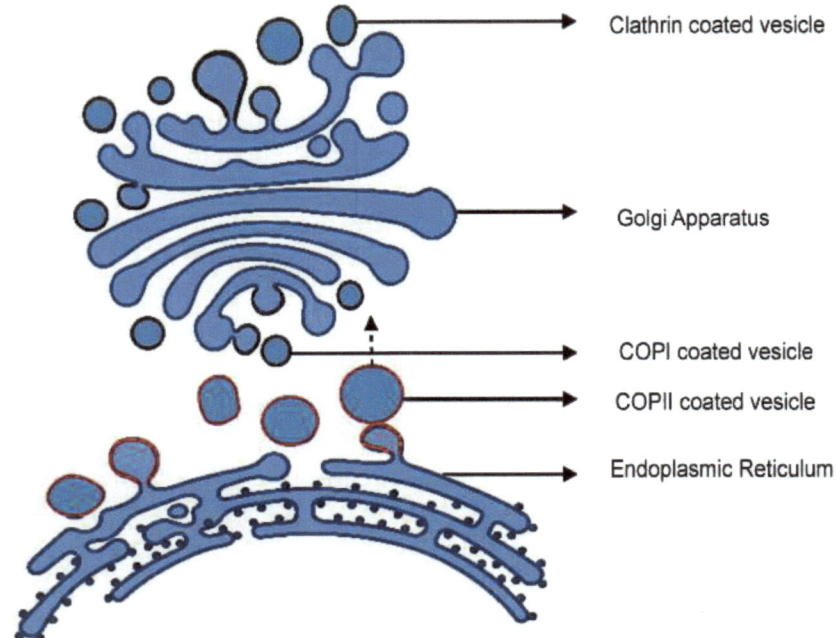

Fig. (2). The Golgi apparatus receives proteins from the endoplasmic reticulum in COPII-coated vesicles and transports them within the Golgi and to their final destinations using COPI and clathrin-coated vesicles.

There are two types of protein glycosylation, N-linked and O-linked glycosylation.

N – Linked Glycosylation: *N*-glycosylation is initiated in the ER by linking a preassembled oligosaccharide structure to specific asparagine residues within a protein sequence. This attachment occurs at asparagine residues that are followed by any amino acid except proline, forming a consensus sequence of Asn---Ser/Thr. Attachment is mediated by the oligosaccharyltransferase (OST) complex. Glucosidases in the ER first trim the precursor glycan, Glc3Man9GlcNAc2, which is an essential process for folding of the proteins and their quality control. Following this step, the glycoproteins continue on to the Golgi for additional modification to process into complex, hybrid, or high-mannose structures. This processing is important for proper protein folding, stability, and transport through the secretory pathway [6].

O – Linked Glycosylation: *O*-glycosylation is a type of glycosylation that involves the attachment of monosaccharides to the hydroxyl group of serine or threonine amino acid residues in a protein. Unlike N-glycosylation, which utilizes a lipid-linked oligosaccharide precursor, O-glycosylation occurs through a stepwise addition of individual monosaccharides, such as N-acetylgalactosamine (GalNAc). The synthesis of O-glycans takes place within the Golgi apparatus, where they undergo further modifications as the protein progresses through its various compartments. The most abundant class of O-linked glycans is the mucin-type glycans, which are of paramount importance regarding the structural integrity and functionality of mucins and proteoglycans [7]. NAA60 is an N-terminal acetyltransferase residing in the Golgi apparatus and, as such, differs from all other known NATs, which generally localize to the cytoplasm and nucleus. The transmembrane domains (TMDs) are presumed to anchor NAA60 to the Golgi membrane stacks. Abrogation of these TMDs causes mislocalization of NAA60 with consequent Golgi fragmentation, demonstrating that such TMDs are indeed essential for maintaining its Golgi localization [8]. Further confirmation of the unusual subcellular distribution of NAA60 was obtained through colocalization analyses with the Golgi markers GM130 and giantin and sensitivity to the Golgi-destroying drug brefeldin A, which induces the dispersal of NAA60 into vesicles. A number of unique regions responsible for its unique Golgi localization, including TMDs and potential endosome/lysosome-targeting sequences, were identified in NAA60 by sequence alignment and structural comparison with NAA50 [8]. It has been observed that NAA60 targets transmembrane proteins of the Golgi apparatus, endoplasmic reticulum, and mitochondria. NAA60 helps in the post-translational modification of the transmembrane protein by N-terminal acetylation (Fig. **3**) [9].

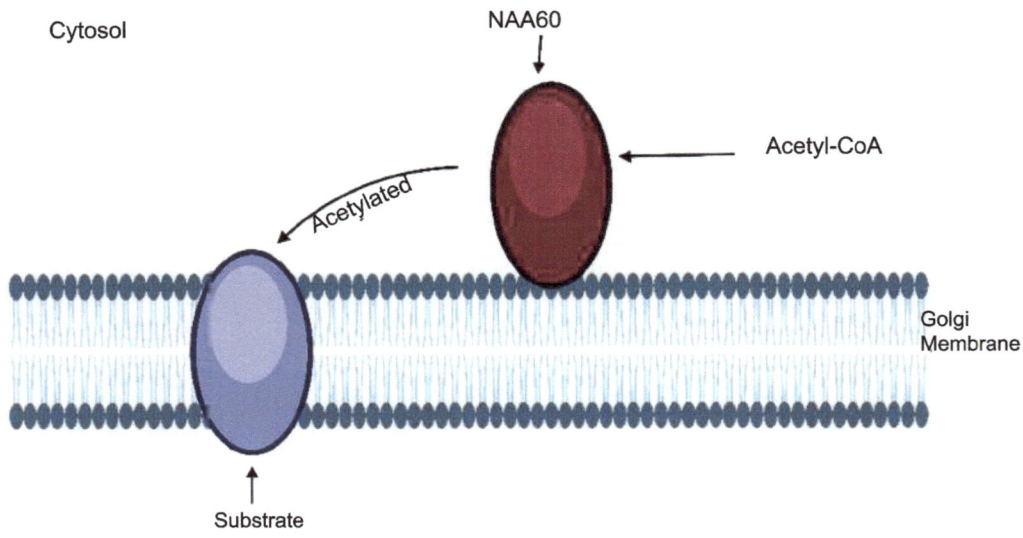

Fig. (3). NAA60 that is being attached to the Golgi membrane through its α-helix (not shown in the image). NAA60 transfers the acetyl group from the Acetyl-CoA to the substrate (transmembrane protein). Thus, the substrate undergoes post-translational modification.

The depletion of NAA60 caused the Golgi ribbon to fragment, suggesting a role of NAA60 in Golgi function, possibly in structural maintenance or in cytoskeleton-dependent processes that enable the Golgi to function. Further support is provided for the role of NAA60 in chromosome segregation during mitosis, connecting its function to the wider cellular process [8].

Lipid and Polysaccharide Metabolism

The glycolipids, along with the oligosaccharide, serve as a receptor in the cell membrane. This glycolipid is a derivative of the ceramide, modified in the Golgi apparatus. The ceramide is first synthesized in the ER and then is transported to the Golgi through a vesicle coated with COPII. On the cytosolic face of the Golgi apparatus, a glucose molecule is attached to the ceramide. The glycosylceramide is then converted into glycolipids on the luminal side. Sphingomyelin is formed when a phosphorylcholine group from phosphatidylcholine is transferred to ceramide on the luminal side of the Golgi apparatus. Sphingomyelin and glycolipids remain localized within the Golgi lumen until they are transported to the plasma membrane in a vesicle coated with COPI [5]. Some of the glycolipids are gangliosides, sulfatides, and cerebrosides.

Protein Sorting and Export

The Trans Golgi Network (TGN) helps in sorting proteins and ensuring that they reach their intended final destinations, like the lysosomes, the plasma membrane, or for secretion outside the cell. The proteins are continuously transported from the Golgi to the plasma membrane for release from the cell. On the other hand, the proteins are also packaged into COPII-coated vesicles in the endoplasmic reticulum (ER) before sending them to the Golgi. Not all proteins are packaged automatically; some require signals that are recognized by receptor proteins in a process called selective packaging. The ER also has a control system that holds onto proteins that are misfolded or not fully assembled, and these are eventually broken down by proteasomes. Once vesicles bud off from the ER, they shed their COPII coat and can undergo fusion, aided by proteins [5].

In certain cells, there's a controlled pathway for secretion where proteins are released in response to signals like hormones from endocrine cells or neurotransmitters from neurons. In polarized cells like epithelial cells, the regular secretion process needs to direct proteins to either the top (apical) or side (basolateral) of the cell membrane. Lysosomal targeting of proteins is mediated by the addition of mannose-6-phosphate residues within the Golgi apparatus. These phosphorylated proteins are subsequently recognized by specific receptors in the TGN, facilitating their delivery to lysosomes. In yeasts and plants, proteins are sent to vacuoles, which have similar functions to lysosomes, using short peptide sequences through a process called vacuolar targeting.

GOLGI APPARATUS DYSFUNCTION AND ASSOCIATED DISEASES

Neurodegenerative Diseases

Neurons have evolved into a complex system to handle their specific needs, such as protein processing and transport for their function and communication. For the processing and transporting of proteins, neurons have different sites, like the ER, Golgi, and ER Exit Sites (ERES, a site from which the COPII-coated vesicle containing folded protein is exported). These are present in the soma region (cell body) of the neuron and also along the neuron's extension (dendrites and axon). There are also some specialized structures like the Dendritic spine apparatus, Golgi outposts, and ER-Golgi Intermediate Compartment (ERGIC) that have a similar function as the Golgi in protein processing.

The fragmentation of the Golgi apparatus is likely to hinder the transportation of proteins necessary for neuronal activity, thus responsible for many neurodegenerative diseases like Parkinson's disease, Huntington's disease, Alzheimer's disease, and Amyotrophic lateral sclerosis. Phosphorylation of the stacking protein GRASP65, a key component of Golgi structure, can induce Golgi fragmentation. This fragmentation may contribute to increased production of Amyloid Precursor Protein (APP), a precursor molecule for amyloid beta (Aβ) peptides. This Aβ is toxic and can cause memory loss, resulting in Alzheimer's disease [10]. High levels of Golgi fragmentation in certain dopamine-producing neurons are linked to the accumulation of alpha-synuclein (α-syn) and the subsequent formation of protein aggregates. These aggregates can disrupt the normal function of the Golgi, leading to impaired protein transport and the buildup of toxic substances that can cause oxidative stress and cell death. α-Synuclein is believed to be a primary factor in the development of Parkinson's disease. The accumulation of α-synuclein aggregates is a hallmark of this neurodegenerative condition [11].

Infectious Disease

It has been identified that a mechanism by which an extracellular bacterial pathogen can disrupt the Golgi apparatus. Streptococcus pneumoniae infection disrupts the Golgi apparatus through the production of hydrogen peroxide (H_2O_2). This disruption can initially hinder bacterial attachment but may also weaken the host cell's defenses in later stages, highlighting a complex interplay between Golgi integrity and infection susceptibility [12]. Similarly, pathogens like Citrobacter rodentium, Herpes simplex virus 1, HIV-1, Measles virus, and S. flexneri target the Golgi-Associated Innate Immune Signalling [13].

Cancer

The Golgi apparatus is essential in various cellular processes, including protein processing, lipid biosynthesis, glycosylation, and intracellular trafficking. However, its morphology and function are often altered in cancer cells. Fragmentation and disassembly of the Golgi apparatus can lead to changes in glycosylation patterns, which in turn can enhance tumor cell survival, proliferation, and migration. Additionally, the Golgi is involved in signalling pathways, regulating kinases, and influencing cellular responses. These alterations enhance the metastatic potential of cancer cells, facilitating their migration and invasion. Moreover, the Golgi apparatus can influence apoptotic signalling, impacting cell survival and potentially contributing to resistance to cancer treatments.

Genetic Mutation

The mutation at the gene level can influence the function of the Golgi. Some examples are

i. Gene LARGE codes for Glycosyltransferases and mutation in this gene can lead to Congenital muscular dystrophy Type 1D [14]
ii. Gene TRAPPC11 codes for Tethering and mutation in this gene can lead to Spondyloepiphyseal dysplasia tarda [15]
iii. Gene RAB33B codes for Rab GTPase, and a mutation in this gene can lead to Smith-McCort dysplasia [16]

THERAPEUTICS

Gaucher Disease (GD)

Gaucher Disease (GD) is a lysosomal storage disorder caused by a genetic deficiency in β-glucocerebrosidase. This enzyme deficiency leads to the accumulation of glucocerebroside (GlcCer) within macrophages. It presents in various forms:

GD1 - This form is marked by symptoms such as hepatosplenomegaly, anemia, thrombocytopenia, and bone disease. Treatment options involve Enzyme Replacement Therapy (ERT) and Substrate Reduction Therapy (SRT). SRT utilizes GlcCer Synthase inhibitors like miglustat and eliglustat to reduce the production of the underlying substrate [17].

GD2 and GD3 - These variants may exhibit neurological symptoms. While eliglustat is not effective for brain-related issues due to its limited ability to penetrate the central nervous system, a new inhibitor called ibiglustat (Genz-682452) is currently being assessed for its potential to cross the blood-brain barrier [17].

- Enzyme Replacement Therapy (ERT) - This method involves providing the missing enzyme to assist in breaking down the accumulated GlcCer.
- Substrate Reduction Therapy (SRT) - This approach focuses on decreasing the production of GlcCer by inhibiting its synthesis, which can be accomplished through the use of GlcCer Synthase inhibitors [17].

Parkinson's Disease

Research shows that there is a link between the improper breakdown of GlcCer and certain types of Parkinson's disease, with 5-10% of patients having mutations in the glucocerebrosidase gene [17]. This suggests a possible connection between lysosomal dysfunction and neurodegenerative disorders.

GTPase Rab1

Role - Rab1 plays a crucial role in the transport of vesicles from the ER to the Golgi.

Therapy Insight - In studies using yeast and mammalian cell models, it has been found that the toxicity of α-synuclein, which hampers ER-to-Golgi transport, can be counteracted by Rab1. This indicates that boosting Rab1 activity may help alleviate Golgi fragmentation and enhance cellular function in Parkinson's Disease (PD).

LRRK2 Mutations

Role - Mutations in LRRK2 (leucine-rich repeat kinase 2), a protein involved in dopamine receptor trafficking, are a common genetic risk factor for Parkinson's disease. LRRK2 interacts with Rab7L1, a GTPase implicated in protein transport.

Therapy Insight - The presence of mutant LRRK2 increases the phosphorylation of Rab7L1, which changes its distribution and affects the morphology of the Trans-Golgi Network (TGN), contributing to Golgi fragmentation. Targeting LRRK2 activity or its interaction with Rab7L1 may present a viable therapeutic approach.

Rab8 and Rab10

Role - Rab8 and Rab10 are substrates of LRRK2, and their phosphorylation is heightened in the presence of mutant LRRK2.

Therapy Insight - Adjusting the activity of Rab8 and Rab10 or their interaction with LRRK2 could be beneficial in addressing Golgi fragmentation in PD [6].

PLA2G6 Mutations

Role: Mutations in PLA2G6 impact phospholipid metabolism, resulting in altered glycosylation patterns and damage to the ER-Golgi intermediate compartment (ERGIC) and Golgi apparatus.

Therapy Insight: Targeting the glycosylation abnormalities or stabilizing the Golgi structure in cells with PLA2G6 mutations may provide therapeutic advantages.

α-Synuclein Aggregates

Role - The aggregation of α-synuclein is a defining feature of PD, leading to Golgi fragmentation by disrupting the levels of transport regulatory proteins such as Rab1, 2, 8, and syntaxin 5. Therapy Insight - Approaches aimed at reducing α-synuclein toxicity or restoring the levels of these transport proteins could potentially reverse Golgi fragmentation [18].

Cancer Therapy

The Golgi apparatus is a key target in cancer photodynamic therapy (PDT) to improve treatment effectiveness. Utilizing photosensitizers such as TPE-PyT-CPS creates localized oxidative stress, which initiates apoptosis in cancer cells [19]. This method results in notable morphological alterations in the Golgi, impairing its function and encouraging cell death, all while reducing harm to nearby healthy tissues.

Inhibitors of Peripheral Golgi-Associated Proteins - TAS-116

TAS-116, an oral inhibitor of Heat-shock protein 90 (HSP90), demonstrates efficacy in reducing KIT protein levels associated with the Golgi apparatus in gastrointestinal stromal tumors (GISTs) that are resistant to imatinib. Studies have revealed that TAS-116 effectively suppresses growth and induces apoptosis in GIST cell lines. Additionally, it has shown promising activity against lung cancers harbouring EGFR mutations that are resistant to gefitinib.

Compounds Affecting GA Protein Localization-2-(Substituted Phenyl)-Benzimidazole (2-PB)

These compounds can disrupt the localization of resident GA proteins, leading to their subsequent degradation. These compounds have demonstrated an ability to decrease cell proliferation in multiple cell lines and have also been shown to inhibit tumor growth in xenograft models of human melanoma.

Nanodelivery Approaches

Recent advancements in nanotechnology are being investigated to enhance the delivery of drugs that specifically target the GA. This includes

- Passive Targeting - Utilizing the Enhanced Permeability and Retention (EPR) effect, which enables nanoparticles to preferentially accumulate in tumor tissues.
- Active Targeting - Conjugating nanoparticles with ligands that selectively bind to receptors that are highly expressed on cancer cells.
- Physical Strategies - Using external stimuli like magnetic fields or ultrasound to guide nanoparticles to the tumor site.

RECENT TRENDS

COVID-19

The SARS-CoV-2 virus relies on the host cell's secretory pathway to replicate, assemble, and spread. The Golgi apparatus, a vital organelle for membrane trafficking, plays a central role in these processes. SARS-CoV-2 infection causes the Golgi apparatus to fragment, which is linked to a decrease in GRASP55, a protein involved in maintaining its structure. This fragmentation accelerates the movement of viral particles within the cell, facilitating their release. Additionally, the infection promotes the production of heparan sulfate, a molecule that enhances the interaction between the virus's spike protein and the ACE2 receptor, a crucial step in viral entry. The disruption of the Golgi apparatus may also affect the synthesis of glycolipids, essential components of the myelin sheath surrounding nerve fibers. This could potentially contribute to the cognitive difficulties, often described as "brain fog," experienced by some COVID-19 patients, as it may impact myelination and neural function [20].

Golgi Dynamics and Regulation

- Golgi Fragmentation: Studies have explored the mechanisms behind Golgi fragmentation and its involvement in various cellular functions, such as cell division, stress responses, and pathological conditions [21].
- Golgi Maturation: Research has explored the molecular pathways involved in Golgi maturation and the factors that regulate this process, such as small GTPases and lipid composition [22].

GOLGI AND DISEASE

Neurodegenerative Diseases: The role of Golgi dysfunction in neurodegenerative conditions like Alzheimer's and Parkinson's disease has been extensively studied [23].

Cancer: Research has explored how Golgi abnormalities contribute to tumorigenesis, metastasis, and drug resistance [24].

Golgi and Cell Signalling

Golgi-Mediated Signalling: The Golgi apparatus has been studied for its involvement in multiple signalling pathways, including those related to cell growth, differentiation, and responses to stress [25].

Golgi and Organelle Communication: Understanding how the Golgi interacts with other organelles, like the ER and mitochondria, has been a focus of research [26].

Golgi and Infectious Diseases

Viral Interactions: Studies have investigated how viruses manipulate the Golgi apparatus to facilitate their replication and spread [23].

Bacterial Interactions: Research has explored the role of the Golgi in host-bacterial interactions and the development of infectious diseases [27].

These are just a few of the recent trends in Golgi biology, and the field continues to evolve with new discoveries and insights.

CONCLUSION

The Golgi apparatus is an organelle that is used to modify, sort, and transport proteins and lipids to other locations within the cell and outside the cell. This chapter talks about how intricate it is with its cis and medial and trans compartments that do important things like glycosylation and lipid metabolism. It also goes into the Golgi's part in many cellular processes like secretion, lysosome synthesis, and its role in diseases like neurodegenerative diseases, cancer, and infections. The recent strides in therapeutic approaches, *i.e.,* targeting Golgi-related mechanisms in diseases like Gaucher and Parkinson's, only underscore its importance in health and disease.

ACKNOWLEDGEMENTS

We would like to acknowledge Ms. Kruthika P, Ms. Raksa Arun, Ms. Srisri SatishKartik, and Mr. Sasitharan T for their contribution in proofreading and editing the work.

REFERENCES

[1] Klute MJ, Melançon P, Dacks JB. Evolution and diversity of the Golgi. Cold Spring Harb Perspect Biol 2011; 3(8): a007849-9.
[http://dx.doi.org/10.1101/cshperspect.a007849] [PMID: 21646379]

[2] Klumperman J. Architecture of the mammalian Golgi. Cold Spring Harb Perspect Biol 2011; 3(7): a005181-1.
[http://dx.doi.org/10.1101/cshperspect.a005181] [PMID: 21502307]

[3] Schmid SL. Clathrin-coated vesicle formation and protein sorting: an integrated process. Annu Rev Biochem 1997; 66(1): 511-48.
[http://dx.doi.org/10.1146/annurev.biochem.66.1.511] [PMID: 9242916]

[4] Adolf F, Rhiel M, Hessling B, *et al.* Proteomic Profiling of Mammalian COPII and COPI Vesicles. Cell Rep 2019; 26(1): 250-265.e5.

[http://dx.doi.org/10.1016/j.celrep.2018.12.041] [PMID: 30605680]

[5] Mohan AG, Calenic B, Ghiurau NA, Duncea-Borca RM, Constantinescu AE, Constantinescu I. The golgi apparatus: A voyage through time, structure, function and implication in neurodegenerative disorders. Cells 2023; 12(15): 1972.
[http://dx.doi.org/10.3390/cells12151972] [PMID: 37566051]

[6] Breitling J, Aebi M. N-linked protein glycosylation in the endoplasmic reticulum. Cold Spring Harb Perspect Biol 2013; 5(8): a013359-9.
[http://dx.doi.org/10.1101/cshperspect.a013359] [PMID: 23751184]

[7] Debets MF, Tastan OY, Wisnovsky SP, *et al.* Metabolic precision labeling enables selective probing of O-linked *N* -acetylgalactosamine glycosylation. Proc Natl Acad Sci USA 2020; 117(41): 25293-301.
[http://dx.doi.org/10.1073/pnas.2007297117] [PMID: 32989128]

[8] Donnarumma F, Tucci V, Ambrosino C, Altucci L, Carafa V. NAA60 (HAT4): the newly discovered bi-functional Golgi member of the acetyltransferase family. Clin Epigenetics 2022; 14(1): 182.
[http://dx.doi.org/10.1186/s13148-022-01402-8] [PMID: 36539894]

[9] Aksnes H, Goris M, Strømland Ø, *et al.* Molecular determinants of the N-terminal acetyltransferase Naa60 anchoring to the Golgi membrane. J Biol Chem 2017; 292(16): 6821-37.
[http://dx.doi.org/10.1074/jbc.M116.770362] [PMID: 28196861]

[10] Joshi G, Bekier ME II, Wang Y. Golgi fragmentation in Alzheimer's disease. Front Neurosci 2015; 9: 340.
[http://dx.doi.org/10.3389/fnins.2015.00340] [PMID: 26441511]

[11] Wei Y. Awan M un N, Bai L, Bai J. The function of Golgi apparatus in LRRK2-associated Parkinson's disease. Front Mol Neurosci 2023; 16: 1097633.

[12] Klabunde B, Wesener A, Bertrams W, *et al.* Streptococcus pneumoniae disrupts the structure of the golgi apparatus and subsequent epithelial cytokine response in an H_2O_2-dependent manner. Cell Commun Signal 2023; 21(1): 208.
[http://dx.doi.org/10.1186/s12964-023-01233-x] [PMID: 37592354]

[13] Tao Y, Yang Y, Zhou R, Gong T. Golgi Apparatus: An Emerging Platform for Innate Immunity. Trends Cell Biol 2020; 30(6): 467-77.
[http://dx.doi.org/10.1016/j.tcb.2020.02.008] [PMID: 32413316]

[14] Grewal PK, McLaughlan JM, Moore CJ, Browning CA, Hewitt JE. Characterization of the LARGE family of putative glycosyltransferases associated with dystroglycanopathies. Glycobiology 2005; 15(10): 912-23.
[http://dx.doi.org/10.1093/glycob/cwi094] [PMID: 15958417]

[15] Larson AA, Baker PR II, Milev MP, *et al.* TRAPPC11 and GOSR2 mutations associate with hypoglycosylation of α-dystroglycan and muscular dystrophy. Skelet Muscle 2018; 8(1): 17.
[http://dx.doi.org/10.1186/s13395-018-0163-0] [PMID: 29855340]

[16] Dupuis N, Lebon S, Kumar M, *et al.* A novel *RAB33B* mutation in Smith-McCort dysplasia. Hum Mutat 2013; 34(2): 283-6.
[http://dx.doi.org/10.1002/humu.22235] [PMID: 23042644]

[17] Zappa F, Failli M, De Matteis MA. The Golgi complex in disease and therapy. Curr Opin Cell Biol 2018; 50: 102-16.
[http://dx.doi.org/10.1016/j.ceb.2018.03.005] [PMID: 29614425]

[18] Martínez-Menárguez JÁ, Tomás M, Martínez-Martínez N, Martínez-Alonso E. Golgi Fragmentation in Neurodegenerative Diseases: Is There a Common Cause? Cells 2019; 8(7): 748.
[http://dx.doi.org/10.3390/cells8070748] [PMID: 31331075]

[19] Liu M, Chen Y, Guo Y, *et al.* Golgi apparatus-targeted aggregation-induced emission luminogens for effective cancer photodynamic therapy. Nat Commun 2022; 13(1): 2179.

[http://dx.doi.org/10.1038/s41467-022-29872-7] [PMID: 35449133]

[20] Wang Y, Gandy S. The Golgi apparatus: Site for convergence of COVID-19 brain fog and Alzheimer's disease? Mol Neurodegener 2022; 17(1): 67.
[http://dx.doi.org/10.1186/s13024-022-00568-2] [PMID: 36271398]

[21] Wortzel, I., Porat, Z. Quantifying Golgi Apparatus Fragmentation Using Imaging Flow Cytometry. In: Barteneva, N.S., Vorobjev, I.A. (eds) Spectral and Imaging Cytometry. Methods in Molecular Biology, vol 2635. Humana, New York, NY, 2023; pp. 173–84.
[http://dx.doi.org/10.1007/978-1-0716-3020-4_10]

[22] Manzer KM, Fromme JC. The Arf-GAP Age2 localizes to the late-Golgi via a conserved amphipathic helix. Mol Biol Cell 2023; 34(12): ar119.
[http://dx.doi.org/10.1091/mbc.E23-07-0283] [PMID: 37672345]

[23] Toader C, Eva L, Covache-Busuioc RA, *et al.* Unraveling the Multifaceted Role of the Golgi Apparatus: Insights into Neuronal Plasticity, Development, Neurogenesis, Alzheimer's Disease, and SARS-CoV-2 Interactions. Brain Sci 2023; 13(10): 1363.
[http://dx.doi.org/10.3390/brainsci13101363] [PMID: 37891732]

[24] Martins M, Vieira J, Pereira-Leite C, Saraiva N, Fernandes AS. The Golgi Apparatus as an Anticancer Therapeutic Target. Biology (Basel) 2023; 13(1): 1.
[http://dx.doi.org/10.3390/biology13010001] [PMID: 38275722]

[25] Khine MN, Sakurai K. Golgi-Targeting Anticancer Natural Products. Cancers (Basel) 2023; 15(7): 2086.
[http://dx.doi.org/10.3390/cancers15072086] [PMID: 37046746]

[26] Yeerken D, Xiao W, Li J, *et al.* Nlp-dependent ER-to-Golgi transport. Int J Biol Sci 2024; 20(8): 2881-903.
[http://dx.doi.org/10.7150/ijbs.91792] [PMID: 38904019]

[27] Read CB, Ali AN, Stephenson DJ, *et al.* Ceramide-1-phosphate is a regulator of Golgi structure and is co-opted by the obligate intracellular bacterial pathogen *Anaplasma phagocytophilum*. MBio 2024; 15(4): e00299-24.
[http://dx.doi.org/10.1128/mbio.00299-24] [PMID: 38415594]

CHAPTER 6

Ribosomes: Engines of Protein Synthesis

Sanjana Dhayalan[1], Shalini Roy[1], K. Kumaran[1] and K.N. Aruljothi[1,*]

[1] *Department of Genetic Engineering, SRM Institute of Science and Technology, Kattankulathur, Chengalpattu, Tamil Nadu, India*

Abstract: Ribosomes are essential macromolecular complexes that translate genetic information from mRNA into proteins. Composed of rRNA and ribosomal proteins, they consist of two subunits, large and small, in both prokaryotic (70S) and eukaryotic (80S) cells, with structural differences reflecting their evolutionary adaptations. Ribosome biogenesis (Ribi) is a tightly regulated process initiated by rDNA transcription and subunit assembly in eukaryotes and involving factors like NusA and NusG in prokaryotes. Ribosomes drive protein synthesis through complex processes of initiation, elongation, termination, and recycling, facilitated by GTPases and elongation factors, ribosomopathies cause diseases such as Diamond-Blackfan Anemia and Shwachman-Diamond Syndrome, linked to mutations in ribosomal proteins. Ribosomal dysfuntion can lead to both hypo- and hyper-proliferation, increasing cancer risk. Recent advances in targeting ribosome-related pathways, such as mTORC1 inhibition, and techniques like ribosome profiling have provided insights into diseases like leukemia and medulloblastoma, revealing non-canonical ORF translation and novel therapeutic targets. These findings highlight the therapeutic potential of modulating ribosome function in disease treatment.

Keywords: ABCE1, Biogenesis, Cancer, Elongation, Eukaryotes, Fingerprinting, GTPases, Prokaryotes, Protein Synthesis, Ribosomal Subunits, Ribosome Profiling, Ribosomopathies, rRNA, Termination, tRNA.

INTRODUCTION

Ribosomes are essential cellular organelles found in all living organisms, serving as the molecular machinery responsible for protein synthesis, a fundamental biological process. These intricate macromolecular complexes comprise ribosomal RNA (rRNA) and ribosomal proteins; they are essential for converting genetic information from messenger RNA (mRNA) into polypeptide chains, which fold into valuable proteins. These indispensable organelles play a central role in synthesizing proteins, which are vital for the structure and function of all

---

[*] **Corresponding author K.N. Aruljothi:** Department of Genetic Engineering, SRM Institute of Science and Technology, Kattankulathur, Chengalpattu, Tamil Nadu, India; E-mail: aruljotn@srmist.edu.in

K.N. Aruljothi, Prakash Gangadaran, Krishnan Anand, Satish Ramalingam, K. Kumaran & Kruthika Prakash (Ed.)
All rights reserved-© 2026 Bentham Science Publishers

living organisms. Their intricate composition and sophisticated mechanisms underscore the complexity of cellular processes and the evolutionary significance of protein synthesis. The ribosomal function provides insight into fundamental biological processes and has implications for medical research, particularly in developing antibiotics that target bacterial ribosomes without affecting eukaryotic cells. Ribosomes are categorized into two primary types based on cellular organization: prokaryotic and eukaryotic ribosomes. Despite their differences, both types share a similar functional architecture. The structural differences between prokaryotic and eukaryotic ribosomes are significant, particularly in the size and composition of the rRNA and protein components [1].

Ribosome in Prokaryotes and Eukaryotes

In eukaryotic cells, ribosomes include two distinct subunits: the large ribosomal subunit (60S) and the small ribosomal subunit (40S). The 60S subunit comprises three rRNA molecules: 5S, 5.8S, 25S/28S rRNA, and 46 associated ribosomal proteins. The small 40S subunit contains a single rRNA molecule, the 18S rRNA, associated with 32 ribosomal proteins. The highly regulated process of ribosomal component assembly takes place in the nucleolus, where mature ribosomal subunits are formed by the combination of ribosomal proteins and rRNA transcription [2]. Prokaryotic ribosomes in organisms such as bacteria are slightly smaller than their eukaryotic counterparts. They consist of a 50S subunit and a 30S subunit. 34 proteins and 5S and 23S rRNA are found in the 50S subunit, whereas 21 proteins and 16S rRNA are found in the 30S subunit.

The protein synthesis, or translation, involves several key steps and the coordinated action of both ribosomal subunits [3]. The small ribosomal subunit first attaches itself to the mRNA molecule and searches for the start codon. The assembly of the ribosomal subunits—40S and 60S in eukaryotes, 30S and 50S in prokaryotes—around the mRNA molecule is the first step in the translation process. In eukaryotes, the initiator tRNA, which carries methionine (Met-tRNAi), and the small 40S ribosomal subunit work together to scan the messenger RNA (mRNA) for the start codon (AUG). The big 60S subunit joins the complex upon identification of the start codon, completing the formation of the 80S ribosome.

The Shine-Dalgarno sequence on the mRNA is bound by the 30S subunit in prokaryotes, and the 50S subunit then joins to form the 70S ribosome. The large and tiny subunits join forces upon identification of the start codon to produce a functional ribosome that can synthesize proteins.

During translation, the ribosome facilitates the decoding of the mRNA sequence into a corresponding amino acid sequence. The primary function of the small subunit is to decode the messenger RNA (mRNA) and guarantee that the appropriate transfer RNA (tRNA) molecules, each containing a distinct amino acid, are introduced into the ribosomal A (aminoacyl) site. Through its peptidyl transferase activity, the larger subunit facilitates the creation of peptide bonds between neighbouring amino acids once the proper tRNA has been positioned. The elongation of the developing polypeptide chain depends on this enzyme action. The tRNA molecules are successively added as the ribosome translocates along the mRNA. When the expanding polypeptide chain reaches a stop codon, it is released from the ribosome, indicating that translation has stopped. The completed polypeptide then undergoes folding and post-translational modifications to achieve its final functional conformation.

In eukaryotic cells, ribosomes are crucial for protein synthesis and can be found in several distinct locations, each serving specific functions. They exist in two primary forms: freely floating in the cytoplasm, referred to as cytoribosomes, or attached to the outer surface of the Rough Endoplasmic Reticulum (RER), thereby forming the rough ER. This association with the RER is significant as it facilitates the co-translational translocation of nascent polypeptide chains into the lumen of the endoplasmic reticulum. Once inside the ER, these polypeptides undergo further modifications, such as glycosylation and folding, before being trafficked to their final destinations within or outside the cell. Additionally, ribosomes are present in the stroma of plastids, such as chloroplasts, and within the mitochondrial matrix. In these organelles, ribosomes synthesize proteins encoded by their respective organellar genomes, which are distinct from nuclear DNA. This unique feature underscores the endosymbiotic theory, suggesting that mitochondria and plastids originated from ancestral prokaryotic cells. In contrast, prokaryotic cells, which lack membrane-bound organelles, exhibit a simpler ribosomal distribution. Ribosomes in prokaryotes are freely distributed throughout the cytoplasm. These ribosomes, typically smaller than their eukaryotic counterparts, are involved in translating mRNA into proteins directly within the cytoplasmic environment. This structural simplicity allows for rapid protein synthesis, which is essential for the adaptability and survival of prokaryotic organisms in diverse environments. The localization of ribosomes in eukaryotic and prokaryotic cells reflects their essential roles in protein synthesis, with distinct adaptations that facilitate cellular functions and processes.

Origin and Evolution of the Ribosome

The origins of ribosomes can be traced back to the early Earth, approximately 4 billion years ago, when the first molecules of life began to form. It is widely accepted that simple RNA molecules, capable of catalyzing chemical reactions, played a crucial role in the emergence of life. This concept is often referred to as the "RNA world hypothesis," which posits that early life forms were based on RNA, which later coevolved with proteins to form more complex biological systems. The ribosome is considered a product of this coevolution, where peptides and RNA are developed together to enhance the stability and reactivity of ribonucleic complexes. The earliest ribosomes likely consisted of simple RNA structures capable of catalyzing peptide bond formation, leading to the synthesis of primitive proteins. Over billions of years, the ribosome's structure has changed significantly [4].

Research has demonstrated that the ribosome's core has not altered much in any of the life domains, including bacteria, archaea, and eukaryotes. The ribosome's ability to convert messenger RNA (mRNA) into proteins depends on this core. Studies show that the ribosome evolved by incorporating new structural components into its pre-existing core without changing its basic structure. As organisms evolved and grew more complex, this process led to the diversification of ribosomal structures. For example, the core of current ribosomes is conserved, but the complex surrounding portions have evolved to meet the unique requirements of many animals. Ribosomal structure has evolved through an accretion process, where new functional elements were added over time, enhancing its dynamics and efficiency.

The ribosome's exit tunnel, crucial for protein folding, evolved from a primitive to a more complex structure, facilitating the transition from simple to complex life forms. The modern ribosome is thought to have largely formed around the time of the Last Universal Common Ancestor (LUCA), approximately 3.5 to 4 billion years ago. At this point, the ribosome had evolved into a more complex structure, incorporating additional rRNA segments and proteins that enhanced its functionality [5]. The core architecture of the ribosome, including the PTC, became conserved across all domains of life, like Bacteria, Archaea, and Eukarya, indicating its fundamental role in the translation process.

The evolution of ribosomes can be described in several key phases: -

- ***Prokaryotic Evolution***: Prokaryotic ribosomes evolved through a series of stages, acquiring capabilities for RNA folding, catalysis, and decoding. This process involved the sequential addition of rRNA expansion segments, which contributed to the ribosome's structural complexity and functional versatility. The evolution of tRNA and mRNA also occurred during this period, allowing for more sophisticated protein synthesis mechanisms.
- ***Eukaryotic Divergence***: With the emergence of eukaryotic cells, ribosomes underwent further modifications, including the addition of tentacle-like rRNA expansions that enhanced their interaction with other cellular components.

Eukaryotic ribosomes also developed unique features, such as the presence of additional RPs that facilitate interactions with the endoplasmic reticulum and other organelles.

- **Structural Conservation and Adaptation:** Throughout evolutionary history, the core structure of the ribosome has remained remarkably conserved, while the outer regions have adapted to meet the needs of increasingly complex organisms. Studies have shown that ribosomes from diverse species, including humans, yeast, and bacteria, share a common core, with variations primarily occurring in the ribosomal surface structures.

The first observations of ribosomes were made in the mid-1950s by Romanian-American cell biologist George Emil Palade. He utilized electron microscopy to identify these dense particles, initially referred to as "Palade granules" due to their granular appearance. In 1958, the term "ribosome" was proposed by Howard M. Dintzis during a symposium as a more suitable designation for these ribonucleoprotein particles involved in protein synthesis. Palade, along with Albert Claude and Christian de Duve, was awarded the Nobel Prize in Physiology or Medicine in 1974 for their contributions to cell biology and the discovery of ribosomes.

Ribosomes Fingerprint

The concept of "fingerprints" in ribosomal structures reveals how new segments were added over time. By digitally reconstructing ancient ribosomes, the ribosome's evolution is characterized by a layering process, where new features are integrated without compromising the existing core structure. Ribosomal fingerprints are primarily composed of the unique sequences and structural motifs present in rRNA and RP. The large ribosomal subunit (LSU) and small ribosomal subunit (SSU) each contribute to the formation of these fingerprints. In eukaryotic cells, the 60S LSU contains 5S, 5.8S, and 25S/28S rRNA, while the 40S SSU contains 18S rRNA. The specific arrangement and interactions of these rRNA molecules, along with the associated RPs, create a unique signature for each species.

Several techniques have been employed to analyze and utilize ribosomal fingerprints:

- **Ribosomal DNA (rDNA) Fingerprinting**: This method involves the use of labeled rDNA probes to hybridize with restriction fragments of chromosomal DNA, creating a unique pattern for each species. For example, a digoxigenin-labeled probe derived from the rrnB rRNA operon of *Escherichia coli* has been used to characterize Listeria monocytogenes isolates [6].

- ***Oligonucleotide Fingerprinting of rRNA Genes (OFRG):*** OFRG is an array-based technique that allows for extensive analysis of microbial community composition. It involves the construction of rDNA clone libraries, followed by hybridization with a series of oligonucleotide probes to create unique fingerprints for each clone [7].
- ***Solid-state Nanopore Fingerprinting:*** This emerging technique utilizes solid-state nanopores to analyze the characteristics of individual ribosomes. By measuring the peak amplitude and dwell time of ribosomes passing through the nanopore, researchers have been able to distinguish between 80S ribosomes and polysomes (complexes of multiple ribosomes translating a single mRNA strand) based on their size and structural properties [8].

Ribosomal fingerprints are not only useful for identification purposes but also provide insights into the evolutionary history of ribosomes. The conservation of certain rRNA and protein components across different domains of life (Bacteria, Archaea, and Eukarya) suggests that these features are essential for the fundamental functions of ribosomes. The variations observed in the outer regions of ribosomes, however, reflect the adaptations that have occurred during the evolution of increasingly complex organisms. Ribosomal fingerprinting is a powerful tool for the identification and characterization of ribosomes from various species. These unique molecular signatures, composed of the distinct arrangements of rRNA and RPs, have applications in fields such as microbial ecology, evolutionary biology, and biotechnology. As research continues to explore the structural and functional dynamics of ribosomes, the concept of ribosomal fingerprinting will likely play an increasingly important role in understanding the complexities of protein synthesis and the evolutionary relationships among diverse life forms.

STRUCTURE OF RIBOSOME

Ribosomes are essential macromolecular machines found in all living cells, playing a critical role in protein synthesis, also known as translation. Composed of rRNA and proteins, ribosomes are classified as ribonucleoproteins and consist of two distinct subunits: large and small. These subunits come together during protein synthesis to form a functional ribosome. The small subunit is responsible for binding to the mRNA and decoding the genetic information, while the large subunit catalyzes the formation of peptide bonds between amino acids, forming the polypeptide chain. Both subunits have their characteristic protuberances and ridges (Fig. **1**).

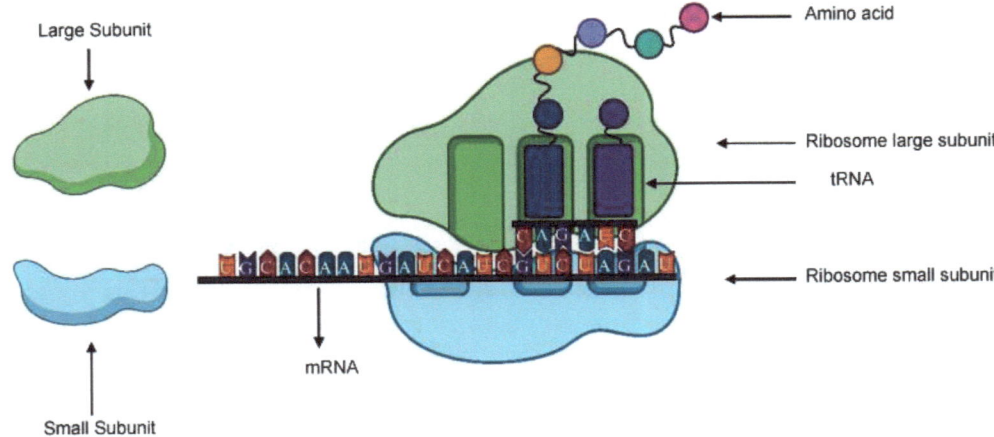

Fig. (1). Structure of the ribosome.

The small subunit consists of the regions, which are the head, platform, and base, whereas the larger subunit consists of the stalk, ridge, and central protuberance. When associated, the head of the small subunit and the central protuberance of the large subunit face each other. The small subunit functions to bind the mRNA and tRNA. The structure of ribosomes contains three key sites for tRNA binding, which are critical for the translation process:

- **A site**: This site is where the incoming A-tRNA binds, bringing the appropriate amino acid for incorporation into the growing polypeptide chain.
- **P site**: The P site holds the tRNA that carries the growing polypeptide chain, facilitating the transfer of the polypeptide to the amino acid at the A site.
- **E site (Exit site)**: This site is where the empty tRNA exits the ribosome after its amino acid has been added to the polypeptide chain.

The coordination of these sites ensures the accurate translation of the mRNA sequence into a functional protein, with the ribosome moving along the mRNA strand, reading codons, and facilitating the addition of amino acids in the correct order. RPs are interspersed throughout the rRNA, stabilizing the ribosome's structure and facilitating its function. The specific interactions between RPs and rRNA are essential for the assembly of the ribosome and its activity during translation. The assembly of ribosomes occurs in the nucleolus of eukaryotic cells, where newly synthesized rRNA combines with RPs to form the subunits, which are then transported to the cytoplasm for protein synthesis. The structural integrity of ribosomes is also significant in the context of Ribosome-Inactivating Proteins (RIPs), such as ricin and Shiga toxin [9]. These proteins can bind to specific RIPs, disrupting the ribosome's function and inhibiting protein synthesis. The binding of

RIPs to the C-terminal regions of P proteins is crucial for their toxic effects, highlighting the vulnerability of ribosomes to certain toxins and the importance of their structure in cellular health. Ribosomes can be visualized using various microscopic techniques, such as electron microscopy. Under an electron microscope, ribosomes appear as small, dense particles scattered throughout the cytoplasm or attached to the RER, forming the rough ER. The size of ribosomes is typically measured using the Svedberg (S) unit, which is a measure of the sedimentation rate during ultracentrifugation. The size of ribosomes varies between prokaryotic and eukaryotic cells, and the size ranges between 20 to 30 nm in diameter [10].

Composition of the Ribosome

Ribosomes are complex molecular machines essential for protein synthesis in all living cells. It is one of the most crucial molecular machineries of cellular life, which assembles amino acids into peptides at the direction of messenger RNAs. The chemical composition of ribosomes consists primarily of rRNA and RP (Fig. 2). The rRNA provides structural support and catalyzes the formation of peptide bonds, while the RPs aid in the proper folding of rRNA and provide binding sites for mRNA and transfer RNA (tRNA) molecules. The two major subunits of a ribosome, along with dozens of RPs and three to four rRNAs, make up a ribosome. The two distinct subunits are a larger subunit and a smaller subunit with densities of 50S and 30S. The 30S subunit has a 16S rRNA with a length of 1540 nucleotides, whereas the 50S subunit has a 5S rRNA with a length of 120 nucleotides and a 23S rRNA with a length of 2900 nucleotides. In the large subunit, 5S rRNA is situated on the central protuberance, while the 3'-end of 23S rRNA is situated below the stalk. The rRNA molecules are partly exposed on the surface. The large subunit contains the P site, A site, and E site. Peptide bond formation occurs at the central protuberance [11]. The 3'-end of 16S rRNA in *E. coli* contains pyrimidine-rich (UCCU) short sequences known as the "SD sequence". The 5' untranslated segment of mRNA has purine-rich short sequences (AGGA).

The initial binding of mRNA to the ribosome occurs by complementary base pairing between these sequences at the 3'-end of 16S rRNA and the 5' untranslated region of mRNA. But in eukaryotes, the binding of mRNA to ribosomes is facilitated by the cap structure at the 5'-end of mRNA. In *E. coli*, they contain 34% protein and 66% RNA, while in eukaryotes, they contain 40% protein and 60% RNA. During protein synthesis, the two subunits join to create a full 70S ribosome. In eukaryotes, the ribosomal subunits possess the 60S and 40S subunits that together make up 80S ribosomes. The 40S subunit in most eukaryotes contains an 18S rRNA with a length of 1900 nucleotides; the 60S subunit contains a 5S rRNA with a length of 120 nucleotides, a 5.8S rRNA with a length of 160 nucleotides, and a 28S rRNA with a length of 4700 nucleotides.

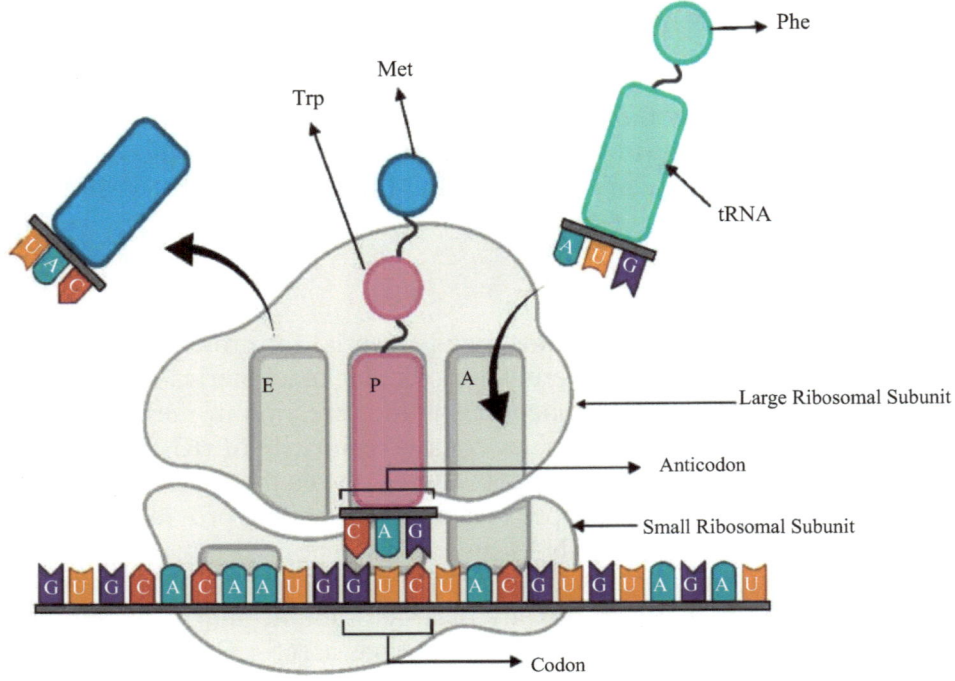

Fig. (2). Composition of the ribosome.

During protein synthesis, the two subunits join to create a full 80S ribosome. In fast-growing cells, rRNAs make up around 80% of all RNAs, while r-proteins make up about 30% of the entire proteome. As a result, a significant portion of the cell's energy budget is required for ribosome manufacturing [12].

FUNCTION OF RIBOSOME

All creatures depend on the ribosome, a ribonucleoprotein complex, for the expression of their genes. Its main role as a molecular machine is to translate the genetic information contained in messenger RNA (mRNA) into certain protein sequences. Through this translation process, the ribosome becomes similar to a molecular CPU by translating a four-letter nucleic acid alphabet into a twenty-letter protein alphabet. Because the residues in its active site are the same in all creatures, rRNA, which is highly conserved across species, is the gold standard for identifying evolutionary relationships. 60% of total transcription is accounted for by rRNA synthesis, whereas ribosome assembly and translation components use up about 80% of a cell's metabolic energy [13, 14]. The process of translation through ribosomes happens in three stages: Initiation, elongation, and termination.

Ribosome Biogenesis (Ribi)

Ribi in Eukaryotic Cells

Ribi in eukaryotic cells is an intricate and highly coordinated process, predominantly occurring within the nucleolus. The process begins with the transcription of rDNA by RNA polymerase (Pol) I, producing a large precursor rRNA (pre-rRNA), and known as 47S pre-rRNA. This is a critical early step in ribosome assembly, regulated by key signalling pathways like MYC and mTOR, which adjust ribosome production based on the cell's needs.

The 47S pre-rRNA is then processed into three smaller rRNAs: 18S, 5.8S, and 28S. At the same time, a separate component, the 5S rRNA, is transcribed by RNA poly III outside of the nucleolus. After transcription, these rRNAs undergo post-transcriptional modifications and assembly with RP, which are transcribed by Pol II and synthesized in the cytoplasm. Around 80 RP join these rRNAs to form the basic building blocks of ribosomes. In the early stages of assembly, these components form a large complex called the 90S processome. This structure eventually splits into two immature ribosomal subunits: pre-40S (which contains 18S rRNA and forms the small ribosomal subunit) and pre-60S (which contains 5.8S, 28S, and 5S rRNAs, forming the large subunit). These subunits are further processed in the nucleoplasm and then exported to the cytoplasm, where they undergo final maturation steps to become fully functional ribosomes, ready for protein synthesis.

Ribi in Prokaryotic Cells

Prokaryotes, including *E. coli*, undergo a highly coordinated, multi-step process, which includes rRNA and RP production, processing, and assembly into functional ribosomes, like in eukaryotes. The rRNA operon is first transcribed by RNA pol, producing a polycistronic pre-rRNA that includes transfer RNAs (tRNAs) and the rRNAs 16S, 23S, and 5S. NusA and NusG are two transcription factors that help control the elongation process of rRNA transcription.

Mature 16S, 23S, and 5S rRNAs are produced when ribonucleases cleave and process the pre-rRNA following transcription. While other enzymes, such as RNase P and RNase E, help with trimming and maturation, RNase III is crucial in bacteria's cleavage of the main transcript. Translation and transcription of RP occur simultaneously with rRNA processing. The 30S subunit contains 21 proteins, while the 50S subunit has 33 proteins. These RPs are produced in the cytoplasm and then imported into the nucleoid region for assembly. They are crucial for the folding and stability of rRNA.

The 16S rRNA and small subunit proteins that make up the 30S subunit are the first to assemble into ribosomes. In a similar manner, big subunit proteins and the 23S and 5S rRNAs combine to form the 50S subunit. In order to assist future rRNA folding and protein binding, specific RPs bind first during the sequential assembly process. Following their formation, the two subunits go through a final maturation process wherein chaperones, assembly factors, and enzymes such as GTPases assist in the proper folding and subunit interaction. This guarantees the formation of functioning 70S ribosomes, which are able to convert mRNA into proteins. The availability of nutrients and the quantity of cellular energy control the process of ribosome In contrast, ribosome synthesis is downregulated in response to stress or dietary constraint.

A number of GTPases, such as Era, RsgA, and ObgE, are involved in the maturation and assembly of ribosomal subunits, which guarantees appropriate RP and rRNA folding [15]. In multicellular organisms, Ribi is more complex than in prokaryotes. Hundreds of additional proteins are involved in human Ribi, with some unique to higher organisms and not found in single-cell organisms. These additional proteins may allow for more sophisticated regulation, enabling cells to adapt ribosome production in response to varying developmental stages and environmental conditions [16]. Ribi is closely linked to therapeutic resistance in cancer. Resistance to chemotherapy and radiation can be enhanced by RP and components involved in rRNA processing. For instance, some RP help cancer cells survive high-dose chemotherapy, while proteins like nucleolin can influence DNA repair to make drugs more sensitive to treatment. NSUN5-induced rRNA modifications enable cancer cells to withstand stress. Furthermore, Ribi factors, such as NOB1, have been shown to be involved in therapeutic resistance, as evidenced by their ability to increase radio sensitivity and suppress tumor development [17].

Role of the Ribosome in Translation – Initiation

Homologous GTPases, like as EF-Tu, IF2, and EF-G, are crucial for translation during its initiation and elongation stages. These three G-proteins, along with elongation factor 4, which appears to be a mimic of EF-G, are among the four closest homologs found by a homology search in Thermus thermophilus. These GTPases engage in homologous interactions with the ribosome, namely with the 50S subunit's Sarcin-Ricin Loop (SRL) and the GTPase-Associated Complex (GAC). These G-proteins bind to the same location on the ribosome as their homologous residues, as shown by structural overlays.

Initially, when the translation process began, the 50S and 30S ribosomal subunits separated. The development of the initiation complex on the 30S subunit before the recruitment of the 50S subunit suggests that the 30S subunit and its decoding core may have an earlier evolutionary origin than the 50S subunit. The steps involved in joining these two subunits include the assembly of initiation factors IF2, IF1, and IF3, the binding of initiator fMet-tRNA (acylated with N-formyl-methionine), and the binding of mRNA to the 30S subunit. During this construction, the ribosome moves dynamically. During initiation and elongation, the 50S subunit spins reversibly relative to the 30S subunit, whereas the 30S subunit's head and beak swivel and tilt in relation to the body.

The fMet-tRNA during initiation and the sequential translocation of tRNAs and mRNA during elongation are both made possible by these movements. The initiation complex can assemble *via* a variety of routes. For example, initiation factors or fMet-tRNA can't be required for an SD sequence on the mRNA to bind directly to the ribosome. The AUG initiation codon is positioned at the P-site of the ribosome by pairing the SD sequence on the mRNA with the Anti-SD (ASD) region close to the 3' end of the 16S rRNA. Through early elongation and the formation of the 50S subunit, this SD-ASD interaction continues.

mRNAs without an SD sequence can still be translated, but they will likely require the assembly of IF1, IF2, IF3, and fMet-tRNA^{fMET} in advance. Preventing the premature assembly of the initiation complex is achieved by IF1 through its blockage of the A-site and assistance in the orientation of IF2 and IF3. The N-Terminal Domain (NTD) and C-Terminal Domain (CTD) of IF3 span the P-site of the ribosome to orient and stabilize fMet-tRNA. The 30S subunit and the IF3 CTD dynamically interact to help position the P-site tRNA. The assembly of the initiation complex promotes the 30S subunit's swivel, and the subsequent assembly of the 50S subunit results in the rotation of the 30S and 50S subunits, stabilizing the P-site tRNA for the formation of peptide bonds. Upon the release of initiation factors, the A-site becomes ready to accept aa-tRNA bound to EF-Tu, allowing peptide bond synthesis to begin.

Role of the ribosome in Translation - Elongation

The ribosome functions as a molecular motor during the elongation phase of translation. It is propelled by a thermal ratchet mechanism that has a step length of roughly 14 angstroms, which is equivalent to the translocation of one codon on mRNA. The ribosome translocates tRNA *via* the A-site, P-site, and E-site in order to read the mRNA in a 5' to 3' orientation. The 30S subunit's reversible rotation in relation to the 50S subunit and the 16S rRNA head's ability to rotate both aid in this process. The force produced by these motions is roughly 13 pN, which places the ribosome in the category of a weak molecular machine. In this process, the elongation factors EF-Tu and EF-G are essential. Elongation factors EF-G and EF-Tu are essential to this process. A-tRNA is delivered by EF-Tu to the ribosome, where it is proofread for accuracy before being incorporated into the A-site to synthesize peptide bonds. By moving the ribosome ahead in preparation for the subsequent cycle of elongation, EF-G promotes translocation. The tRNA molecules fold into an L shape and function as stiff adapters during translation. The 3'-CCA end and anticodon loop of tRNA molecules are crucial for amino acid addition and codon recognition. Minor distortions occur as tRNA navigates the ribosome. The accuracy of translation is ensured by a codon-anticodon latch, formed by interactions between rRNA and RPs, which prevents incorrect tRNA pairing by not allowing the latch to close when an error occurs. Large rotations from the A/T-site to the A/A-site are required for tRNA accommodation in order to reject both near-cognate and non-cognate tRNAs.

The ribosome's P Transferase Center (PTC) catalyzes the synthesis of peptide bonds by positioning the 3' ends of tRNA for P-transfer. This center facilitates bond formation through a proton shuttle mechanism. The ribosome's thermal ratchet mechanism propels translocation, which includes rotation between the 30S and 50S subunits as well as swiveling and tilting of the 30S head and beak. These movements are propelled forward by EF-G·GTP hydrolysis, which guarantees correct tRNA and mRNA transport. The dynamics of the ribosome are controlled during this process by intersubunit bridges, such as h44 and h69, that connect the 30S and 50S subunits. Pawls, such as codon-anticodon interactions and 3'-CCA tRNA attachments, regulate the ribosome's translocation by preventing backward slippage and preserving forward motion. To guarantee precise and effective translocation during translation, smaller kink-turns in the rRNA, such as Kt-42 and Kt-38, further control the ribosome's dynamics in addition to greater ratcheting and swiveling motions [18, 19].

Role of the Ribosome in Translation – Termination

The ternary complex plays a critical role in the interactions between eukaryotic release factors eRF1 and eRF3, which orchestrate the termination of protein synthesis. Through its amino-terminal domain (N), eRF1 may identify stop codons in the ribosomal A site. Meanwhile, its middle domain (M) has a GGQ motif that causes the P-site P-tRNA to release the developing polypeptide. eRF1's carboxy-terminal domain (C) has a mini-domain that affects stop codon selectivity and interacts with both eRF3 and ABCE1.

The three stop codons in the canonical genetic code are UAG, UAA, and UGA. The sequence context, in particular the existence of a purine at the +4 position and, in the case of a pyrimidine at the +4 nucleotide, a purine at the +5 position, affects the translation termination efficiency. Some creatures, though, show variations to this universal code. For example, UGA is reallocated to encode an amino acid (sense codon) in some ciliate protists, green algae, and diplomonads, but UAG and UAA continue to act as stop codons. In other instances, UGA is the only stop codon, and UAG and UAA are reallocated as sense codons. Additionally, in some organisms, like Condylostoma magnum and a trypanosomatid from the Blastocrithidia clade, all three stop codons act as sense codons in internal positions but still function as termination signals near the 3'-end of mRNA.

When a stop codon is recognized, eRF3, a GTPase connected to translation factors like EF-Tu, interacts with eRF1 and hydrolyzes GTP. Switch I and Switch II components in eRF3's G domain control how it interacts with eRF1 and the ribosome. There are two different types of eRF3: eRF3b, which is only expressed in the brain, and eRF3a, which is found everywhere. For precise stop codon recognition and termination, GTP binding must be stabilized by the complex made up of eRF1 and eRF3.

In order to recognize stop codons, eRF1 must interact with the ribosomal decoding center. It is also sensitive to the context of the nucleotide sequence, especially if a +4 purine is present, as this increases termination efficiency. eRF1 changes its conformation in the post-hydrolysis stage, placing its GGQ motif in the ribosome's PTC to sever the ester bond and release the developing protein. The structural framework for eRF1/eRF3-mediated termination has been established through various studies, including cryo-electron microscopy, providing insights into the conformational changes that enable eRF1 to recognize stop codons and activate the catalytic release of the polypeptide [20].

Ribosome Recycling

The highly conserved protein ABCE1, which is essential for dissociating stalled elongation complexes and post-termination complexes (post-TCs), is involved in ribosome recycling following translation termination. The way that ABCE1 and eRF1 interact at the ribosomal A site determines how recycling works. The termination and ribosome recycling phases of protein synthesis depend on this connection. Two Nucleotide-Binding Domains (NBDs) and an iron-sulfur (FeS) domain are present in ABCE1. Through cycles of ATP binding and hydrolysis, ABCE1's ATPase activity produces conformational changes that result in ribosomal subunit splitting. Notably, in order for ABCE1 to attach itself to the ribosome, eRF3 needs to separate.

Though the specifics of how ABCE1 mediates this process are still unknown, during recycling, ATP hydrolysis is necessary for the release of ABCE1 from the small ribosomal subunit following the dissociation of the large subunit. There is evidence, for example, that eRF1-bound ribosomes markedly increase the ATPase activity of ABCE1, but this connection is not as strong with the 40S subunit alone. Starter factors, including eIF1A, eIF1, and eIF3, help displace remaining components on the 40S subunit, facilitating the release of mRNA and deacylated tRNA from the subunit. Furthermore, though to differing degrees of efficiency, ligatin (eIF2D) and proteins like MCT-1 and DENR can also mediate this release.

Regulation of Termination

Ribosomes play a central role in the regulation of translation termination, ensuring the precise synthesis of proteins by interpreting mRNA codons. The end of translation is indicated by stop codons (UAA, UAG, and UGA) in mRNA, which are recognized by ribosomes. To stop translation, the ribosome enlists release factors, including eRF1 and eRF3. Premature or ineffective stop codon recognition may lead to abnormal protein synthesis, which is why accurate recognition and termination are crucial [21]. The nascent polypeptide is released from the tRNA in the P site by ribosomes working with release factors. Because the big ribosomal subunit's ribosomal peptidyl transferase center catalyzes this activity, ribosomes are necessary for the last stage of translation [22].

In cases where stop codon read-through occurs (either naturally or in response to specific viral or eukaryotic signals), the ribosome continues elongating the polypeptide chain beyond the stop codon. This flexibility allows the production of extended protein isoforms, but it must be carefully regulated to avoid detrimental effects, such as non-functional proteins [22, 23]. Ribosomes need to be recycled for further translation cycles after termination. The process of ribosome recycling is crucial for preserving the effectiveness of cellular translation because it releases tRNAs and mRNA when ribosomal subunits separate, which is caused by factors such as ABCE1. At stop codons, some emerging peptides can cause ribosome stalling, which can impact the expression of genes downstream. Therefore, ribosomes not only catalyze peptide synthesis but also regulate translation by reacting to the structure of the developing peptide [22, 24].

DYSFUNCTION AND DISEASE OF THE RIBOSOME

Dysfunctions in ribosomes, collectively referred to as "ribosomopathies," can lead to a range of diseases due to their critical role in protein synthesis. These dysfunctions may arise from mutations in RPs, rRNA, or factors involved in Ribi and function. The following are a few noteworthy instances of ribosome dysfunction-related diseases:

Diamond-Blackfan Anemia (DBA): DBA was first identified by Josephs in 1936 in a review of anemia in infancy and children. Diamond and Blackfan later classified DBA as a congenital hypoplastic anemia. Diamond-Blackfan Mutations in RP genes, including RPS19, RPL5, and RPL11, result in abnormalities in Ribi, which is the cause of anemia. DBA is characterized by a failure to produce sufficient red blood cells (anemia), craniofacial abnormalities, and an increased risk of cancer. Anemia is the outcome of erythroid progenitor cells' inability to divide and proliferate. Currently, Diamond-Blackfan Anemia (DBA) treatments begin with transfusions and corticosteroids. For steroid-resistant patients, Hematopoietic Stem Cell Transplantation (HSCT) is used, especially when sibling donors are available. Novel therapeutic approaches include L-leucine for mRNA translation, Sotatercept for erythropoiesis, and SMER28 for autophagy modulation, among others.

Shwachman-Diamond Syndrome (SDS): Similar to DBA, ribosome assembly is impaired in SDS, which is brought on by mutations in the SBDS gene. Shwachman *et al.* initially discovered SDS in 1964 in a group of five children who were being monitored in a Harvard University cystic fibrosis clinic. Skeletal deformities, pancreatic insufficiency, and bone marrow failure are caused by ribosome dysfunction. It has been discovered that SDS stem and multipotent progenitors have active TGF-β signalling, which inhibits hematopoiesis. In SDS bone marrow cells, inhibition of this system with drugs such as AVID200 and SD208 has promoted the development of hematopoietic colonies. Furthermore, 50% of individuals with SDS have Premature Termination Codons (PTCs) due to mutations such as c.183–184 TA > CT in the SBDS gene. It has been demonstrated that ataluren, a medication that permits the ribosome to evade PTCs, restores SBDS expression in patient myeloid cells, improving differentiation and lowering apoptosis in patient-derived PBMCs.

Myelodysplastic syndrome (MDS) with chromosome 5q deletion or 5q-Syndrome: 5q-Syndrome is a subtype of myelodysplastic syndrome (MDS). The deletion of part of chromosome 5q, which includes the RPS14 gene, disrupts Ribi. Van Den Berghe initially reported the 5q− condition in 1974 with three cases. This results in dysplastic megakaryocytes and macrocytic anemia because erythroid progenitor cell activity is compromised. The 5q syndrome is managed with a range of treatments, including red blood cell transfusions, recombinant erythropoietin, thalidomide, and retinoid injections. Lenalidomide, a derivative of thalidomide, is currently the gold standard for managing transfusion-dependent patients with 5q syndrome. It works by promoting p53 degradation, which is typically overactivated in these patients due to nucleolar stress. Lenalidomide achieves this by inhibiting the auto-ubiquitination of MDM2 *via* decreased expression of PPP2Acα, which is linked to lenalidomide resistance in some patients.

Treacher Collins Syndrome (TDS): British ophthalmologist Edward Treacher Collins outlined the key components of a condition in 1900. TDS results from mutations in genes related to ribosome synthesis in the early stages of development, such as TCOF1, POLR1C, or POLR1D. Hearing loss, anomalies in the eyes, and craniofacial deformities are caused by defective Ribi in neural crest cells. TCS is currently managed by addressing each patient's specific clinical needs. In terms of therapeutic alternatives, proteasome inhibitors such as MG132 and Bortezomib, used in multiple myeloma, have shown potential in reducing craniofacial malformations in zebrafish models of TCS. These inhibitors help prevent the degradation of the cellular nucleic acid-binding protein (Cnbp), which is essential for proper craniofacial development.

Cartilage-hair Hypoplasia - Anauxetic Dysplasia (CHH-AD): McKusick *et al.* originally identified CHH in numerous Amish families in 1965 as a type of short-limbed dwarfism brought on by skeletal dysplasia. Mutations in the RMRP gene, which encodes an RNA component involved in rRNA processing, result in CHH. This defect hinders the proliferation of cells, which increases the risk of cancer, anemia, immunological deficits, and short stature. Mutations in the DKC1 gene cause X-linked Dyskeratosis Congenita (X-DC), a disorder that impairs telomerase activity and ribosome synthesis. This results in aberrant skin pigmentation, accelerated cellular aging, bone marrow failure, and an increased risk of cancer. Neutropenia in CHH is typically managed with granulocyte colony-stimulating factor (G-CSF) injections to stimulate the production of neutrophils, while bone marrow transplantation is used to correct immunodeficiency. However, these treatments do not address the chondrodysplasia characteristic of CHH. In terms of therapeutic perspectives, recombinant growth hormone injections have been used experimentally in a CHH patient to increase height, though this approach carries potential risks and requires further validation through larger studies. Additionally, XAV939, an inhibitor of Wnt/β-catenin signalling, has shown promise in zebrafish models for alleviating chondrodysplasia and improving bone mineralization.

North American Indian Childhood Cirrhosis (NAIC): Mutations in the *CIRH1A* gene are associated with NAIC, a condition that causes progressive liver cirrhosis and liver failure in children by compromising ribosome assembly and hepatocyte function. Ribosomal dysfunction in cancer is facilitated by enhanced protein synthesis, which is brought on by mutations or dysregulation of RPs such as RPL5 and RPL11. Furthermore, p53, a tumor suppressor protein, can be activated by abnormal Ribi, establishing a connection between ribosome dysfunction and oncogenesis.

X-linked- Dyskeratosis Congenita (DKC): Oral leukoplakia, nail dystrophy, and aberrant skin pigmentation are the hallmarks of dyskeratosis congenita (DKC). DKC was first identified in 1910 and subsequently connected to bone marrow failure. Cytopenia affects over 90% of patients, and bone marrow failure is the main cause of death. Patients with DKC are at a heightened risk for solid tumors, leukemia, and pulmonary fibrosis. Abnormalities in the DKC1 gene, which affects rRNA modification, produce X-linked DKC, a condition caused by abnormalities in components of the telomerase complex. The sole curative option is still stem cell transplantation, yet side effects, including lung fibrosis, make it less successful. In DKC, cytopenia is treated similarly to DBA, with chronic transfusions and HSCT for transfusion-dependent patients. Due to the pleiotropic nature of DKC, patients are monitored across several medical specialties. Emerging treatments for DKC, such as eltrombopag (EPAG) and Danazol, are being explored to improve hematologic outcomes. Danazol, in particular, has shown the ability to up-regulate telomerase reverse transcriptase (TERT), leading to telomere elongation and improving blood counts in DKC patients.

Bowen-Conradi Syndrome (BCS): A rare autosomal recessive developmental condition (OMIM #211180). Significant prenatal and postnatal psychomotor abnormalities, growth retardation, microcephaly, micrognathia (a small jaw), and congenital vertical talus are its defining characteristics. Usually, these symptoms lead to an early demise. The EMG1 protein has a single amino acid alteration that results in BCS: at position 86, glycine takes the place of aspartate. Nucleolar RNA methyltransferase EMG1 is essential for the small Ribi. This mutation interferes with the processing of 18S rRNA by preventing EMG1 from accessing its pre-ribosomal binding sites in the nucleus. As a result of the severity and fatality linked to BCS, there aren't any medications available right now to address symptoms.

Ribosomal Haploinsufficiency: Mutations in the genes producing RPs result in ribosomal haploinsufficiency, which causes specific symptoms such as anemia, developmental abnormalities, and congenital abnormalities like craniofacial deformities. Even though ribosomes play a widely recognized role in protein synthesis, ribosomal malfunction can affect specific cell types, notably those that are rapidly proliferating, like erythroid cells. Disrupted Ribi, which causes an increase of free RPs that bind to MDM2, a p53 repressor, and cause p53 activation, is one of the primary mechanisms of ribosomal haploinsufficiency. This leads to cell cycle halt and apoptosis, which causes anemia. A further theory proposes that the inadequate development of ribosomal subunits causes a delay in the translation of globin genes, resulting in an overabundance of free heme and erythroid cell death, ultimately leading to anemia. The most often mutated gene in DBA is RPS19, the first ribosomal gene connected to human disease; mutations in this gene cause haploinsufficiency. Research has demonstrated that the DBA phenotype, which is observed in hematopoietic progenitor cells with impaired proliferation and differentiation, is replicated when RPS19 expression is knocked down in vitro. Similarly, in vivo models that reflect the essential characteristics of ribosomopathies have been established, including zebrafish and mice. These models are vital for creating new treatment strategies and offer important insights into the biology of ribosomal malfunction [25].

Ribosomes and Cancer: Ribosomopathies, which are disorders caused by defects in Ribi, show an increased susceptibility to cancer despite the early hypo-proliferative phenotype often seen in patients. It is well-known that this contradictory transition from inhibited cell proliferation to an eventual neoplastic hyper-proliferation is known as "Dameshek's riddle." This paradox is caused by aberrations in Ribi, which at first suppress cell proliferation. However, over time, these abnormalities interfere with important tumor-suppressive pathways, such as p53 activation, and encourage the development of cancer by means of translation reprogramming and the generation of "onco-ribosomes."

RP Mutations and Cancer

Mutations in RPs, such as RPL5, RPL11, RPL10, and RPS15, are increasingly recognized in both hematologic and solid tumors. These mutations frequently alter mRNA translation and impede nucleolar stress responses, tipping the scales from normal cell development to malignant transformation. RPL5 mutations have been linked to multiple myeloma, melanoma, and breast cancer, suggesting a broad function for RPL5 in the genesis of cancer. RPL10 is frequently mutated in pediatric T-cell acute lymphoblastic leukemia (T-ALL); this mutation affects proper translation and contributes to oncogenesis. RPS15 RP is significantly mutated in Chronic Lymphocytic Leukemia (CLL), that are aggressive, and RPS15A is implicated in a number of malignancies, including liver, stomach, and lung cancers. In addition to impairing ribosome function, these mutations directly cause cancer by increasing the translation of oncogenic proteins and blocking p53-mediated tumor-suppressive pathways.

Onco-Ribosomes and Cancer Development

Research on the development of "onco-ribosomes," in which growth-promoting and survival proteins are preferentially translated by ribosomes with changed composition and function, is fascinating. Furthermore, some RPs' extra-ribosomal activities—such as those related to stress responses and transcriptional regulation—have an impact on the development of cancer. Many malignancies have increased Pol I transcription, which results in hyperactive ribosome synthesis. This promotes greater protein synthesis, which causes cancer cells to develop a "translation addiction" in which they depend on higher protein synthesis for growth and survival [26-29].

DIAGNOSTIC AND THERAPEUTIC PROPERTY OF RIBOSOMES

Some therapies for genetic diseases involve targeting ribosomes to alter their function, particularly in promoting stop codon read-through. For instance, aminoglycosides and ataluren target ribosomes to allow them to bypass premature stop codons, offering a potential treatment for genetic disorders like cystic fibrosis [25].

The discovery of multifunctional or moonlighting proteins (MLPs) in prokaryotes and eukaryotes has greatly expanded our understanding of protein functionality beyond their primary roles. Wool proposed in 1996 that many RP could have additional roles beyond translation. These MLPs have a wide range of activities in eukaryotic cells, such as DNA repair, translation self-regulation, and blocking MDM2-mediated p53 ubiquitination and destruction. These proteins also have the ability to stop intron removal and block the splicing of their own RNA transcripts. Additionally, they bind to the HDM2 protein, which is a crucial negative regulator of the tumor suppressor p53, and block it while also encouraging p53 translation following DNA damage. Additional roles include decreasing their own mRNA half-life, binding to DNA to promote Rep helicase protein unwinding, and taking part in the NSP-Interacting Kinase 1 (NIK1) receptor-mediated defense pathway to guard against the Gemini virus substrate. They may also serve as NIK1's binding partners.

RP as Anti-microbial

An important discovery is that several RPs also function as antimicrobial peptides (AMPs), offering protection against a wide range of pathogens. For example, in 1999, Pütsep *et al.* found the first RP, uL1 from Helicobacter pylori, which contains two amphipathic α-helices connected by a hinge. The N-terminal region of RP uL1 from *H. pylori* forms a well-structured amphipathic helix, a feature not observed in other bacterial counterparts. They synthesized a peptide based on this N-terminal segment, known as Hp (2–20), which includes only the first α-helix. This peptide exhibited antimicrobial activity against *E. coli* and Bacillus megaterium, including enterohemorrhagic strains of *E. coli* responsible for severe foodborne illnesses, such as bloody diarrhea and potentially life-threatening Hemolytic Uremic Syndrome (HUS).

Building on Pütsep's work, further studies have explored the homology of RPs across various organisms, shedding light on the conserved nature of their antimicrobial activity. In 2010, Khairulina *et al.* identified a homology between the eukaryotic RP uS15 and the prokaryotic RP uS19. Later, Qu *et al.* demonstrated that RP uS15 from *Branchiostoma japonicum* (amphioxus) displayed antimicrobial activity against pathogens such as *Aeromonas hydrophila, E. coli, Staphylococcus aureus,* and *Bacillus subtilis*, pathogens known to cause food poisoning, diarrhea, and skin infections. RPs derived from Lactic Acid Bacteria (LAB), such as RP bS21 from *Lactobacillus sakei* and RP bL36 from Pediococcus acidilactici, have also shown antimicrobial activity against critical foodborne pathogens like *Listeria monocytogenes* and *Escherichia coli*. This widespread conservation of antimicrobial activity among RPs suggests these functions may have ancient evolutionary origins. As a result, synthetic RPs based on these conserved core regions have shown great promise as potent antimicrobial agents.

Therapeutic Targeting of Ribi and mTORC1 in Cancer

Nowadays, targeting the mTORC1 pathway and Ribi has emerged as a viable treatment approach. Tumor growth has been slowed by inhibiting mTORC1, which is a crucial regulator of Ribi and protein synthesis. One way to do this is by using medications such as rapamycin. On the other hand, compensatory activation of other pathways frequently results in resistance to mTOR inhibitors. To overcome resistance and improve therapeutic efficacy, combination treatments targeting several pathways (*e.g.*, PI3K/Akt and Myc pathways) are presently under investigation.

RECENT TRENDS

Novel targets that stabilize mTOR mRNA, including LARP1, have the potential to open up new treatment options. Furthermore, in order to develop more potent treatments that interfere with the metabolic and biosynthetic requirements of cancer cells, it is imperative to comprehend the molecular role that Ribi plays in cancer [30].

Advancements in Ribosome Profiling and Cancer

Ribosome profiling enables the detection of ribosome positions on mRNAs, revealing critical information on translation initiation, elongation, termination, and ribosome stalling. Its benefit is that it makes it possible for researchers to examine translational events with never-before-seen accuracy, including non-AUG start, stop codon read-through, and the translation of previously thought-to-be untranslated small ORFs and non-coding RNAs.

The study of translation in disease and therapy has been revolutionized by ribosome profiling and its advanced techniques, including Ribosome-Nascen--Chain Complex Sequencing (RNC-seq) and Single-Cell Ribosome Sequencing (scRibo-seq). Whereas scRibo-seq delivers single-cell resolution insights into cellular heterogeneity, RNC-seq offers a holistic picture by isolating full-length translating mRNAs, assisting in the discovery of RNA splice variants and alternative ORFs. These techniques have proven invaluable in the study of cancer, exposing translational changes in diseases like as leukemia, medulloblastoma, glioblastoma, and Triple-Negative Breast Cancer (TNBC). For example, ribosome profiling has revealed the involvement of circRNA translational regulation in glioblastoma and the relevance of mutations in genes like DDX3X in medulloblastoma.

RPL10 mutations and METTL3 activity have been connected to survival pathways in leukemia, and the peptide CIP2A-BP has been demonstrated to control metastasis in TNBC through the PI3K/AKT pathway. Additionally, ribosome profiling has brought attention to the oncogenic activation of pathways like mTORC1 and the function of RNA methylation in translation regulation in malignancies of the prostate and digestive system. These discoveries not only broaden our knowledge of cancer biology but also point to possible treatment targets for future research. Although the method has drawbacks, ribosome profiling offers precise positional data. Over-accumulation at start and stop codons can result from experiment-induced distortions that disturb ribosome mobility, such as the use of elongation inhibitors like cycloheximide. Furthermore, errors in footprint length and data interpretation may result from nuclease digestion processes, ribosome conformation, and RNA secondary structures.

Ribosome Profiling Reveals Non-Canonical ORF Translation in High-Risk Childhood Medulloblastoma

A study on childhood medulloblastoma highlighted the dysregulation of RNA translation, particularly focusing on non-canonical open reading frames (ORFs). Using ribosome profiling of 32 medulloblastoma samples, researchers identified widespread translation of these non-canonical ORFs. They employed CRISPR-Cas9 screens to uncover the roles of various lncRNA-ORFs and upstream ORFs in promoting cell survival. Notably, the microprotein ASNSD1-uORF was found to be upregulated, linked to MYC-family oncogenes, and crucial for cell survival by interacting with the prefoldin-like chaperone complex. This work emphasizes the significance of non-canonical ORF translation in medulloblastoma and suggests the need for further investigation into these ORFs as potential cancer targets [31].

CONCLUSION

Ribosomopathies arise from mutations affecting RPs, rRNAs, and factors involved in Ribi, leading to a spectrum of disorders, including Diamond-Blackfan anemia, Shwachman-Diamond syndrome, and various cancers, which reflect ribosomal dysfunction's impact on cellular proliferation, differentiation, and survival. These disorders reveal a paradox in ribosome biology, as hypo-proliferative defects in early disease stages often transition to hyper-proliferative phenotypes, implicating altered Ribi pathways and defective nucleolar stress responses in oncogenesis. Recent advances in ribosome profiling have expanded the understanding of translation dynamics and the role of non-canonical ORFs, especially in malignancies, revealing potential targets for cancer therapies. Therapies targeting ribosomes, mTORC1, and translation initiation combined with precision techniques like ribosome profiling, RNC-seq, and scRibo-seq provide insights into translation regulation in health and disease. Future research should prioritize optimizing these advanced techniques, developing ribosome-targeted therapies, and investigating the mechanistic role of Ribi in cancer cell metabolism to facilitate new, targeted treatments. Understanding the dual role of ribosomes as both protein synthesizers and regulators of cellular stress responses may offer novel intervention points across ribosomopathies and malignancies, highlighting the importance of investigating both canonical and extra-ribosomal functions of RPs for broader therapeutic application.

ACKNOWLEDGEMENTS

We would like to acknowledge Ms. Kruthika P, Ms. Raksa Arun, Ms. Srisri SatishKartik, and Mr. Sasitharan T for their contribution in proofreading and editing the work.

REFERENCES

[1] STRUCTURE AND FUNCTION OF PROKARYOTIC AND EUKARYOTIC RIBOSOMES. Progress in Biophysics and Molecular Biology. Elsevier 1978; pp. 193-231.
[http://dx.doi.org/10.1016/B978-0-08-020295-2.50007-6]

[2] Myasnikov AG, Simonetti A, Marzi S, Klaholz BP. Structure–function insights into prokaryotic and eukaryotic translation initiation. Curr Opin Struct Biol 2009; 19(3): 300-9.
[http://dx.doi.org/10.1016/j.sbi.2009.04.010] [PMID: 19493673]

[3] Mahoney SJ, Dempsey JM, Blenis J. Cell signaling in protein synthesis ribosome biogenesis and translation initiation and elongation. Prog Mol Biol Transl Sci 2009; 90: 53–107.

[4] Petrov AS, Gulen B, Norris AM, et al. History of the ribosome and the origin of translation. Proc Natl Acad Sci USA 2015; 112(50): 15396-401.
[http://dx.doi.org/10.1073/pnas.1509761112] [PMID: 26621738]

[5] Fox GE. Origins and Early Evolution of the Ribosome. Evolution of the Protein Synthesis Machinery and Its Regulation. Cham: Springer International Publishing 2016; pp. 31-60.
[http://dx.doi.org/10.1007/978-3-319-39468-8_3]

[6] Pimentel T, Marcelino J, Ricardo F, Soares AMVM, Calado R. Bacterial communities 16S rDNA fingerprinting as a potential tracing tool for cultured seabass Dicentrarchus labrax. Sci Rep 2017; 7(1): 11862.
[http://dx.doi.org/10.1038/s41598-017-11552-y] [PMID: 28928412]

[7] Severgnini M, Cremonesi P, Consolandi C, Caredda G, De Bellis G, Castiglioni B. ORMA: a tool for identification of species-specific variations in 16S rRNA gene and oligonucleotides design. Nucleic Acids Res 2009; 37(16): e109-9.
[http://dx.doi.org/10.1093/nar/gkp499] [PMID: 19531738]

[8] Raveendran M, Leach AR, Hopes T, Aspden JL, Actis P. Ribosome Fingerprinting with a Solid-State Nanopore. ACS Sens 2020; 5(11): 3533-9.
[http://dx.doi.org/10.1021/acssensors.0c01642] [PMID: 33111519]

[9] Wilson DN, Doudna Cate JH. The structure and function of the eukaryotic ribosome. Cold Spring Harb Perspect Biol 2012; 4(5): a011536-6.
[http://dx.doi.org/10.1101/cshperspect.a011536] [PMID: 22550233]

[10] Korostelev A, Ermolenko DN, Noller HF. Structural dynamics of the ribosome. Curr Opin Chem Biol 2008; 12(6): 674-83.
[http://dx.doi.org/10.1016/j.cbpa.2008.08.037] [PMID: 18848900]

[11] Yusupova G, Yusupov M. High-resolution structure of the eukaryotic 80S ribosome. Annu Rev Biochem 2014; 83(1): 467-86.
[http://dx.doi.org/10.1146/annurev-biochem-060713-035445] [PMID: 24580643]

[12] Siodmak A, Martinez-Seidel F, Rayapuram N, *et al.* Dynamics of ribosome composition and ribosomal protein phosphorylation in immune signaling in *Arabidopsis thaliana*. Nucleic Acids Res 2023; 51(21): 11876-92.
[http://dx.doi.org/10.1093/nar/gkad827] [PMID: 37823590]

[13] Sanbonmatsu KY, Blanchard SC, Whitford PC. 9.5 Dynamics of Very Large Systems: The Ribosome. Comprehensive Biophysics. Elsevier 2012; pp. 76-85.
[http://dx.doi.org/10.1016/B978-0-12-374920-8.00907-3]

[14] Bhagavan NV, Ha CE. RNA and Protein Synthesis. Essentials of Medical Biochemistry. Elsevier 2011; pp. 301-20.
[http://dx.doi.org/10.1016/B978-0-12-095461-2.00023-0]

[15] Panse VG. Getting ready to translate: cytoplasmic maturation of eukaryotic ribosomes. Chimia (Aarau) 2011; 65(10): 765-9.
[http://dx.doi.org/10.2533/chimia.2011.765] [PMID: 22054128]

[16] Ni C, Buszczak M. Ribosome biogenesis and function in development and disease. Development 2023; 150(5): dev201187.
[http://dx.doi.org/10.1242/dev.201187] [PMID: 36897354]

[17] Elhamamsy AR, Metge BJ, Alsheikh HA, Shevde LA, Samant RS. Ribosome Biogenesis: A Central Player in Cancer Metastasis and Therapeutic Resistance. Cancer Res 2022; 82(13): 2344-53.
[http://dx.doi.org/10.1158/0008-5472.CAN-21-4087] [PMID: 35303060]

[18] Hurtado-Rios JJ, Carrasco-Navarro U, Almanza-Pérez JC, Ponce-Alquicira E. Ribosomes: The New Role of Ribosomal Proteins as Natural Antimicrobials. Int J Mol Sci 2022; 23(16): 9123.
[http://dx.doi.org/10.3390/ijms23169123] [PMID: 36012387]

[19] Opron K, Burton ZF. Ribosome Structure, Function, and Early Evolution. Int J Mol Sci 2018; 20(1): 40.
[http://dx.doi.org/10.3390/ijms20010040] [PMID: 30583477]

[20] Hellen CUT. Translation Termination and Ribosome Recycling in Eukaryotes. Cold Spring Harb Perspect Biol 2018; 10(10): a032656.
[http://dx.doi.org/10.1101/cshperspect.a032656] [PMID: 29735640]

[21] He F, Jacobson A. Nonsense-mediated mRNA decay: Degradation of defective transcripts is only part of the story. Annu Rev Genet 2015; 49(1): 339-66.
[http://dx.doi.org/10.1146/annurev-genet-112414-054639] [PMID: 26436458]

[22] Shoemaker CJ, Green R. Kinetic analysis reveals the ordered coupling of translation termination and ribosome recycling in yeast. Proc Natl Acad Sci USA 2011; 108(51): E1392-8.
[http://dx.doi.org/10.1073/pnas.1113956108] [PMID: 22143755]

[23] Houck-Loomis B, Durney MA, Salguero C, *et al.* An equilibrium-dependent retroviral mRNA switch regulates translational recoding. Nature 2011; 480(7378): 561-4.
[http://dx.doi.org/10.1038/nature10657] [PMID: 22121021]

[24] Wang Z, Sachs MS. Ribosome stalling is responsible for arginine-specific translational attenuation in *Neurospora crassa.* Mol Cell Biol 1997; 17(9): 4904-13.
[http://dx.doi.org/10.1128/MCB.17.9.4904] [PMID: 9271370]

[25] Keeling KM, Xue X, Gunn G, Bedwell DM. Therapeutics based on stop codon readthrough. Annu Rev Genomics Hum Genet 2014; 15(1): 371-94.
[http://dx.doi.org/10.1146/annurev-genom-091212-153527] [PMID: 24773318]

[26] Gentilella A, Kozma SC, Thomas G. A liaison between mTOR signalling, ribosome biogenesis and cancer. Biochimica et Biophysica Acta (BBA) -. Gene Regulatory Mechanisms 2015; 1849(7): 812-20.

[27] Narla A, Ebert BL. Ribosomopathies: human disorders of ribosome dysfunction. Blood 2010; 115(16): 3196-205.
[http://dx.doi.org/10.1182/blood-2009-10-178129] [PMID: 20194897]

[28] Kang J, Brajanovski N, Chan KT, Xuan J, Pearson RB, Sanij E. Ribosomal proteins and human diseases: molecular mechanisms and targeted therapy. Signal Transduct Target Ther 2021; 6(1): 323.
[http://dx.doi.org/10.1038/s41392-021-00728-8] [PMID: 34462428]

[29] Orgebin E, Lamoureux F, Isidor B, *et al.* Ribosomopathies: New Therapeutic Perspectives. Cells 2020; 9(9): 2080.
[http://dx.doi.org/10.3390/cells9092080] [PMID: 32932838]

[30] Su D, Ding C, Qiu J, *et al.* Ribosome profiling: a powerful tool in oncological research. Biomark Res 2024; 12(1): 11.
[http://dx.doi.org/10.1186/s40364-024-00562-4] [PMID: 38273337]

[31] Hofman DA, Ruiz-Orera J, Yannuzzi I, *et al.* Translation of non-canonical open reading frames as a cancer cell survival mechanism in childhood medulloblastoma. Mol Cell 2024; 84(2): 261-276.e18.
[http://dx.doi.org/10.1016/j.molcel.2023.12.003] [PMID: 38176414]

CHAPTER 7

Lysosomes: The Cell's Digestive System

B. Surya[1], Faizaan Khan[1], Abhinav Roy[1], Sanjana Dhayalan[1], K. Kumaran[1], Krishnan Anand[2] and K.N. Aruljothi[1,*]

[1] *Department of Genetic Engineering, SRM Institute of Science and Technology, Kattankulathur, Chengalpattu, Tamil Nadu, India*

[2] *Department of Chemical Pathology, School of Pathology, Faculty of Health Sciences, University of the Free State, Bloemfontein, South Africa*

Abstract: Lysosomes serve as essential organelles in eukaryotic cells, facilitating the recycling and degradation of cellular waste through the action of hydrolytic enzymes. This acidic organelle engages with various vesicles, such as phagosomes and endosomes, to dismantle biomolecules encompassing proteins, lipids, and nucleic acids. Recent studies have illuminated the function of lysosomes in autophagy, a process wherein they contribute to the degradation of cellular constituents in response to stress or periods of starvation. Three main types of autophagic processes, macroautophagy, microautophagy, and chaperone-mediated autophagy, use central mechanisms that help the trafficking of their selective cargo. The activity of lysosomes is connected with several diseases: lysosomal storage diseases, Parkinson's disease, and muscular dystrophy are generally caused by either a lack of efficiency of the autophagosome and lysosome fusion or impairment of lysosome digestion. Mechanisms of lysosomal regeneration involve Autophagic Lysosome Reformation (ALR) and Endocytic Lysosome Reformation (ELR), with the presence of essential functions to maintain lysosomal function. Further knowledge about such processes may allow for the creation of therapies for neurodegenerative and muscular disorders characterized by a significant contribution of lysosomal dysfunction.

Keywords: Acidification, Autophagosome, Autophagy, Endocytosis, Endosomes, Hydrolytic Enzymes, Lysosomal Dysfunction, Lysosomes, Lysosomal Membrane Proteins, Neurodegeneration, Phagocytosis, SNARE Proteins..

INTRODUCTION

Commonly known as suicide bags, they are cell organelles in Eukaryotic cells that help break down and recycle waste products from the cell. Lysosomes have hydrolytic enzymes that help break down biomolecules. Some of those hydrolytic

---

[*] Corresponding author K.N. Aruljothi: Department of Genetic Engineering, SRM Institute of Science and Technology, Kattankulathur, Chengalpattu, Tamil Nadu, India; E-mail: aruljotn@srmist.edu.in

K.N. Aruljothi, Prakash Gangadaran, Krishnan Anand, Satish Ramalingam, K. Kumaran & Kruthika Prakash (Ed.)
All rights reserved-© 2026 Bentham Science Publishers

enzymes include Proteases, Lipases, Nucleases, and Glycosidases. Due to these enzymes, the lysosome has an acidic pH of about 5, and the pH of the cytoplasm is neutral (7.2 to 7.8); the lysosome's membrane separates both. We currently understand lysosomes as a very varied group of organelles that vary in location, acidity, functions, and the hybrid organelles they create through fusion/fission and "kiss and run" processes. Endolysosomes, phagolysosomes, autolysosomes, and other terms are used to identify different lysosomal compartments. For example, endolysosomes are intermediate compartments formed during the fusion of endosomes and lysosomes, and they are typically acidic.

Phagolysosomes form when phagosomes fuse with lysosomes to degrade engulfed particles. Autolysosomes result from the fusion of autophagosomes with lysosomes during autophagy. There's also a distinction between acidic lysosomes (like endolysosomes, which actively degrade material) and non-acidic compartments (such as terminal lysosomes), which have completed their digestive function [1, 2].

Discovery of Lysosomes

During the 1950s, de Duve *et al.* from the University of Louvain in Belgium wanted to know how insulin acts in liver cells. They wanted to find where the glucose-6-phosphatase enzyme resides, which regulates blood sugar. They added rat liver debris to distilled water and centrifuged it. They found the enzyme with high activity, but when they purified the enzyme from the cell extracts, they could not precipitate it; however, they could not dissolve it again [3]. The strategy they used was a more advanced differential centrifugation that separates cell constituents based on size and density. They lysed the liver cells, fractioned them in a sucrose medium, and assayed glucose-6-phosphatase activity in the microsomal fraction. Differential centrifugation led them to the conclusion that the activity of the control enzyme, acid phosphatase, was only 10% of what it should have been. One of the scientists kept some of the fractioned samples in the fridge for five days (Fig. 1). On the other hand, when assaying the phosphatase activity, it was found to be equal to that which was expected. De Duve and his co-workers then hypothesized that a membrane could limit the access of an enzyme to its substrate, and resting the samples would give the enzymes time to become accessible. To these membrane-bound sacs containing acid phosphatase along with other lytic enzymes, de Duve gave the name "lysosomes" [4]. An acceptance of the discovery of lysosomes came when Christian de Duve was awarded the Nobel Prize in Physiology or Medicine in 1974. That same year, Alex Novikoff of the University of Vermont paid a visit to de Duve's laboratory. As an expert microscopist, Novikoff took the first electron micrographs of this newly discovered organelle from partially purified samples of the lysosome. The vital role of lysosomes, for instance, was discovered by the work of Werner Strauss and his laboratory [2]. Strauss sought to determine how cells can draw molecules from outside their boundary wall through a so-called endocytosis process. He tagged proteins, followed their movement within cells, and found that protein fragments are engulfed in lysosomes. He concluded that lysosomes break down proteins.

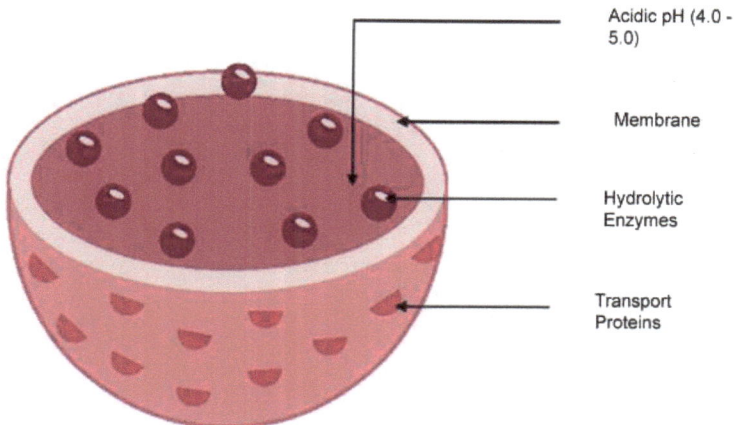

Fig. (1). Lysosome structure showing its membrane, digestive enzymes, and organelles involved in cellular degradation and recycling.

In other experiments, Zanvil Cohn incubated radiolabeled bacteria with macrophages and found that the different components of bacteria, including lipids, carbohydrates, and amino acids, were concentrated in lysosomes. Cohn deduced that lysosomes are the digestive organs of the cell, disposing of materials not produced by the cell itself and cellular products. Thus, lysosomes can be regarded as recycling stations that aid in cellular waste disposal and recycling of material ends [5].

Evolution of Lysosomes

Lysosomes are formed when vesicles from the Golgi Complex fuse with endosomes. Endocytosis is a process in which a section of the plasma membrane pinches off to form vesicles known as endosomes (Fig. **2**) [5].

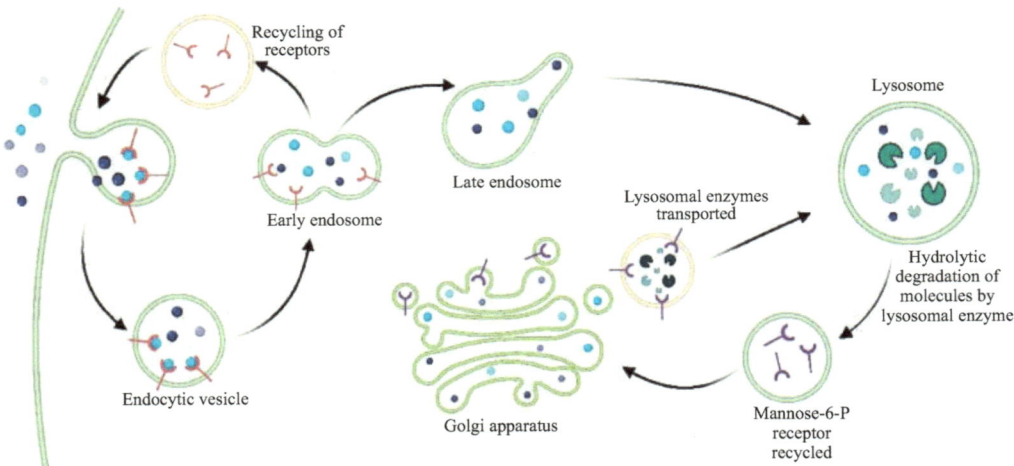

Fig. (2). Formation of lysosomes through the fusion of late endosomes with lysosomal enzymes produced by the Golgi apparatus, leading to the degradation of cellular components.

Endosomes mature in 3 stages:

Early Endosomes

The endosomes formed through endocytosis at the plasma membrane form the early endosomes. They differentiate between substances that are destined for degradation and those that are meant to be recycled back to the plasma membrane.

Recycling Endosomes

The molecules, like cell receptors, are subsequently sent to recycling endosomal pathways and returned to the plasma membrane.

Late Endosomes

Late endosomes are vesicles that have undergone maturation. During this stage, the internal pH drops to around 5, which facilitates the delivery of lysosomal acid hydrolases. Mannose-6-Phosphate (M6P) residues are recognized by mannose--phosphate receptors in the trans-Golgi network, and these receptors help package lysosomal proteins into clathrin-coated vesicles that target these proteins to late endosomes [5]. Once inside the late endosome, the acidic pH triggers the release of M6P receptors from their ligands. This enables the enzymes to be released back into the endosome, while the M6P receptors are recycled back to the Golgi.

FUNCTION

Autophagy (Process of Self-destruction)

Out of the two major degradative pathways in eukaryotes, in addition to the proteasome pathway, is autophagy or "cellular self-eating." Proteasomes can only degrade specific ubiquitinated proteins, so autophagy is used to degrade enormous amounts of proteins, lipids, and glycogen, as well as organelles. Autophagic degradation is not exclusively catabolic, as the resulting monomers are recycled in biosynthetic and energy-generating activities after degradation. This is a critical point to consider. A variety of autophagic pathways have been discovered, all aimed at directing cellular cargo to the lysosomal lumen. This classification of various autophagic mechanisms is based on the methods the cell employs to achieve that [6].

The most prevalent form of autophagy is macroautophagy, in which a phagophore, a membrane cistern, forms around a specific region of the cytoplasm, converting it into a double membrane autophagosome. This organelle conveys the material sequestered to the lysosomal compartment by fusing its outer membrane with the lysosomal limiting membrane. The outcome is an autolysosome that degrades both the inner autophagosomal membrane and the autophagic contents. It makes no difference whether the autophagosome fuses with late endosomes or creates an amphisome, as amphisomes will eventually fuse with lysosomes. An amphisome is a hybrid vesicle formed when an autophagosome fuses with a late endosome. In metazoans, it has even been suggested that late endosome-autophagosome fusion constitutes a requisite step that needs to precede lysosomal fusion [7].

In a process that is topologically analogous to the formation of multivesicular bodies during endosome maturation, the lysosomal membrane may engulf smaller volumes of cytoplasmic cargo through lysosomal membrane invaginations and engulfment. Microautophagy refers to this specific type of autophagy. Lamp-2A is a protein component of lysosomal membranes found in mammals, birds, and all vertebrate cells. It can form a homohexameric channel with the assistance of Hsc70 during chaperone-mediated autophagy [8].

Phagocytosis [Function of Specialized Cells (macrophages, Neutrophils)]

By engulfing large extracellular particles such as bacteria and directing them towards lysosomal breakdown, specialized cells (*i.e.*, macrophages) carry out phagocytosis.

Lysosomal Membrane Proteins

Multiple clathrin-dependent and independent pathways facilitate the transport of lysosomal membrane proteins synthesised in the endoplasmic reticulum to the lysosome. Following synthesis in the endoplasmic reticulum and subsequent glycosylation, proteins enter the Trans Golgi Network (TGN). From there, they may proceed *via* the secretory pathway to the plasma membrane, where they can later be reinternalized *via* endocytic processes, or they may be directed through endosomal compartments to the lysosomes (Fig. **3**) [9].

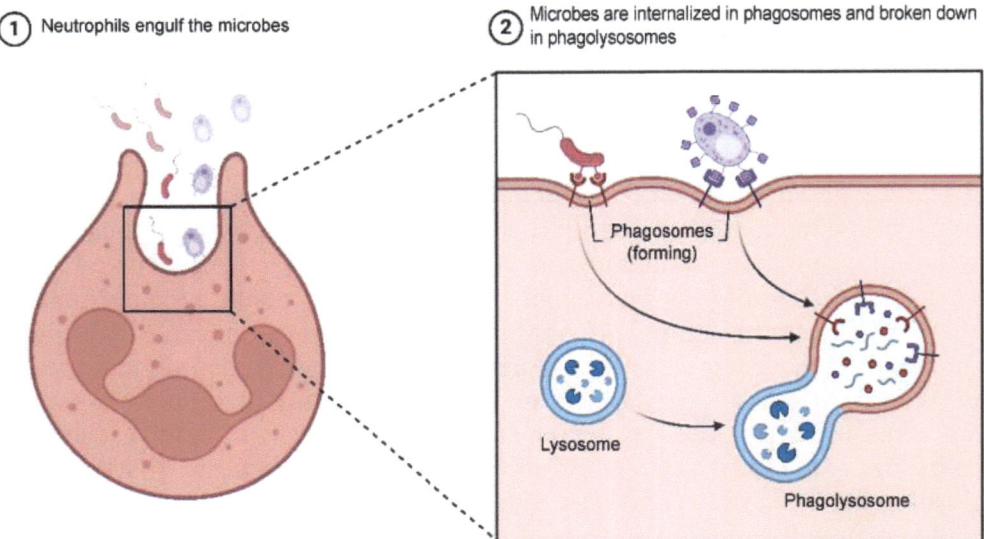

Fig. (3). Formation of lysosomes occurs through the fusion of late endosomes with lysosomal enzyme-containing vesicles derived from the Golgi apparatus, enabling the degradation of internalized molecules.

To date, proteomics has identified more than 100 lysosomal membrane proteins. Type 1 transmembrane proteins that are expressed in abundance encompass Lysosome-Associated Membrane Proteins (LAMP)-1 and -2, Lysosomal Integral Membrane Protein (LIMP)-2, as well as CD63 (LIMP1). Their essential roles in the mechanisms of lysosomal membrane transport have been elucidated. The removal of LAMP proteins results in a notable buildup of cholesterol within lysosomes. The brief cytosolic sequence of lysosomal membrane proteins possesses the capability to engage with cytosolic proteins and/or proteins situated on alternative organelles. The transmembrane domains of these proteins frequently play a crucial role in the transport processes that occur across the membrane. LAMP proteins, known as Lysosome-Associated Membrane Proteins, have been demonstrated to play a crucial role in stabilising a polypeptide translocation mechanism referred to as TAPL. TAPL plays a crucial role in facilitating the transport of cytosolic peptides into lysosomes, thereby ensuring their effective degradation.

In addition to these functions, LAMP-2A, one of the three LAMP-2 isoforms, is involved in a specialised autophagic pathway known as Chaperone-Mediated Autophagy (CMA). This pathway is responsible for the breakdown of around thirty percent of all cytoplasmic soluble protein molecules that contain the KFERQ sorting motif for lysosomal breakdown. LAMP proteins also facilitate lysosome motility, lysosomal exocytosis, and the fusion of lysosomes with the plasma membrane. Likewise, LAMP-1 is essential for the preservation of lysosomal structure and equilibrium, which in turn contributes to the biogenesis and function of lysosomes [10]. Additional lysosomal membrane proteins comprise a range of transporters and exchangers, such as cystinonin and sugar channels like spindling (SPIN). Amino acid transporters include SLC38A9, LYAAT-1, LAAT-1, and SNAT7. The Ion channels, such as mucolipin 1 (MCOLN1), also referred to as TRPML1, play a crucial role in the regulation of calcium levels and the storage of amino acids, sugars, as well as ions, including Cu^+ and Fe^{2+}. In addition to the lysosomal transmembrane proteins, the lysosomal membrane also contains additional proteins, such as SNARE proteins and tethering factors, which play an essential role in lysosomal fission and fusion. A parallel four-helix bundle of four SNARE proteins, known as a SNAREpin or trans-SNARE complex, facilitates lysosomal fusion actions at target organelles [11].

The assembly of an R-SNARE, such as VAMP7 or VAMP8, at the lysosomal membrane involves three Q-SNAREs, including syntaxin 7, syntaxin 8, and VTI1B, located on late endosomes or similar components on autophagosomes, such as SNAP27 and syntaxin 17, facilitating fusion with late endosomes. After its assembly, the SNAREpin undergoes a conformational change that is reliant on calcium. The binding of calcium ions initiates a structural alteration in the SNARE proteins, thereby improving their capacity to facilitate membrane fusion. The conformational change aids in overcoming the repulsive forces between lipid bilayers, thereby facilitating the formation of a fusion pore. Following membrane fusion, the AAA ATPase N-ethylmaleimide-Sensitive Factor (NSF) disassembles the SNAREpin. The BLOC-1-related complex (BORC) is a newly identified multi-subunit protein complex. This complex localizes to the lysosomal membrane and regulates motility by interacting with kinesin-5 *via* ARL8 [12].

Lysosome Reformation

The cell's supply of lysosomes is quickly depleted by cargo breakdown. Thus, an important homeostatic mechanism for preserving the lysosome pool is lysosomal regeneration following destruction. Cells from patients with Lysosomal Storage Diseases (LSDs) have been found to have defective lysosome reformation. This defect has also been linked to Parkinson's disease. We will go more in-depth about the involvement of lysosomal defects that lead to Parkinson's disease in a later topic [13].

Endocytic Lysosome Reformation (ELR)

ELR is an ATP-dependent mechanism, according to earlier research. Since lysosome reformation was prevented by inhibiting lysosomal acidification, the need for ATP is believed to maintain the proton gradient. Another crucial element is intra-lysosomal Ca^{2+}. Later, it was discovered that the lysosomal calcium channel TRPML1 and the PtdIns3P 5-kinase PIKfyve are both necessary for ELR. PtdIns(3,5) P2, which is produced by PIKfyve and activates TRPML1 to regulate lysosomal Ca^{2+} efflux. PIKfyve, TRPML1, and mTOR govern the shrinking of phagosomes and entotic vacuoles in addition to regulating lysosome renewal [14].

During autophagy, the phagophore elongates to form the autophagosome, which creates the autolysosomes through fusion with lysosomes. Tubulation events occur with the assistance of the KIF5B and VPS34-UVRAG complex. PIP5K1B aids in converting PI(4)P to PI(4,5)P2, and the resulting budded tubules are fragmented by DNM2 and PIP5K1A, forming Proto lysosomes. These proto-lysosomes mature into functional lysosomes, ready to participate in further fusion events and degradation pathways.

Autophagic Lysosome Reformation (ALR)

ALR, or lysosome reconstruction from autolysosomes, was first seen when autophagy was induced through starvation. MTOR is a critical kinase regulating cell growth and is inhibited early in starvation-induced autophagy, and all lysosomes are converted into autolysosomes. Through lysosomal degradation of macromolecules upon prolonged fasting, mTOR is reactivated. Consequently, autolysosomes can generate lysosomal tubules that also give rise to empty proto-lysosomes. The mechanisms for how proto-lysosomes mature into functional lysosomes are still unknown. ALR is essential to replenish the lysosome pool during autophagy. Fibroblasts from patients suffering from different LSDs harbor dysfunctional mTOR reactivation and ALR.

One theory is that ALR is produced by phosphatidylinositol 4,5-bisphosphate (PtdIns(4,5) P2), clathrin, and the adaptor protein complexes AP2 and AP4. PtdIns4P 5-kinase 1B converts the membrane-ubiquitous phosphatidylinositol 4-phosphate (PtdIns4P) to PtdIns(4,5) P2 on specific microdomains of the autolysosome. Clathrin is further recruited by PtdIns(4,5) P2 *via* the AP2 complex

[6, 14]. Clathrin drives the formation of PtdIns(4,5) P2-enriched microdomains, in which proto-lysosomal formation begins. The kinesin heavy chain KIF5B, interacting with PtdIns(4,5) P2, is recruited to the microdomains and promotes the expansion of proto-lysosomal tubules along microtubules. Lysosomal Membrane proteins such as LAMP1 are recruited to these microdomains through interaction with AP4.

In addition, PtdIns(4,5)P2 interacts with WHAMM, a member of the WASP (Wiskott-Aldrich syndrome protein) family that activates Arp2/3 for actin nucleation. The WASP family is significant because it plays a crucial role in regulating actin dynamics, which are essential for various cellular processes such as cell shape rigidity, motility, and intracellular transport. WHAMM specifically promotes proto-lysosome tubulation by facilitating the formation of branched actin networks on autolysosomes. This branching structure is vital for lysosomal trafficking and fusion events, highlighting the importance of actin remodeling in lysosomal function [14].

The second PtdIns4P 5-kinase, PIP5K1A, generates PtdIns(4,5) P2 on the proto lysosomal tubules, which causes clathrin rebinding and proto lysosome scission that is dependent on dynamin 2. ALR requires the balance between PtdIns4P and PtdIns(4,5) P2 on autolysosomal microdomains. Muscular dystrophy is a consequence of the inositol polyphosphate-5-phosphatase K deletion, which degrades PtdIns(4,5) P2 to PtdIns4P and blocks the repopulation of lysosomes. ALR involves more detail. For example, gene products for two of the most prevalent autosomal-recessive hereditary forms of spastic paraplegia, spastizin/SPG15 and spatacsin/SPG11, assemble into a complex and localize to lysosomes through the spastizin FYVE domain that promotes initiation of tubules on the autolysosome. Whether the PtdIns(4,5) P2-clathrin axis interacts with the spastin-spatacsin complex to regulate ALR remains unknown [12, 13].

Phagocytic Lysosome Reformation (PLR)

Depending upon whether the cell is still living or apoptotic, the procedure of phagocytosis results in the formation of phagosomes that are eventually destined to fuse with lysosomes. It was found that nondegradable content, such as erythrocytes, permits the shriveling of phagolysosomes in addition to the reforming of the tubuloreticular network of lysosomes in bone marrow-derived macrophages, although other forms of nondegradable content, such as latex beads, impede the reforming of the tubular network. Researchers were able to prove that PIKfyve, TRPML1, and SLC-36.1 are essential in regulating PLR in human cells and C. elegans. PIKfyve inhibition in mammalian cells compromises the resolution of the phagolysosome and the entotic vacuole but does not affect processes that happen before phagolysosome formation.

Inhibiting the lysosomal Ca^{2+} channel TRPML1 has the same impact on phagolysosome shrinkage. The deficiency in phagolysosome shrinkage induced by PIKfyve loss may be circumvented by overexpressing TRPML1. Therefore, PIKfyve partly regulates phagolysosome shrinkage *via* TRPML1. During the genetic screen, important regulators of PLR in C. elegans were found. These included SLC-36.1, a homologue of the neutral amino acid transporters SLC36A1-4/PAT1-4 in mammals, and PIKfyve PPK-3, an orthologue in worms. In embryos that are genetically modified, tubulation takes place on phagolysosomes that contain cell corpses. Nevertheless, ppk-3 and slc-36.1 mutants show an increase in phagolysosomal vacuoles that do not produce lysosomal tubules. As part of its efforts to control PLR, SLC-36.1 unites with PPK-3 and attaches to lysosomes. It is possible that PLR plays a crucial role in preserving a normal amount of lysosomes, as there are fewer free lysosomes in the engulfing cells of ppk-3 and slc-36.1 mutant embryos [15].

In the adult hypodermis of C. elegans, PIKfyve/PPK-3 and SLC-36.1 play an essential role in ALR. When these proteins are missing, a small number of bigger autolysosomes build up and are unable to form lysosome reformation tubules. Evidence from the identification of PIKfyve, TRMPL1, and SLC-36.1/SLC30A--4 shows that comparable regulators are present for the different types of gastric lysosomes involved in lysosome reformation. To ensure cell viability, embryonic development, and lysosome rebuilding, amino acids must be exported *via* phagosomes and autolysosomes. During lysosome reconstruction, three lysosomal transporters—SLC-36.1/SLC30A1-4, Spinster, and LAAT-1/PQLC2—stimulate PIKfyve-dependent nutrient export from phagolysosomes. We still don't know whether additional lysosomal exporters are involved in lysosome reformation or if PIKfyve controls these transporters *via* PtdIns(3,5) P2.

Key Players of Membrane Fusion

Organelles attached to membranes aid in the compartmentalisation of different biochemical processes, and eukaryotic cells are bounded by a plasma membrane. This compartmentalisation, enabled by membrane fusion, allows the cell to meticulously regulate various activities and transport routes among organelles. With certain exceptions, such as mitochondrial fusion, the majority of membrane fusions depend on analogous components, as shown below. Each cellular membrane has a distinct mix of lipids and proteins, with individual components significantly influencing the regulation of mobility and fusion of particular organelles. This encompasses proteins from the small GTPase families, including Arf (ADP-ribosylation factor) and Rab (Ras-associated binding). These proteins can attract effector proteins when in their GTP-bound (active) form, acting as molecular identifiers that determine the fate of the organelle.

Motor proteins, SM (Sec1/Munc-18) proteins, tethering elements, and SNAREs (Soluble NSF attachment protein) receptors are all part of this complex. Autophagosome formation-essential Atg (autophagy-related) proteins were first discovered in Saccharomyces cerevisiae. Similar proteins in higher metazoans are still active and serve their original purposes. The activation and maturation of the autophagosome and the phagophore are controlled by Atg proteins. Two systems similar to ubiquitin that play a role in autophagosome formation are the Atg12 and Atg8 conjugating systems. Atg8 and Atg12 are ubiquitin-like proteins, and these enzymatic mechanisms try to lipidate and bind them to the expanding phagophore membrane. Atg8 family proteins in mammals are categorised into two subgroups: LC3 and GABARAP. AbaraPL1, GABARAP, and GABARAP2 are all components of GABARAP. All six of these proteins seem to play a role in autophagosome formation, but Atg8/LC3s is particularly important since it helps the phagophore elongate at the beginning.

Samples also pointed out that conjugation-deficient cells degraded the inner membrane more slowly. The autophagosomal SNARE Syntaxin17/ STX17 (described below) might still be acquired by these imperfect, open autophagosomes, this is important. This demonstrates that Atg8 lipid conjugation plays a role in autophagosome completion but not in fusion and in the targeting of STX17. However, as stated in other studies, members of the Atg8 family are not required to generate autophagosomes but are important in the fusion of autophagosomes with lysosomes by recruiting PLEKHM1 to autophagosomes. Lastly, it has been suggested that GABARAP proteins regulate the lipid composition of the autophagosomes for the subsequent fusion of the autophagosome with the lysosome organelle. The Atg8 proteins that conjugate with the pre-autophagosomal membranes could have fusogenic activity in the context of the enlarging phagophores. Furthermore, it has been shown that the LC3 homologue from C. elegans binds to the HOPS tethering complex to promote autophagosome-lysosome fusion (but not GABARAP).

Atg4, a cysteine protease, plays a key role in autophagy by recycling Atg8 and removing it from the outer membrane of autophagosomes. In yeast, this process is essential because it prevents autophagosomes from fusing with vacuoles prematurely. Atg8 must be removed for the fusion to happen properly, as its presence can hinder the interaction of other proteins required for membrane fusion. This clearance ensures that the necessary fusion machinery can effectively merge the membranes. This suggests that Atg8 has two important roles in the fusion process: first, it helps attract proteins that facilitate fusion, and second, it needs to be cleared away for the final step of fusion to occur.

Besides Atg8, other Atg proteins also assist in vesicle fusion. For example, in yeast, the SNARE protein Vam7 is drawn to autophagosomes by the Atg17-Atg31-Atg29 complex and Atg11. In mammals, Atg14 might act as a tether to help bring autophagosomes and lysosomes together. The ESCRT (endosomal sorting complexes required for transport) system was thought to help close the phagophore, a structure like the multivesicular bodies. Research has shown that ESCRT helps seal autophagosomes in both yeast and mammalian cells, potentially involving Vps21/Rab5 in yeast. In human cells lacking ESCRT, open autophagosomes pile up but can still recruit STX17 and eventually fuse with lysosomes, though at a slower rate [7, 9, 16].

Interestingly, the issues caused by losing ESCRT are like those seen in cells that cannot perform the ATG conjugation process. In both cases, the inner membrane of the unsealed autophagosomes still gathers LAMP1 (Lysosome-Associated Membrane Protein 1) after fusion with lysosomes—something that does not usually happen with completely closed autophagosomes. This also disrupts the breakdown of the inner membrane. Since lysosomal enzymes usually digest membranes, LAMP proteins are heavily glycosylated to protect lysosomal membranes from damage. In cells lacking ESCRT or ATG conjugation, this delayed degradation may be due to an accumulation of lysosomal membrane proteins or lipids, which could explain the buildup of autophagosomes in ESCRT-deficient cells [9].

One powerful inhibitor of vacuolar ATPase (V-ATPase) is bafilomycin A1, which stops lysosomes from acidifying and degrading their contents. Further, it prevents autophagosomes from merging with lysosomes, which prompted researchers to conclude that acidity in lysosomes is essential for fusion. Consistent with the idea that acidification and fusion are distinct mechanisms, further research demonstrated that V-ATPase-deficient cells may still produce non-degrading autolysosomes in the absence of acidification. This discovery provides further evidence that the pH of the lysosome is not a determinant in autophagosome-lysosome fusion. Unlike pH, the lysosomal lumen's ion composition seems to regulate the fusion of autophagosomes and lysosomes. It is believed that the lysosomal voltage-gated calcium channel and Ca^{2+} efflux from lysosomes promote fusion, as the lack of this channel leads to an increase in autophagosomal-lysosomal fusion intermediates in both fly and mouse cells. Nevertheless, autophagic flux is restricted in this instance due to elevated intracellular Ca^{2+} levels caused by heavy metal-induced ER stress. Keep in mind that elevated Ca2+ levels might have an inhibitory effect as well. The data suggest that autophagosome completion, Atg8 conjugation proteins, or lysosomal acidification are not necessary for STX17 recruitment and lysosome fusion. In addition to facilitating phagophore formation and edge sealing, Atg8 proteins may also play a role in timing autophagosome-lysosome fusion. Autophagic flow is reduced when open autophagosomes fuse prematurely because the inner membrane of the autophagosome does not degrade immediately. When lysosomal material reaches the gap between the autophagosomes' outer and inner membranes, the inner membrane often ruptures. Atg8 conjugation and ESCRT-deficient cells show that the inner membrane may generate lysosomal membrane proteins that restrict the access of lysosomal hydrolases to it if the closure hole is left open. Thus, it is crucial to comprehend how cargo may be efficiently broken down without endangering the autolysosome's integrity. This emphasises the flexibility and intricacy of autophagic processes.

LYSOSOMAL DYSFUNCTION

"Lysosomal dysfunction" describes the state in which lysosomes lack the ability to carry out their typical roles. Lipids, carbohydrates, and proteins are only some of the waste products that lysosomes break down and recycle. Cellular damage and instability may result from lysosome failure and the accumulation of cellular waste products. Lysosomal dysfunction may be caused by a number of factors, including ageing, exposure to environmental pollutants, and genetic disorders. Lysosomal dysfunction is a hallmark of a number of hereditary illnesses, including Tay-Sachs and Niemann-Pick. Depending on the severity and location of the defect, lysosomal dysfunction may manifest in several ways. Some malfunctions may not produce any symptoms, while others might lead to a variety of symptoms such as neurodegeneration, muscular weakness, or organ failure. The root cause of lysosomal dysfunction will dictate the course of therapy. Medication may help with symptoms, but in certain situations, treatments like stem cell transplants or enzyme replacement therapy may be required. Researchers are always looking for new lysosomal dysfunction treatments, including pharmaceutical and gene therapy methods.

A number of diseases may be brought on by problems with autophagy, which is essential for maintaining cellular homeostasis. Therefore, many diseases have been associated with the loss of genes that code for proteins that are involved in the process of autophagosome-lysosome fusion. Mutations in the gene that codes for the tethering factor/adaptor EPG-5 produce a variety of symptoms in the Vici syndrome, including agenesis of the corpus callosum, cataracts, retinitis pigmentosa, cardiomyopathy, and immunodeficiency. Charcot-Marie-Tooth type 2B neuropathy is caused by mutations in the Rab7 gene. Diseases like osteopetrosis, which may be caused by mutations in Rab7 interactors like PLEKHM1, are possible. Mutations in mucopolysaccharidosis-plus syndrome are caused by mutations in the VPS33A gene, while mutations in the VPS11 gene cause neuroaxonal dystrophy in dogs and autosomal recessive leukoencephalopathy in humans. Extremely severe CEDNIK syndrome is caused by SNAP-29 mutations that include one base pair deletion or insertion, which cause premature termination. Lastly, PIKfyve mutations may induce François-Neetens mouchetée fleck corneal dystrophy. Lysosomal malfunction or insufficient endosomal traffic might potentially contribute to these diseases, since all gene products participate in endosomal-lysosomal activities apart from autophagy. Dysfunctional locomotion and ongoing neurodegeneration in the mutant flies show that Syx17 isn't necessary for endocytic degradation in fruit flies. Therefore, changes in autophagy and endocytic traffic/degradation occur concurrently and lead to the development of various illnesses [1, 3, 5].

Having said that, there is one outlier, Joubert syndrome 1, characterised by a wide range of neurodevelopmental abnormalities, is caused by mutations in the INPP5E gene. Endosomal-lysosomal fusion and lysosomal integrity are not regulated by INPP5E. The idea that the condition is caused by insufficient autophagic flux was supported when the Joubert syndrome 1 mutant version of this protein was engineered in INPP5E knockdown neurons, but it was unable to restore autophagy. As the cost of next-generation sequencing drops, large-scale studies of rare genetic diseases and cancer subtypes will find more patients whose symptoms are partially driven by problems with autophagosome-lysosome fusion or lysosomal degradation.

Lysosome Storage Disorders (LSD)

Mutations in the genes that create lysosomal hydrolases, which are crucial elements of the lysosomal degradation process, are usually the cause of lysosome storage disorders (LSDs), a class of genetically diverse, multisystem illnesses. LSDs can lead to significant disability, mortality, and progressive neurological decline. The buildup of stored substances in lysosomes, resulting from reduced degradative activity and autophagy inhibition, is a characteristic histological feature of Lysosomal Storage Disorders (LSDs).

ALR levels are reduced in the fibroblasts of individuals with lysosomal storage disorders, including Scheie syndrome (mucopolysaccharidosis type I), Fabry disease, and aspartylglucosaminuria. Scheie syndrome results from a deficiency in alpha-L-iduronidase (IDUA), leading to the accumulation of glycosaminoglycans (GAGs) within lysosomes. This condition is a multisystem disorder characterised by bone abnormalities, cardiac complications, and cognitive deficits. Fabry disease involves alterations in the lysosomal hydrolase GLA (galactosidase alpha), leading to episodic or chronic pain and potentially life-threatening complications affecting the kidneys, heart, and cerebrovascular system. Aspartylglucosaminidase (AGA) is a lysosomal enzyme that is inactive, leading to aspartylglucosuria.

This condition initiates in childhood, characterised by learning and behavioural difficulties, and advances swiftly into adulthood, manifesting as skeletal abnormalities, compromised dental health, epileptic seizures, and psychological disorders. It is hypothesised that fibroblasts from individuals with lysosomal storage diseases will exhibit ALR inhibition due to impaired mTOR activation, given the detrimental effects of these diseases on lysosomal hydrolases. Beneficial deterioration. The failure of lysosomal hydrolases to degrade autolysosome contents during autophagy obstructs amino acid synthesis and efflux, thereby hindering mTOR reactivation and the commencement of ALR. Approximately 50 distinct forms of LSD exist, with numerous additional variants potentially associated with ALR inhibition. Spin (spinster), a sugar transporter, is essential for mTOR activation during autophagy, thereby initiating ALR. Disruption of spin in Drosophila or its zebrafish equivalent (SPNS1/NRS, spinster homolog 1 "Drosophila") leads to progressive neurotoxicity resembling that of lysosomal storage disorders.

Fibroblasts from individuals with Niemann-Pick type C disease-1 (NPC) exhibit reduced expression of the human ortholog SPNS1, alongside abnormalities characteristic of an ALR deficit, such as autophagy suppression and the accumulation of larger autolysosomes/lysosomes. Rectification of these deficiencies can be achieved by enhancing SPNS1 expression or supplementing cells with L-leucine, an amino acid known to activate mTOR. While some lysosomal storage disorders exhibit signs associated with mTOR hyperactivation that restrict autophagy and TFEB-mediated lysosome formation, this condition is not prevalent. The implications of prolonged mTOR stimulation on ALR remain ambiguous, presenting a compelling opportunity for additional investigation. Prolonged mTOR activation inhibits autophagy, thereby restricting the production of autolysosomes that contain ALR.

Parkinson's Disease

About one percent of the population over 60 suffers from Parkinson's disease (PD), a neurodegenerative illness that is both common and severe, and which is linked to the inhibition of autophagy. Glucosyl ceramidase beta 1 (GBA1) and cyclin G-Associated Kinase (GAK), two genes necessary for ALR, are related to 5-10% of Parkinson's disease patients who have genetic risk factors. Reduced lysosome activity and autophagy suppression are associated with neurodegeneration, and they are caused by GBA1 loss. Despite GBA1's need for ALR, its deletion impairs lysosome homeostasis maintenance by blocking MITF/TFEB-dependent de novo lysosome production. Reduced autophagic mTOR reactivation and lysosomal acidification/degradative function cause GBA1 mutations in patient-derived fibroblasts to have low ALR. Concurrent with -synuclein accumulation is ALR suppression. While autophagy is generally responsible for removing synuclein, a major pathogenic characteristic of Parkinson's Disease (PD) is its buildup in Lewy bodies in neurons, making this finding more significant (Fig. **4**).

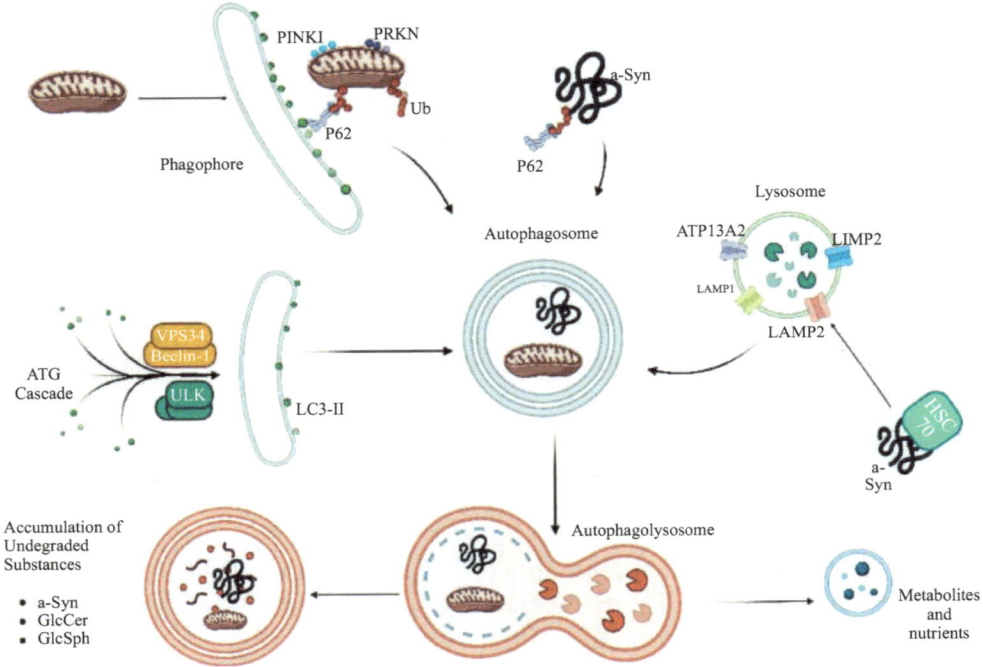

Fig. (4). Proper function of the Autophagy-Lysosomal Pathway (ALP) is essential for cellular homeostasis, preventing the accumulation of damaged proteins and organelles. Dysfunctional ALP in Parkinson's disease leads to the buildup of protein aggregates and defective mitochondria. Key proteins involved include α-SYN, PRKN, PINK1, LRRK2, and various lysosomal and autophagy-related factors

New research suggests that ALR dysfunction is linked to GAK and other risk loci for Parkinson's disease. Both Parkinson's disease and LSD Gaucher disease are impacted by the GBA1-dependent ALR pathway. GBA1 mutations are also a prominent cause of LSD Gaucher disease. Inflammation, hepatosplenomegaly, poor haematopoiesis, bone abnormalities, and, in very rare instances, severe neurological problems are symptoms of Gaucher disease, which is caused by the accumulation of a glycosphingolipid (glucocerebroside) that is mostly cleared in macrophage lysosomes. One of the ALR's key components and an activator of the GBA1 enzyme, PSAP (prosaposin), is cleaved to create Sap C (saposin C). Alterations in PSAP lead to an extremely uncommon form of Gaucher's disease. In GBA1-deficient MEFs, the enzyme replacement therapy iglucerase (Cerezyme) successfully restores mTOR activation and lysosome homeostasis. Similarly, patient-derived fibroblasts defective in Sap C show an improvement in lysosomal enzyme activity and restoration of ALR after administration of (2-hydroxypropyl)-cyclodextrin (HP-CD), a medication now being tested in LSD clinical trials. Consequently, ALR-targeting drugs could potentially be used to treat LSD and Parkinson's disease in the future [6].

Muscular Dystrophy

The skeletal muscle exhibits a heightened degree of basal autophagy, which it utilises for cytoprotection. Muscle sickness arises from the suppression of autophagy, whereas inherited skeletal muscle disorders, referred to as muscular dystrophies or myopathies, result from abnormalities in crucial autophagy regulatory genes. The activation of autophagy in skeletal muscle represents a significant physiological response to both fasting and exercise. During fasting, the body engages in an accelerated process of autophagic breakdown of muscle proteins, resulting in the release of amino acids that can be utilised in other areas when nutritional intake is insufficient. Enhanced autophagy activation during post-exercise recovery is essential for the regulation of mitochondrial quality. For the enhancement and sustained function of autophagy, it is essential that these adaptive autophagy processes receive sufficient lysosome replenishment. ALR dysfunction facilitates muscular degeneration and yields significant repercussions for skeletal muscle. Clinically, Marinesco-Sjögren syndrome bears resemblance to the multisystem disorder resulting from INPP5K mutations. Individuals impacted by this condition exhibit congenital muscular dystrophy, cataracts, and a range of complications, including developmental delays, cognitive impairments, microcephaly, and reduced stature. Muscle wasting is a condition that unfolds gradually, initiating in childhood with a decline in mobility and advancing into adulthood. Mutations in INPP5K may disrupt the membrane-targeting SKIP C-terminal Homology (SKICH) domain or the 5-phosphatase domain, resulting in a diminished catalytic activity of the enzyme towards PtdIns(4,5)P2. Preliminary investigations indicated that autophagy malfunction does not play a role in INPP5K congenital muscular dystrophy; nonetheless, the SKICH domain, which was first identified in INPP5K, is also present in CALCOCO2/NDP52 (calcium binding and coiled-coil domain 2), TAX1BP1 (Tax1 binding protein 1), and

CALCOCO1 (calcium binding and coiled-coil domain 1), all of which exhibit this characteristic.

Moreover, the observation of rimmed vacuoles and SQSTM1/p62 within the muscle tissue of individuals harbouring INPP5K mutations suggests a significant role of autophagy dysregulation. Rimmed vacuoles hold considerable importance as they signify disrupted autophagic processes and are frequently linked to muscle disorders, acting as a histological marker for autophagic dysfunction. Mice exhibiting a conditional, skeletal muscle-specific deletion of Inpp5k manifest a profound muscular disorder characterised by early onset and marked autophagy inhibition attributable to ALR suppression. INPP5K is brought into autolysosomes during ALR in muscle cells, where it dephosphorylates PtdIns(4,5)P2 to produce PtdIns4P. The absence of INPP5K results in an elevation of PtdIns(4,5)P2 on autolysosomes, which in turn causes a reduction in PtdIns4P. This alteration disrupts the dynamic connection of clathrin to the remodelling tubules, ultimately diminishing the formation of lysosomes. It is noteworthy that muscle cells harbouring INPP5K disease mutants, which demonstrate diminished enzymatic activity towards PtdIns(4,5)P2, do not display ALR. This represents the inaugural occurrence of muscular dystrophy associated with ALR suppression, presenting a persuasive case for additional exploration into the influence of other muscular dystrophy/myopathy genes on ALR activity [2].

FUTURE DIRECTIONS FOR LYSOSOME RESEARCH

Essential concepts concerning the function and life cycle of lysosomes were just recently uncovered, despite being found more than 70 years ago. Recent research has shown that lysosomes play a significant role in cellular aging, tissue homeostasis, and cellular rejuvenation. According to a study done by Butsch *et al*, lysosome-aging cells have reduced lysosomal and autophagic activity, which leads to cellular degradation by allowing damaged organelles to accumulate. Inducing autophagy in experimental models by genetic modification or pharmacological treatments demonstrates enhanced cellular health and increases in diverse organisms such as yeast and mammals. Another study (Hu *et al*., 2020) found that tiny extracellular vesicles produced from embryonic stem cells (ESC-sEVs) can regenerate senescent brain stem cells. The ESC-sEVs provide particular miRNAs that block mTORC1 activity, a recognized repressor of lysosomal formation and function. This pathway enables the return of lysosomal activity, resulting in enhanced NSC proliferation, neurogenesis, and cognitive function in VD models [6].

CONCLUSION

There is a long history that establishes that lysosomes are not merely "suicide bags" but rather complex, multifunctional organelles important for cell health and function. While these organelles are active participants in the degradation and recycling of cellular waste, their acidic environment and hydrolytic enzymes help them to maintain cellular homeostasis. Since Christian de Duve discovered lysosomes, we have gained significant insights into their role in so many cellular activities, such as autophagy and phagocytosis, and other key signalling pathways. Such involvement in multiple physiological and pathological processes, such as cellular aging, neurodegeneration, and LSDs-makes the lysosome a forefront of modern research. These diseases underscore an important need for lysosome function: genetic mutations affecting the enzymes or membrane proteins in lysosomes lead to catastrophic cellular dysfunction and systemic disease. Lysosome research is the future therapeutic application that is likely to introduce novel approaches to treating neurodegenerative diseases or even aging. Other critical future interventions involve lysosome modulation, enzyme replacement, and gene therapy, restoring or compensating malfunctioning lysosomes, further pointing out the importance of these organelles in maintaining cellular well-being and longevity.

ACKNOWLEDGEMENTS

We would like to acknowledge Ms. Kruthika P, Ms. Raksa Arun, Ms. Srisri Satish Kartik, Ms. Sanjana Dhayalan, and Mr. Sasitharan T for their contribution in proofreading and editing the work.

REFERENCES

[1] Butsch TJ, GB, & B ert, KA. Organelle-Specific Autophagy in Cellular Aging and Rejuvenation. Adv Geriatr Med Res [Internet]. 2021. Available from: https://agmr.hapres.com/htmls/AGMR_1386_Detail.html

[2] Wartosch L, Bright NA, Luzio JP. Lysosomes. Curr Biol 2015; 25(8): R315-6.
[http://dx.doi.org/10.1016/j.cub.2015.02.027] [PMID: 25898096]

[3] Cao W, Li J, Yang K, Cao D. An overview of autophagy: Mechanism, regulation and research progress. Bull Cancer 2021; 108(3): 304-22.
[http://dx.doi.org/10.1016/j.bulcan.2020.11.004] [PMID: 33423775]

[4] Sabatini DD, Adesnik M. Christian de Duve: Explorer of the cell who discovered new organelles by using a centrifuge. Proc Natl Acad Sci USA 2013; 110(33): 13234-5.
[http://dx.doi.org/10.1073/pnas.1312084110] [PMID: 23924611]

[5] Goldenring JR. Recycling endosomes. Curr Opin Cell Biol 2015; 35: 117-22.
[http://dx.doi.org/10.1016/j.ceb.2015.04.018] [PMID: 26022676]

[6] Wolfe DM, Lee J, Kumar A, Lee S, Orenstein SJ, Nixon RA. Autophagy failure in A lzheimer's disease and the role of defective lysosomal acidification. Eur J Neurosci 2013; 37(12): 1949-61.
[http://dx.doi.org/10.1111/ejn.12169] [PMID: 23773064]

[7] Hu G, Xia Y, Zhang J, et al. ESC-sEVs rejuvenate senescent hippocampal NSCs by activating lysosomes to improve cognitive dysfunction in vascular dementia. Adv Sci (Weinh) 2020; 7(10): 1903330.
[http://dx.doi.org/10.1002/advs.201903330] [PMID: 32440476]

[8] Uribe-Querol E, Rosales C. Phagocytosis: Our Current Understanding of a Universal Biological Process. Front Immunol 2020; 11: 1066.
[http://dx.doi.org/10.3389/fimmu.2020.01066] [PMID: 32582172]

[9] Hurley JH. ESCRT s are everywhere. EMBO J 2015; 34(19): 2398-407.
[http://dx.doi.org/10.15252/embj.201592484] [PMID: 26311197]

[10] Jung J, Genau HM, Behrends C. Amino acid-dependent mTORC1 regulation by the lysosomal membrane protein SLC38A9. Mol Cell Biol 2015; 35(14): 2479-94.
[http://dx.doi.org/10.1128/MCB.00125-15] [PMID: 25963655]

[11] Luzio JP, Hackmann Y, Dieckmann NMG, Griffiths GM. The biogenesis of lysosomes and lysosome-related organelles. Cold Spring Harb Perspect Biol 2014; 6(9): a016840-0.
[http://dx.doi.org/10.1101/cshperspect.a016840] [PMID: 25183830]

[12] Margiotta A. Membrane Fusion and SNAREs: Interaction with Ras Proteins. Int J Mol Sci 2022; 23(15): 8067.
[http://dx.doi.org/10.3390/ijms23158067] [PMID: 35897641]

[13] Mauvezin C, Nagy P, Juhász G, Neufeld TP. Autophagosome–lysosome fusion is independent of V-ATPase-mediated acidification. Nat Commun 2015; 6(1): 7007.
[http://dx.doi.org/10.1038/ncomms8007] [PMID: 25959678]

[14] Mizushima N. The ATG conjugation systems in autophagy. Curr Opin Cell Biol 2020; 63: 1-10.
[http://dx.doi.org/10.1016/j.ceb.2019.12.001] [PMID: 31901645]

[15] Scott CC, Vacca F, Gruenberg J. Endosome maturation, transport and functions. Semin Cell Dev Biol 2014; 31: 2-10.
[http://dx.doi.org/10.1016/j.semcdb.2014.03.034] [PMID: 24709024]

[16] Pols MS, van Meel E, Oorschot V, et al. hVps41 and VAMP7 function in direct TGN to late endosome transport of lysosomal membrane proteins. Nat Commun 2013; 4(1): 1361.
[http://dx.doi.org/10.1038/ncomms2360] [PMID: 23322049]

CHAPTER 8

Vacuoles: The Storage Vaults of the Cell

S. Sahasra[1], S. Aswini[1], K. Kumaran[1], Sanjana Dhayalan[1] and K.N. Aruljothi[1,*]

[1] *Department of Genetic Engineering, SRM Institute of Science and Technology, Kattankulathur, Chengalpattu, Tamil Nadu, India*

Abstract: Vacuoles are cell structures enclosed in membranes that play roles in various cellular functions within plant cells, some animal cells, and fungi cells. This section delves into the array of duties that vacuoles perform, like storing water and nutrients and aiding in maintaining turgor pressure, pH balance, and detoxification within cells. It also discusses the variations seen among organisms with a focus on specific vacuole types, including the central vacuole found in plants, the autophagic variety seen in animals, and the versatile fungal vacuoles. Furthermore, the chapter explores how vacuoles play a role in autophagy and defense mechanisms and the importance of vacuole-specific proteins such as Tonoplast Intrinsic Proteins (TIPs). It also covers progress in vacuole research, including their potential for use in drug delivery systems, enhancing plants' ability to withstand stress.

Keywords: Abscisic Acid, Autophagy, Cellular Detoxification, Glycerol, Fungal Vacuole, Ion Regulation, Membrane, Nutrients, Osmoregulation, Plant Vacuole, Tonoplast, Tonoplast Intrinsic Proteins (TIPs), Turgor Pressure, Vacuolar pH..

INTRODUCTION

Vacuoles are membrane-bound organelles found primarily in plant and fungal cells and some animal cells. In animal cells, vacuoles are relatively small, typically ranging from 50 to 500 nanometers in size. In contrast, plant cells contain a large central vacuole that can occupy between 30% and 90% of the cell's total volume, depending on the cell type and developmental stage. They act as storage compartments that can hold a variety of substances, including water, ions, nutrients, waste products, and pigments.

Vacuoles are essential for various cellular functions. They store water, nutrients, and waste products. Their ability to regulate water content helps maintain cell shape and rigidity, contributing to turgor pressure, the internal pressure exerted by

---

[*] **Corresponding author K.N. Aruljothi:** Department of Genetic Engineering, SRM Institute of Science and Technology, Kattankulathur, Chengalpattu, Tamil Nadu, India; E-mail: aruljotn@srmist.edu.in

K.N. Aruljothi, Prakash Gangadaran, Krishnan Anand, Satish Ramalingam, K. Kumaran & Kruthika Prakash (Ed.)
All rights reserved-© 2026 Bentham Science Publishers

the vacuole in/within the cell. Additionally, vacuoles are crucial for maintaining the pH of the cell's internal environment through proton pumps and ion exchange, ensuring optimal conditions for cellular processes like enzyme activity and metabolism. They also serve as detoxification centers, storing harmful substances. Furthermore, vacuoles are responsible for pigmentation by storing pigments like anthocyanins and carotenoids that provide color to flowers, fruits, and leaves. Lastly, they play a role in autophagy, a cellular process that involves breaking down cellular components.

The discovery of vacuoles dates back to the early days of microscopy. While Anton van Leeuwenhoek is often credited in the late 17th century for his observations of single-celled organisms, earlier scientists likely observed vacuoles using basic microscopes. As microscopy techniques advanced, scientists began to appreciate the diversity and significance of vacuoles in different cell types. While the term "vacuole" comes from the Latin word "vacuoles," which means "empty," the organelle is not truly empty space but instead contains a variety of substances. Research on vacuoles remains a vital area in cell biology, with scientists investigating their roles in plant growth, development, and responses to stress.

Types of Vacuoles

Plant Vacuoles

In plant cells, the central vacuole is the most prominent type of vacuole. This large, membrane-bound organelle takes up a significant portion of the cell's volume. The central vacuole is surrounded by a membrane called tonoplast, which is also known as the vacuolar membrane. The central vacuole serves various functions, including:

- Storage: It stores water, nutrients, ions, toxins, and waste products.
- Turgor pressure: The pressure exerted by the fluid inside the vacuole maintains the cell's rigidity and the cell's structure.
- pH regulation: It regulates the pH of the cell's internal environment.
- Detoxification: It stores harmful substances.
- Pigmentation: It contains pigments that provide colour to flowers, fruits, and leaves.
- Autophagy: It breaks down cellular components.

There are primarily two types of plant vacuoles.

- Lytic Vacuole (LV): LVs are developed early during embryogenesis. These vacuoles help in breaking down unwanted cellular substances using hydrolase enzyme. Therefore, hydrolase enzyme are very crucial in the degradation and recycling of cellular substances [1].

- Protein Storage Vacuole (PSV): PSVs are formed after LVs during embryogenesis. In addition, in maturing seeds, LVs are transformed to PSVs, but during germination, this process is reversed. These vacuoles are used to store nutrients and proteins, which may be needed at different stages of growth or under stress conditions. It's possible for protein storage vacuoles and lytic vacuoles to combine to form a large central vacuole [1].

Animal Vacuoles

Animal cells typically lack the large, central vacuoles that are characteristic of plant cells. Nevertheless, they possess various smaller vacuoles that fulfill specific roles. These vacuoles are generally less prominent than the central vacuoles found in plants.

Different Types of Vacuoles Present in Animals

Food Vacuoles - These are membrane-bound structures that form when cells take in food particles through a process called phagocytosis. They merge with lysosomes, which contain enzymes for digestion, to break down the ingested food.

Autophagic Vacuoles - These are created when cells need to recycle their own materials. They engulf damaged organelles or cellular waste and transport them to lysosomes for breakdown.

Secretory Vacuoles - These vacuoles are responsible for storing and releasing substances like hormones or neurotransmitters.

Contractile Vacuoles - Although more commonly found in protists, these vacuoles can sometimes be present in animal cells, particularly those in hypotonic environments. They play a role in osmoregulation by maintaining the cell's water balance and expelling excess water.

Fungal Vacuoles

Fungal vacuoles are organelles that carry out a variety of essential functions for the growth, differentiation, symbiosis, and pathogenesis of fungi. These acidic storage compartments are involved in protein degradation, maintaining cellular homeostasis, membrane trafficking, signalling processes, and nutrient storage. They play a role in glycoprotein turnover and hydrolysis, store phosphate, calcium, and amino acids, and help regulate pH, ion homeostasis, osmotic balance, and cytoplasmic detoxification. Additionally, they are crucial for maintaining an acidic lumen, which is important for various transport and storage activities [2].

Vacuoles are vital for the fungal lifestyle, participating in long-distance nutrient transport, hyphal extension and branching, cell-cycle timing, and the initiation of morphogenetic processes like appressorium formation and pseudohyphal growth. In pathogenic fungi, vacuoles support growth in nutrient-limited environments and are essential for forming appressoria, specialized structures that facilitate host invasion.

Differences from Plant and Animal Vacuoles

Although fungal vacuoles have similarities with plant vacuoles and mammalian lysosomes, they differ significantly in their functions and roles within the organism. Plant vacuoles are generally larger and take up a more substantial portion of the cell volume, often acting as a central storage area for ions, water, and nutrients. They also help maintain turgor pressure, which is essential for a plant cell's rigidity and structural support.

In contrast, mammalian lysosomes mainly focus on degrading macromolecules and detoxifying harmful substances. They possess various hydrolytic enzymes that can break down proteins, carbohydrates, nucleic acids, and lipids. Fungal vacuoles, while also involved in detoxification and macromolecule degradation, have a broader range of functions that are closely linked to the ecological niche and lifestyle of fungi.

Protist Vacuole

Protists, which are single-celled organisms, have evolved unique adaptations that allow them to thrive in a variety of environments. One such adaptation is the presence of vacuoles, which are membrane-bound organelles that are essential for various cellular functions. Living in freshwater or soil, protists encounter significant osmotic challenges because of the differences in water concentration between their internal environment and the surrounding medium. As a result, osmoregulation is necessary for these organisms to maintain their internal water balance stable.

To tackle these challenges, many protists have evolved a specialized organelle known as the contractile vacuole. This endomembrane organelle is vital for osmoregulation, as it collects excess water and expels it from the cell. By managing water content, contractile vacuoles help maintain the cell's osmolarity, preventing it from either bursting or shrinking due to osmotic imbalances. This function is crucial for the survival of protists in environments where water concentration can vary widely [3].

Structure

The structure of plant vacuoles is intricate and adaptable, featuring several important characteristics.

Tonoplast

The tonoplast is a selectively permeable phospholipid bilayer that encloses the vacuole in plant cells. It plays a crucial role in regulating the movement of ions, nutrients, and water, and helping to maintain cellular homeostasis. The tonoplast is dynamic and flexible, allowing it to interact with other cellular components and undergo fusion under certain conditions [4].

Embedded within the tonoplast are Tonoplast Intrinsic Proteins (TIPs), a group of aquaporins that function as water channels. These proteins facilitate the transport of water and small solutes across the membrane and are involved in various physiological processes, such as stomatal movement, vacuolar sequestration, and responses to environmental stress. The selective permeability of the tonoplast enables efficient compartmentalization within plant cells, ensuring the vacuole's distinct role in storage, detoxification, and osmoregulation.

TIPs are regulated by gating mechanisms that respond to various environmental cues, enabling plants to precisely control the movement of solutes across the tonoplast membrane and adapt to changing conditions. Additionally, the tonoplast contains receptors that can bind to signalling molecules, including hormones like abscisic acid. This binding initiates signalling cascades that can modify cellular metabolism or gene expression. For instance, when abscisic acid binds to its receptor, it can influence the activity of TIPs, thereby affecting the cell's water balance and ion concentrations [5].

To further illustrate the significance of Tonoplast Intrinsic Proteins (TIPs), the example of Oryza sativa, commonly known as rice, is particularly relevant. OsTIP1;2, OsTIP2;2, OsTIP4;1, and OsTIP5;1 are the TIPs of Oryza sativa that are capable of transporting water, while the TIPs OsTIP1;2, OsTIP3;2, and OsTIP4;1 can also facilitate the transport of glycerol (Fig. **1**) [5].

Nicotiana tabacum NtTIPa, an aquaporin, is a protein found in tobacco vacuoles that facilitates the transport of both water and specific non-electrolyte molecules like glycerol and urea [6]. The tonoplast is a dynamic membrane essential for regulating water and solute transport in plant cells. It contains TIPs and receptors that enable plants to effectively respond to environmental changes and maintain cellular homeostasis.

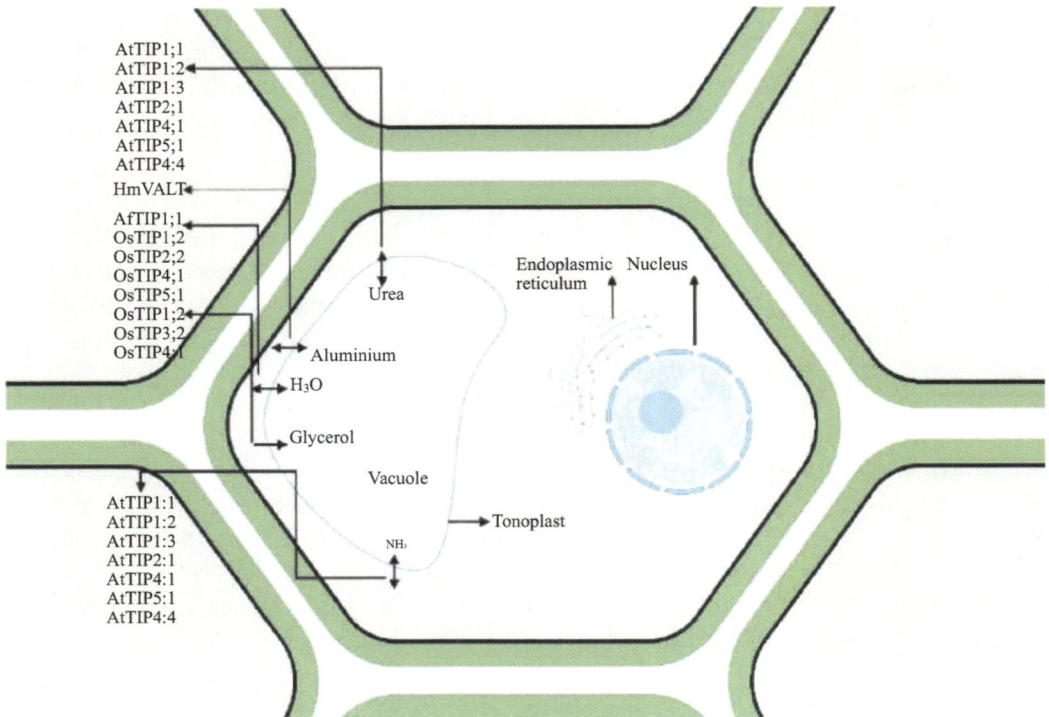

Fig. (1). The various Tonoplast Intrinsic Proteins (TIP) aiding in the transport of different molecules in a plant cell.

Morphological Changes

Vacuoles can alter in size and shape in response to different environmental factors. For instance, during stomatal movement, vacuoles in guard cells experience notable morphological changes, where smaller vacuoles can merge to create larger ones during stomatal opening, and then divide back into smaller vacuoles when closing [4].

Dynamic Structures

Vacuoles' shape and size change in response to the external factor. The vacuole can house various structures, including transvascular strands and spherical bodies known as "bulbs." These structures are often connected to the vacuole membrane and can frequently change shape. Trans-Vacuolar Strands (TVS) are flexible structures that link various areas of the cytoplasm inside the vacuole. They help in moving organelles and metabolites, which helps maintain the cell's structural organization. The actin cytoskeleton dynamically shapes and reorganizes TVS, enabling efficient cell-to-cell communication and adaptability to environmental fluctuations [7]. In tonoplast, there are double membrane invaginations, protruding into the vacuole lumen. It has been hypothesized that it may be the degrading factor of specific TIP [8]. The frequency of these bulbs decreases under starvation and as cell development progresses [4].

Vacuoles are indispensable for plant growth, as they regulate turgor pressure, a driving force behind cell expansion [9]. They also play a role in storing and breaking down cellular components, aiding the plant's response to both biotic and abiotic stresses. This is essential in the function of plant cells, enabling them to adjust to changing conditions and fulfill multiple roles within the cell.

Components

The sap within the Contractile Vacuole (CV) of Paramecium is vital for osmoregulation [9]. It helps the cell maintain its water balance by collecting excess water, which prevents the cell from bursting under osmotic pressure. The sap may also contain various ions that aid in maintaining the overall ionic balance within the cytosol. Research has concentrated on the ion content of the CV, especially sodium (Na^+) and potassium (K^+). In C. carolinensis, the concentration of Na^+ was found to be higher in the CV than in the cytosol, while K^+ levels were lower. Additionally, the sap may include bicarbonate and ammonium ions, which are byproducts of cellular metabolism.

The spongiome, a network of tubules and vacuoles, plays a role in water transport to the contractile vacuole. Known as the contractile vacuole complex, it contains V-ATPases (Vacuolar-type H^+-ATPases), proton-pumping enzymes that generate the membrane potential of the complex (CVC). These enzymes are essential for creating ionic gradients that drive water into the vacuole [10]. The sap also contains Ca^{2+} for osmoregulation. Under the conditions of calcium stress combined with a lack of external potassium (K^+), the CV accumulates Ca^{2+} to maintain K^+ hypertonicity relative to the cytosol, as the balance of Ca^{2+} and K^+ is crucial for osmoregulation. This accumulation is facilitated by transport processes, potentially including Ca^{2+} pumps or exchangers, that regulate the concentration of Ca^{2+} in the CV fluid [10]. The dynamics of the contractile vacuole's membranes, including the smooth and decorated tubules. The smooth membranes store bending energy, which is released during water expulsion [9].

Inclusion Bodies in the Vacuole

While inclusion bodies can be present in the cytoplasm, they are often seen within vacuoles as well. The size, shape, and composition of these structures can differ based on the type of cell and its specific functions. Common inclusion bodies found in vacuoles include:

Pigments

Pigments are substances that provide color to plants, flowers, and fruits. Examples are anthocyanins (red, purple, and blue pigments), carotenoids (orange, yellow, and red pigments), and chlorophyll (the green pigment). Crystals - These are solid, crystalline formations that can develop within vacuoles. They may consist of various materials, such as calcium oxalate, calcium carbonate, and silica. Crystals can fulfill multiple roles, including mineral storage, protection against herbivores, and providing structural support.

Vacuoles vs. Vesicles

A vesicle is a small, membrane-enclosed organelle within a cell that contains various types of fluids. On the other hand, a vacuole is a type of vesicle that primarily contains water. Vacuoles are larger than vesicles and are crucial for maintaining cellular structure and storage.

Structure

Both vacuoles and vesicles are membrane-bound organelles, but they vary in size and composition. Vesicles are usually smaller and can contain a mix of substances such as water, nutrients, enzymes, waste products, harmful compounds, and ions. In contrast, vacuoles are generally larger and mainly consist of water, acting as storage compartments.

Function

Vacuoles and vesicles have different roles in cellular metabolism. Vesicles are involved in various metabolic activities, providing temporary storage for food and enzymes while assisting in the transport of molecules. They also function as reaction chambers for chemical processes. Vacuoles, on the other hand, are mainly tasked with storing substances and play a crucial role in maintaining the structural integrity of cells, especially in plants.

Though both vesicles and vacuoles are essential for cellular functions, they are quite different in terms of size, composition, and purpose. Vesicles are smaller and multifunctional, whereas vacuoles are larger and primarily dedicated to storage and structural support for cells.

FUNCTION OF VACUOLES

Water Storage - Vacuoles hold significant amounts of water, which is essential for maintaining turgor pressure in plant cells. This pressure is what keeps the plant upright and supports its overall structure.

Nutrient Storage - The vacuole serves as a critical storage compartment for various nutrients that are essential for the cell's survival and function. For instance, vacuoles can store sugar, which can be used as an energy source by the cell; they can store protein like phaseolin, which is broken down and used to support the growth of the seedling, during germination, and can also store other molecules like ions for osmotic balance.

Waste Storage - Vacuoles temporarily hold waste products and harmful substances, keeping them separate from the cytoplasm. This includes detoxifying materials that could potentially damage the cell.

Ion Regulation - They play a key role in regulating internal ion concentrations, for regulating plant stress. Using Na^+/H^+ antiporters located on the vacuolar membrane stores Na^+ ions in response to salt stress. Small vesicles around the vacuole also likely store Na^+ and fuse with the main vacuole, increasing its capacity to sequester Na^+ [11].

Storage of Secondary Metabolites - Vacuoles are involved in storing secondary metabolites, like nicotine in certain plants, which can serve various ecological roles, including deterring herbivores.

Turgor Maintenance - By managing water content, vacuoles help maintain turgor pressure, which is vital for cell growth and structural integrity.

Support for Structural Rigidity - In plant cells, vacuoles contribute to the rigidity of the cell wall by maintaining pressure, which aids in the overall structural support of the plant.

Facilitating Transport - Vacuoles can assist in transporting stored materials to other parts of the plant during growth or stress responses, such as breaking down into smaller vesicles that can merge with the plasma membrane to release nutrients.

Temporary Storage of Sugars - During processes like nectar production, vacuoles serve as a source of sugars, directly participating in the secretion needed to attract pollinators.

Pigmentation – Vacuoles contain water-soluble pigments like anthocyanins and fat–soluble pigments like carotenoids and xanthophylls, which give colour to the fruits and vegetables in plants.

pH Regulation –Vacuoles help in maintaining the pH of the cell. Vacuolar ATPases (V-ATPases) play a vital role in keeping cellular organelles acidic. These proton pumps harness energy from ATP to move protons against their concentration gradient, thereby establishing an acidic environment. The activity of V-ATPases is influenced by both internal and external factors, such as the dissociation of the V1 domain and the presence of counterion channels. The efficiency of proton transport depends on the quantity of V-ATPases, their operational status, and the equilibrium between proton pumping and leakage. Maintaining proper pH levels is crucial for various cellular activities, including membrane trafficking and the elimination of pathogens. Any disruption in pH balance can lead to significant issues in these essential processes [12].

Autophagy – Autophagic vacuoles encapsulate the cytoplasmic components, like damaged organelles and proteins, and then fuse with the lytic vacuole to degrade the components [13].

Signalling – Vacuole can release and receive signals by releasing and receiving Ca^{2+} ions from the cytosol. It is aided by a Ca^{2+} permeable membrane called TPC1 (Two–Pore Channel 1)

Defense mechanism - The plant cell lacks immune cells, like those in animal cells. The lytic vacuole can defend the plant cell using hydrolytic enzymes. There are two types of defence systems used by vacuoles.

Vacuolar Membrane-collapse System

This system helps in defending against the virus. When a virus infects the plant cell, the enzyme, Vacuolar Processing Enzyme (VPE), is activated. VPE lyses the vacuolar membrane, leading to the release of vacuolar contents, including

nucleases and other enzymes, into the cytoplasm. The released nucleases target and degrade viral nucleic acids, thus preventing the replication of the virus inside the cell. VPE accumulation prevents the formation of visible lesions [14].

Vacuole-plasma Membrane-fusion System

This system helps in defending against bacteria. When a virulent bacterium infects a plant cell, cell shrinkage and cytoplasmic aggregation occur. The vacuolar membrane would then merge with the plasma membrane, thus connecting the vacuole to the extracellular region. The vacuolar defence proteins are then released into the extracellular region, where the bacterial cells are proliferating. The defence protein exhibits antibacterial properties, thus inhibiting the bacterial growth in the extracellular region (Fig. **2**) [14].

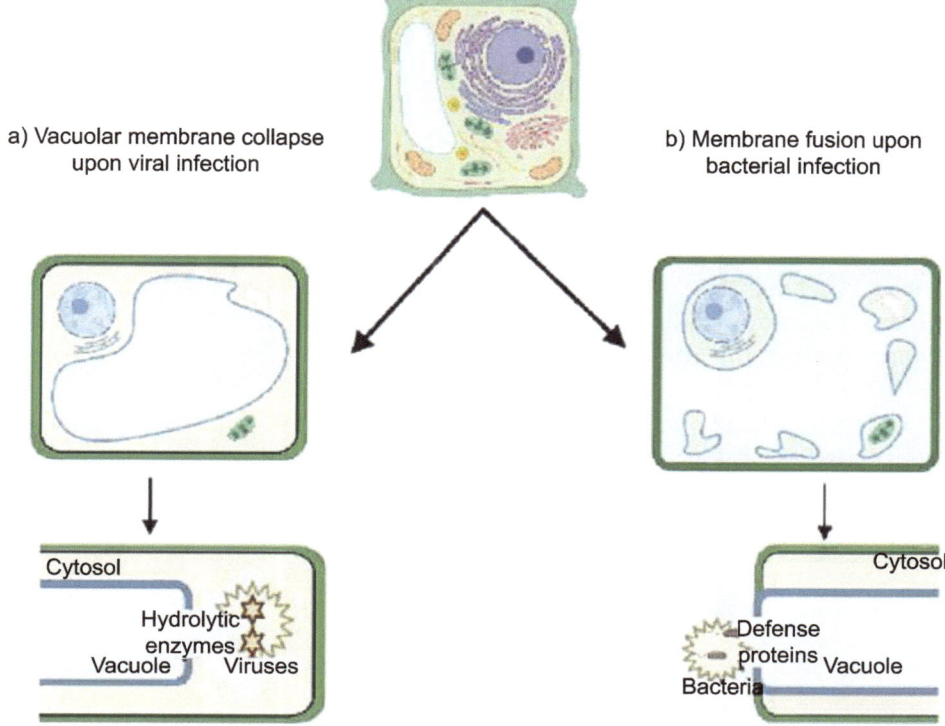

Fig. (2). Vacuolar defence mechanism in a plant cell. a) Vacuolar membrane collapse upon viral infection, where the lysed vacuole releases the hydrolytic enzyme in the cytosol to kill viruses. b) Membrane fusion upon bacterial infection, where the vacuolar membrane fuses with the plasma membrane to release defence proteins to inhibit bacterial growth.

Autophagosome-lysosome Fusion

Autophagosome-lysosome fusion is a process by which autolysosomes are formed in eukaryotes. Autolysosomes help in breaking down cellular components like misfolded proteins and other macromolecules that are no longer needed.

Formation of Autolysosomes

- Phagophore Formation
    - A double-membrane structure known as a phagophore begins to form, surrounding various cellular components.
- Autophagosome Maturation

- The phagophore then closes, resulting in the creation of an autophagosome.
- The fusion of the autophagosome with the lysosome is a critical step in the autophagic process and requires the involvement of specific SNARE (Soluble N-ethylmaleimide-sensitive factor Attachment Protein Receptor) proteins.
- SNARE proteins, including STX17, SNAP29, and VAMP7, bind to the autophagosome.
* Fusion with Late Endosome
    - HOPS and EPG-5 help in facilitating the fusion of the autophagosome and late endosome.
    - SNARE proteins also play a crucial role in facilitating the fusion between the autophagosome and the late endosome.
* Formation of Amphisome
    - The outer membrane of the autophagosome and the membrane of the late endosome merge, forming an amphisome.
* Fusion with Lysosome
    - The amphisome and the lysosome fuse with one another, leading to the formation of an autolysosome.
    - The materials captured inside the autolysosome are broken down by lysosomal enzymes (Fig. 3) [15].

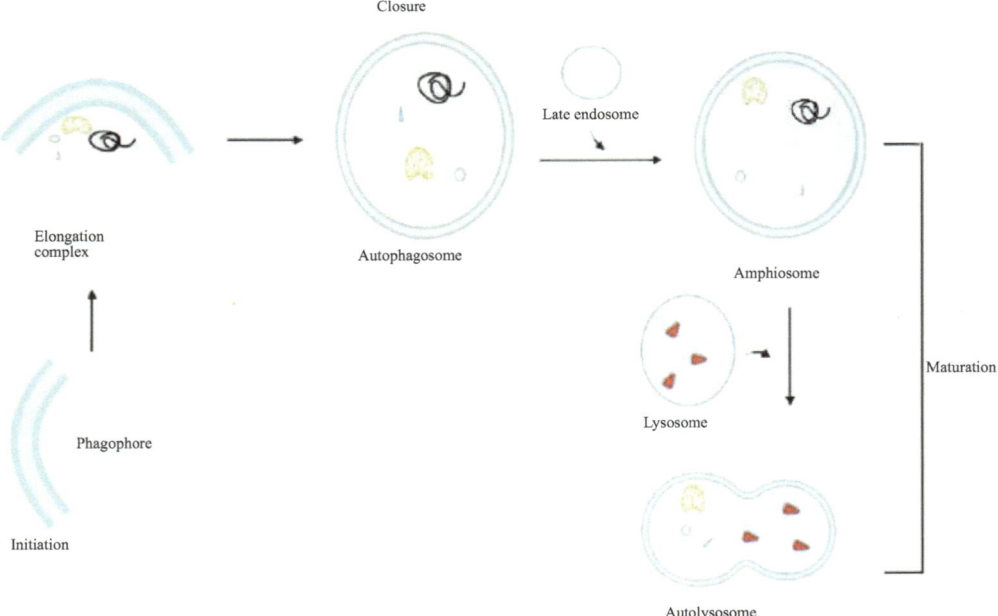

Fig. (3). Steps involved in the formation of an autolysosome.

DIAGNOSTIC AND THERAPEUTIC APPLICATION OF VACUOLES

In Tay-Sachs disease, lysosomes accumulate GM2 gangliosides due to a deficiency in the enzyme hexosaminidase A, leading to the presence of numerous electron-lucent membrane-bound vacuoles in various cell types, including cultured fibroblasts and peripheral blood lymphocytes. These lysosomal abnormalities serve as key diagnostic indicators and can be detected through biochemical assays or imaging techniques. Such diagnostic approaches use vascular abnormalities as key indicators of disease presence and severity [16, 17].

In neurodegenerative diseases such as Alzheimer's disease, vacuoles in neurons can reveal pathological changes. Vacuolar changes are associated with the accumulation of amyloid-beta plaques and tau protein tangles. Diagnosing tau accumulation inside and outside neurons primarily reflects later stages of disease, when significant neuronal damage has already occurred. In contrast, vacuolar analysis can reveal early signs of cellular dysfunction, such as impaired autophagy and protein clearance, before tau tangles fully form. This makes vacuolar changes a more sensitive and dynamic marker for early detection and monitoring of neurodegenerative diseases, by analysing the vacuolar proteome or using imaging techniques to observe these changes, diagnosis and monitor the progression of neurodegenerative diseases more effectively [18].

Maintaining vacuolar pH is essential for cancer cell survival. A higher vacuolar pH helps cancer cells evade acid-induced cell death, considering the fact that they are generally more tolerant to acidic environments. A higher vacuolar pH allows cancer cells to maintain key cellular functions like nutrient storage, protein degradation, and resistance to stress, which helps them evade certain forms of cell death. Additionally, altered vacuolar pH can disrupt ion homeostasis, impair protein degradation, and affect cell-cell adhesion—conditions that make cells more vulnerable to entosis. Inhibiting vacuolar acidification disrupts these processes, making cells more susceptible to treatment. Entosis, while not directly pH-dependent, can still be influenced by these altered cellular conditions, contributing to cell death under specific circumstances. Inhibiting vacuolar acidification has been shown to increase cell death, suggesting that manipulating vacuolar pH could enhance cancer treatment efficacy [19].

RECENT TRENDS

The enhancement of imaging technologies, such as super-resolution microscopy, specifically STED (Stimulated Emission Depletion) microscopy. This cutting-edge technology has provided unprecedented insights into vacuole dynamics, revealing intricate details about how vacuoles interact with other cellular components, particularly during stress responses. Studies focused on model plants, such as *Arabidopsis thaliana*, where researchers have explored the modification of vacuoles to enhance their ability to manage ion balance and osmotic pressure. These studies aim to bolster the plants' resilience to extreme environmental conditions such as drought and high salinity [20].

A notable example is the successful use of vacuoles in yeast cells to encapsulate and deliver anticancer drugs. This innovative method allows for targeted drug delivery, reducing side effects and improving therapeutic effectiveness. The vacuole-mediated drug delivery system works by encapsulating anticancer drugs within the vacuoles of yeast cells, which naturally possess an efficient system for storing and transporting various substances. These yeast cells can be engineered to target specific cancer cells, where the vacuoles release their drug payloads in response to changes in the microenvironment, such as pH or enzymatic activity. This targeted release allows for higher concentrations of the drug to be delivered directly to cancer cells, reducing systemic exposure and minimizing side effects, while enhancing the overall therapeutic efficacy [21].

CONCLUSION

This chapter has explored the diverse roles of vacuoles across different organisms, including plants, animals, fungi, and protists. Vacuoles are essential for various cellular functions such as water storage, pH regulation, detoxification, and autophagy. In plants, vacuoles play a crucial role in maintaining turgor pressure, storing nutrients, and regulating ion balance. In animal cells, vacuoles are smaller and primarily involved in intracellular transport, waste management, and storage. The structural and functional differences between plant and animal vacuoles highlight their specialized adaptations to cellular needs. Additionally, vacuole-specific proteins, such as Tonoplast Intrinsic Proteins (TIPs), contribute to selective permeability and efficient transport across the membrane. Recent research has expanded our understanding of vacuoles, demonstrating their potential applications in drug delivery, stress tolerance in plants, and cellular homeostasis. These findings highlight the significance of vacuoles beyond their traditional roles, opening new avenues in both medical and agricultural research.

ACKNOWLEDGEMENTS

We would like to acknowledge Ms. Kruthika P, Ms. Raksa Arun, Ms. Srisri Satish Kartik, Ms. Sanjana Dhayalan, and Mr. Sasitharan T for their contributions in proofreading and editing the work.

REFERENCES

[1] Tan X, Li K, Wang Z, Zhu K, Tan X, Cao J. A Review of Plant Vacuoles: Formation, Located Proteins, and Functions. Plants 2019; 8(9): 327.
[http://dx.doi.org/10.3390/plants8090327] [PMID: 31491897]

[2] Veses V, Richards A, Gow NAR. Vacuoles and fungal biology. Curr Opin Microbiol 2008; 11(6): 503-10.
[http://dx.doi.org/10.1016/j.mib.2008.09.017] [PMID: 18935977]

[3] More KJ, Kaur H, Simpson AGB, Spiegel FW, Dacks JB. Contractile vacuoles: a rapidly expanding (and occasionally diminishing?) understanding. Eur J Protistol 2024; 94: 126078.
[http://dx.doi.org/10.1016/j.ejop.2024.126078] [PMID: 38688044]

[4] Zhang C, Hicks GR, Raikhel NV. Plant vacuole morphology and vacuolar trafficking. Front Plant Sci 2014; 5: 476.
[http://dx.doi.org/10.3389/fpls.2014.00476] [PMID: 25309565]

[5] Sudhakaran S, Thakral V, Padalkar G, et al. Significance of solute specificity, expression, and gating mechanism of tonoplast intrinsic protein during development and stress response in plants. Physiol Plant 2021; 172(1): 258-74.
[http://dx.doi.org/10.1111/ppl.13386] [PMID: 33723851]

[6] Gerbeau P, Güçlü J, Ripoche P, Maurel C. Aquaporin Nt-TIPa can account for the high permeability of tobacco cell vacuolar membrane to small neutral solutes. Plant J 1999; 18(6): 577-87.
[http://dx.doi.org/10.1046/j.1365-313x.1999.00481.x] [PMID: 10417709]

[7] Hoffmann A, Nebenführ A. Dynamic rearrangements of transvacuolar strands in BY-2 cells imply a role of myosin in remodeling the plant actin cytoskeleton. Protoplasma 2004; 224(3-4): 201-10.
[http://dx.doi.org/10.1007/s00709-004-0068-0] [PMID: 15614481]

[8] Maîtrejean M, Vitale A. How are tonoplast proteins degraded? Plant Signal Behav 2011; 6(11): 1809-12.
[http://dx.doi.org/10.4161/psb.6.11.17867] [PMID: 22057339]

[9] Allen RD. The contractile vacuole and its membrane dynamics. BioEssays 2000; 22(11): 1035-42.
[http://dx.doi.org/10.1002/1521-1878(200011)22:11<1035::AID-BIES10>3.0.CO;2-A] [PMID: 11056480]

[10] Stock C, Grønlien HK, Allen RD. The ionic composition of the contractile vacuole fluid of Paramecium mirrors ion transport across the plasma membrane. Eur J Cell Biol 2002; 81(9): 505-15.
[http://dx.doi.org/10.1078/0171-9335-00272] [PMID: 12416727]

[11] Hamaji K, Nagira M, Yoshida K, et al. Dynamic aspects of ion accumulation by vesicle traffic under salt stress in Arabidopsis. Plant Cell Physiol 2009; 50(12): 2023-33.
[http://dx.doi.org/10.1093/pcp/pcp143] [PMID: 19880402]

[12] Huynh KK, Grinstein S. Regulation of vacuolar pH and its modulation by some microbial species. Microbiol Mol Biol Rev 2007; 71(3): 452-62.
[http://dx.doi.org/10.1128/MMBR.00003-07] [PMID: 17804666]

[13] Eskelinen EL. Maturation of autophagic vacuoles in Mammalian cells. Autophagy 2005; 1(1): 1-10.
[http://dx.doi.org/10.4161/auto.1.1.1270] [PMID: 16874026]

[14] Shimada T, Takagi J, Ichino T, Shirakawa M, Hara-Nishimura I. Plant Vacuoles. Annu Rev Plant Biol 2018; 69(1): 123-45.

[15] Lőrincz P, Juhász G. Autophagosome-Lysosome Fusion. J Mol Biol 2020; 432(8): 2462-82.
[http://dx.doi.org/10.1016/j.jmb.2019.10.028] [PMID: 31682838]

[16] Ferreira CR, Gahl WA. Lysosomal storage diseases. Transl Sci Rare Dis 2017; 2(1-2): 1-71.
[http://dx.doi.org/10.3233/TRD-160005] [PMID: 29152458]

[17] Lopez Vasquez K. Tay-Sachs disease. J Neonatal Nurs 2020; 26(6): 316-8.
[http://dx.doi.org/10.1016/j.jnn.2020.02.001]

[18] Hondius DC, Koopmans F, Leistner C, *et al.* The proteome of granulovacuolar degeneration and neurofibrillary tangles in Alzheimer's disease. Acta Neuropathol 2021; 141(3): 341-58.
[http://dx.doi.org/10.1007/s00401-020-02261-4] [PMID: 33492460]

[19] Su Y, Ren H, Tang M, *et al.* Role and dynamics of vacuolar pH during cell-in-cell mediated death. Cell Death Dis 2021; 12(1): 119.
[http://dx.doi.org/10.1038/s41419-021-03396-2] [PMID: 33483474]

[20] Scheuring D, Löfke C, Krüger F, *et al.* Actin-dependent vacuolar occupancy of the cell determines auxin-induced growth repression. Proc Natl Acad Sci USA 2016; 113(2): 452-7.
[http://dx.doi.org/10.1073/pnas.1517445113] [PMID: 26715743]

[21] Choi W, Heo MY, Kim SY, Wee JH, Kim YH, Min J. Encapsulation of daunorubicin into Saccharomyces cerevisiae-derived lysosome as drug delivery vehicles for acute myeloid leukemia (AML) treatment. J Biotechnol 2020; 308: 118-23.
[http://dx.doi.org/10.1016/j.jbiotec.2019.12.008] [PMID: 31846628]

CHAPTER 9

Plasma Membrane: Gateway and Sentinel of Cellular Exchange

S. Aswini[1], K. Kumaran[1], Sanjana Dhayalan[1], Prakash Gangadaran[2,3,4,*] and K.N. Aruljothi[1]

[1] *Department of Genetic Engineering, SRM Institute of Science and Technology, Kattankulathur, Chengalpattu, Tamil Nadu, India*

[2] *Department of Nuclear Medicine, School of Medicine, Kyungpook National University, Daegu, Republic of Korea*

[3] *Cardiovascular Research Institute, Kyungpook National University, Daegu, Republic of Korea*

[4] *BK21 FOUR KNU Convergence Educational Program of Biomedical Sciences for Creative Future Talents, School of Medicine, Kyungpook National University, Daegu, Republic of Korea*

Abstract: The plasma membrane is a critical cellular structure that acts as a selective barrier, controlling the movement of molecules in and out of the cell while enabling communication with the external environment. This chapter provides an in-depth exploration of the plasma membrane, beginning with its structure and the fluid-mosaic model, which describes the dynamic organization of lipids, proteins, and carbohydrates. It then examines the various functions of the membrane, including maintaining cellular integrity, supporting signal transduction, and regulating the movement of substances through mechanisms like passive diffusion, facilitated diffusion, active transport, and group translocation. Key transport proteins such as glucose transporters and ion channels are discussed in detail, highlighting their roles in maintaining cellular homeostasis. The chapter also addresses the pathological significance of plasma membrane dysfunction, linking abnormalities to diseases such as hypertension, sphingolipid-related disorders, and neuronal vulnerabilities. Further, it explores the role of the plasma membrane in neurotrophic signaling and insulin regulation. Finally, recent advances in therapeutic approaches, including CAR-T cell therapy and liposomal drug delivery systems, are examined for their potential in disease treatment.

Keywords: Active Transport, Aquaporin, Bilayer, Diffusion, Endocytosis, Exocytosis, Facilitated Diffusion, Insulin, Lipid Raft, Membrane Potential, Neurotransmitter, Phospholipid, Potassium Channel, Sphingolipidoses..

---

[*] **Corresponding author Prakash Gangadaran:** Department of Nuclear Medicine, School of Medicine, Kyungpook National University, Daegu, Republic of Korea; E-mail: prakashg@knu.ac.kr

K.N. Aruljothi, Prakash Gangadaran, Krishnan Anand, Satish Ramalingam, K. Kumaran & Kruthika Prakash (Ed.)
All rights reserved-© 2026 Bentham Science Publishers

INTRODUCTION

The plasma membrane is a lipid bilayer with embedded proteins that control the movement of substances in and out of the cell while enabling interactions. Organellar membranes have a similar structure but differ in composition and function based on the organelle's role. Membrane-bound organelles include the mitochondria, endoplasmic reticulum, Golgi apparatus, lysosomes, and peroxisomes. Non-membranous organelles include ribosomes, centrioles, cilia, and flagella [1].

STRUCTURE

Gorter and Grendel Membrane Theory (1920)

This theory proposes that the plasma membrane is made up of a lipid bilayer. An experimental investigation on erythrocytes from different sources was carried out, and the experimental results supported their hypothesis. Gorter and Grendel failed to explain the membrane functions and could not explain the other membrane components other than lipids. This model set the stage for later models by highlighting the importance of lipid organization (Fig. **1**) [2].

Fig. (1). Depiction of the difference between the garter and Grendel membrane model and the Davson-Danielli model.

Pauci-molecular Theory

In 1935, Davson and Danielli came up with a model for plasma membrane called the Pauci-molecular theory. According to this theory, the membranes have a central "lipoid" region, which in general is surrounded by monolayers of lipids and is covered by protein sheets. This model is like a sandwich model made of protein-lipid-protein. In the year 1959, J. David Robertson proposed that all the cellular membranes possess a similar underlying structure, the unit membrane. He used extensive metal staining for the visibility of plasma membranes under an electron microscope. When observed under a microscope, the cell membrane exhibits a trilaminar appearance, featuring a lighter inner region and two darker outer regions. Robertson noted that the membrane consists of a lipid bilayer with sheets of mucoproteins on both sides, which supports the pauci-molecular theory [3]. The pauci-molecular theory proposed that all the membranes have equivalent structure, the same thickness, and the same lipid-protein ratio in the membrane surface. The Pauci-molecular theory couldn't explain the presence of membrane proteins outside the membrane or the active transport of macromolecules. It also failed to account for the dynamic nature and heterogeneity of membrane structures. Until 1972, the most accepted plasma membrane model was the pauci-molecular theory. After 1972, the scenario completely changed [4].

Fluid Mosaic Model (1972)

S. Jonathan Singer and Garth Nicolson proposed the fluid-mosaic model. This theory says that the plasma membrane is made up of a lipid bilayer where proteins are embedded in it. The phospholipid bilayer illustrates the effects of hydrophilic and hydrophobic interactions in lipids. The non-polar fatty acid chains of the phospholipid stand away from being in contact with water by increasing the hydrophobic interaction. The ionic groups are in direct contact with the aqueous phase, thereby giving rise to hydrophilic interactions. The hydrophilic and hydrophobic interactions are crucial for maintaining the membrane's integrity and functionality, as they help create a stable barrier that controls the movement of substances in and out of the cell. The dipole-dipole interaction between ionic pairs of phosphatidylcholines (zwitterionic phospholipid) is the reason for the stable lipid bilayer structure. According to this theory, there are two types of membrane proteins-peripheral and integral (Fig. **2**).

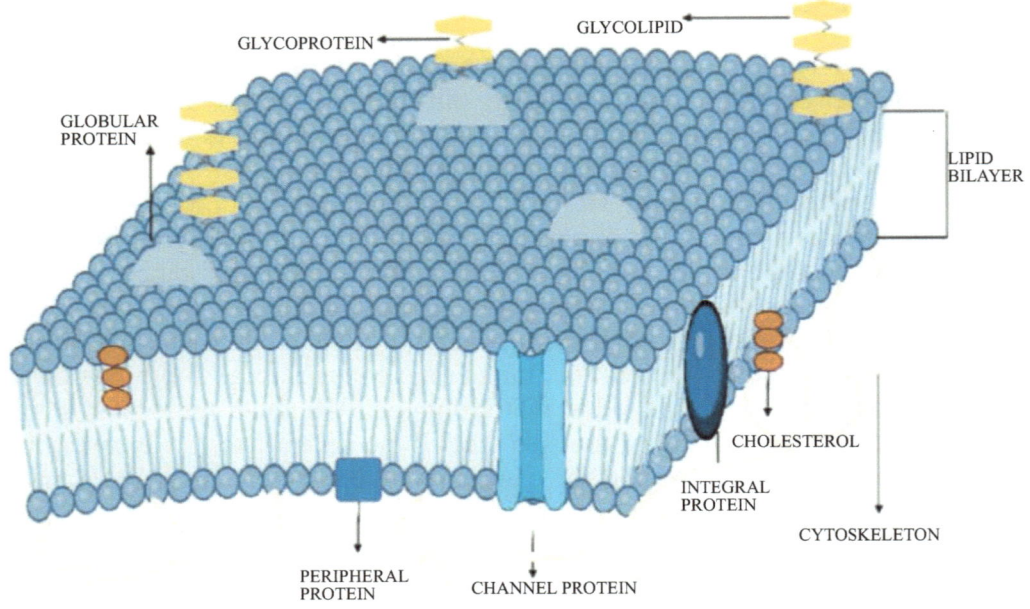

Fig. (2). Fluid mosaic model.

Peripheral Proteins (Extrinsic Proteins)

These coat the outer and inner surfaces of the phospholipid bilayer of the plasma membrane. The association with the hydrophilic parts of lipids or intrinsic proteins is caused by ionic charges. These proteins are fully soluble in water. These require mild treatments like the addition of chelating agents to dissociate from the membrane. In their dissociated state, the proteins are fairly soluble in neutral buffers, indicating that they are attached to the membrane through weak covalent interactions. The cytochrome c, which plays a major role in the electron transport chain by mitochondria, gets dissociated from the phospholipid layer under high electrolyte concentration is a good example of a peripheral protein [5].

Integral Proteins (Intrinsic Proteins)

These proteins directly insert into the hydrophobic part of the membrane. They require extreme treatment with organic solvents, protein denaturants, bile acids, and reagents to dissociate from the membrane. In most cases, they remain associated with the membrane even after the isolation. They are highly insoluble in neutral buffers. On the basis of topology, there are three classes for intrinsic proteins-monotopic, bitopic, and polytopic.

- Monotopic- These proteins are embedded in the membrane and do not span through the membrane. Their hydrophilic domain is exposed to hydrophilic surroundings.
- Bitopic- These proteins span the membrane once. Examples include spike glycoproteins of enveloped viruses like influenza, vaccinia virus, and HIV. In HIV, the protruding protein gp160 is responsible for the recognition and attack of leucocytes. Glycophorin, which helps the erythrocyte to carry glucose, is also a bitopic protein.
- Polytopic- They have multiple spanning domains in the membrane. Bacteriorhodopsin is a protein present in some photosynthetic bacteria, is a light-driven proton pump that helps in the production of ATP by creating a proton gradient.
- Lipid raft- The plasma membrane of animal cells contains mainly four phospholipids (phosphatidylcholine, phosphatidylethanolamine, phosphatidylserine, and sphingomyelin), glycolipids, and cholesterol. The outer leaflet of the plasma membrane consists of phosphatidylcholine, glycolipids, and sphingomyelin, while the inner leaflet mainly consists of phosphatidylethanolamine and phosphatidylserine. Phosphatidylinositol is also present in the inner leaflet and is involved in various cell signalling pathways. Cholesterol and sphingolipids create membrane domains known as lipid rafts, which allow for free diffusion within the plasma membrane. These lipid rafts can move laterally and interact with membrane proteins, highlighting their importance in cell movement, endocytosis, and signalling [6]. The Fluid Mosaic Model is the most accepted for plasma membranes, depicting them as a flexible, dynamic bilayer with embedded proteins and carbohydrates. It explains the membrane's fluidity, asymmetry, and selective permeability, which are crucial for cellular processes and functions [7].

Handerson and Unwin's Membrane Theory

In 1975, Henderson and Unwin elucidated the three-dimensional structure of the purple membrane protein bacteriorhodopsin, marking the first time the α-helix structure of a membrane protein was demonstrated. This addresses the rotational movement and mobility of membrane proteins. The molecules are arranged around a 3-fold axis with a 2nm-wide space filled with lipids. It is found that the protein extends from both sides of the lipid bilayer and is composed of seven α-helices packed together running perpendicular to the plane of the membrane. Henderson and Unwin's model contributed to understanding the structural role of proteins in membranes, emphasizing their involvement in transport and signalling, and refining the concept of membrane permeability and function [8].

Evolutionary Aspects of Plasma Membrane

Simpler amphiphilic molecules (having both hydrophilic and hydrophobic parts), which are simpler to make than phospholipids and are capable of integrating and accumulating other amphiphilic molecules, are likely to have created membranous boundaries during early prebiotic evolution.

FUNCTIONS OF PLASMA MEMBRANE

Acting as a Physical Barrier

The plasma membrane separates the cell's interior contents from the external environment. Playing a key role in maintaining homeostasis by controlling the movement of ions, nutrients, and waste, ensuring a stable internal environment. The cytoskeleton interacts with the plasma membrane through multiple structural and functional connections that help maintain cell shape, enable movement, and regulate intracellular signaling. This organises the cell shape by using a dynamic framework of several polymer networks.

Selective Permeability

It has the capability to balance the concentration of substances inside the cell. The substances include: ions (Ca^{2+}, Na^+, K^+, and Cl^-), sugar, fatty acids, amino acids, and CO_2 (leaving the cell as waste). The phospholipid bilayer of the plasma membrane only permits substances meeting certain criteria to pass through it. Fat-soluble drugs and Fat-soluble vitamins like A, E, D, and K pass through the plasma membrane in the digestive tract. Oxygen and carbon dioxide pass through the plasma membrane through diffusion.

Endocytosis by the Plasma Membrane

The plasma membrane surrounds foreign particles or substances to be engulfed, which then bud off to form a vesicle, which acts as a transporter. The term endocytosis was coined by the scientist Christian deDuve in the year 1963. Endocytosis includes:

- Phagocytosis (or eating): It involves the identification and ingestion of particles larger than $0.5\mu m$ into the plasma membrane through a derived vesicle called a phagosome. Phagocytes engulf the microbial pathogens like bacteria and apoptotic cells-apoptosis is getting rid of the cells in the body that have been damaged beyond repair. Most phagocytic cells, such as macrophages and neutrophils, are selective in what they digest due to factors like pH differences (dead cells tend to have a higher pH than living ones) and surface recognition

markers that help identify harmful or foreign particles. This selectivity allows them to target pathogens or damaged cells while avoiding healthy tissue.
- Pinocytosis (or drinking): The name Pinocytosis was found and given by a scientist called Warren Lewis. It is the vesicular uptake of small particles (like lipoproteins, immune complexes, colloids), soluble macromolecules (hormones, antibodies, toxins, enzymes).

Exocytosis

The process of secreting stored macromolecules in the vesicles by the fusion of vesicles to the plasma membrane is nothing but exocytosis. In plants, exocytosis is controlled by the changes in the cytosolic Ca^{2+} ions. Exocytosis also plays a crucial role in neurotransmitter release in neurons, where the fusion of vesicles containing neurotransmitters with the plasma membrane allows these chemical signals to be released into the synaptic cleft, transmitting signals between neurons.

Exocytosis is of two types,

- Constitutive exocytosis- This is when vesicles or granules fuse with the plasma membrane impulsively.
- Regulated exocytosis- This is when vesicles or granules fuse with the plasma membrane in response to cell stimulation, like an increase in intracellular calcium ions (Ca^{2+}), changes in membrane potential, or the binding of signalling molecules like hormones or neurotransmitters to cell surface receptors.

These fusions are followed by the discharge of vesicles to the extracellular space.

Facilitating Communication and Signalling among Cells

Cells receive signals from other cells in the form of chemicals. Individual cells receive many signals at the same time. These chemical signals can be growth factors, hormones, or neurotransmitters. Among these, neurotransmitters are the short-range signalling molecules. Signalling molecules bind to the proteins called receptors present in the plasma membrane and initiate a physiological response [7].

Providing Shape to the Cytoskeleton and Maintaining Cell Potential

The cells depend on adhesive proteins to maintain physical contact with the neighbouring cells. The epithelial cells, cells of the gastrointestinal tract, are some of the good examples of cell-to-cell recognition. Various types of cell-cell adhering molecules include cadherins, nectins. Cell-cell adhesion decides the polarity of the cells within tissues. Alteration in cell adhesions may lead to tumorigenesis and metastasis [9].

Membrane Transport Mechanism

A biological membrane is semipermeable, indicating that it is mostly impermeable to most solutes (including various biomolecules and salts) in the surrounding solution, while permitting the passage of other molecules, mainly water. Charles Overton's groundbreaking research in the 1890s contributed to the development of the important concept of asymmetrical transmembrane distribution and, as a result, permeability differences between water and other solutes. The lipid bilayer's hydrophobic interior, a very thin layer measuring about 40 Å in thickness, primarily serves as a barrier to the movement of solutes. Within different membranes, there is a change in the permeability; permeability is lower in the case of closely packed lipids that make up the lipid bilayer. The permeability of solutes varies, with sodium ions (Na^+) at approximately 10^{-12} cm/s and water (H_2O) at about 0.2×10^{-2} cm/s [10].

Fick's First Law

In 1855, physiologist Adolf Fick suggested that solutes tend to move from areas of higher concentration to areas of lower concentration. This idea later became known as Fick's laws of diffusion. Both free solution and diffusion across membranes are governed by the same principles. By measuring the concentration and flow of salt diffusing between two connected reservoirs connected by water reservoirs, he was able to formulate his proposal.

Diffusion rate = -DA(dc/dx)

D = diffusion coefficient (Smaller molecules have a higher diffusion coefficient (D value))

A = the area of cross-section through which diffusion occurs.

dc/dx = solute concentration gradient

Small solutes have high diffusion coefficients, allowing them to diffuse rapidly. Under a specific given physiological condition, the diffusion rate remains constant. This equation can explain the limitation of the enzymatic reaction rate [11].

Osmosis

Water cannot pass through the semipermeable barrier without the force of osmotic pressure. Until its potential is zero, its net movement continues. William Hewson's pioneering studies on the hemolysis of red blood cells in the 1770s provided an initial demonstration of the basic principles of osmosis. A reflection coefficient (σ) describes deviation from ideality. When a solute is completely impermeable (ideal semipermeable barrier), σ = 1. σ = 0 if the solute is completely permeable (permeability equals water). With σ=0.75 to 1, biological membranes make a good semipermeable membrane. Hypotonic solutions have lower solute concentrations than cells, causing them to swell. Hypertonic solutions have higher solute concentrations, causing cells to shrink. Isotonic solutions have equal solute concentrations inside and outside cells, so cells remain the same size (Fig. **3**) [12].

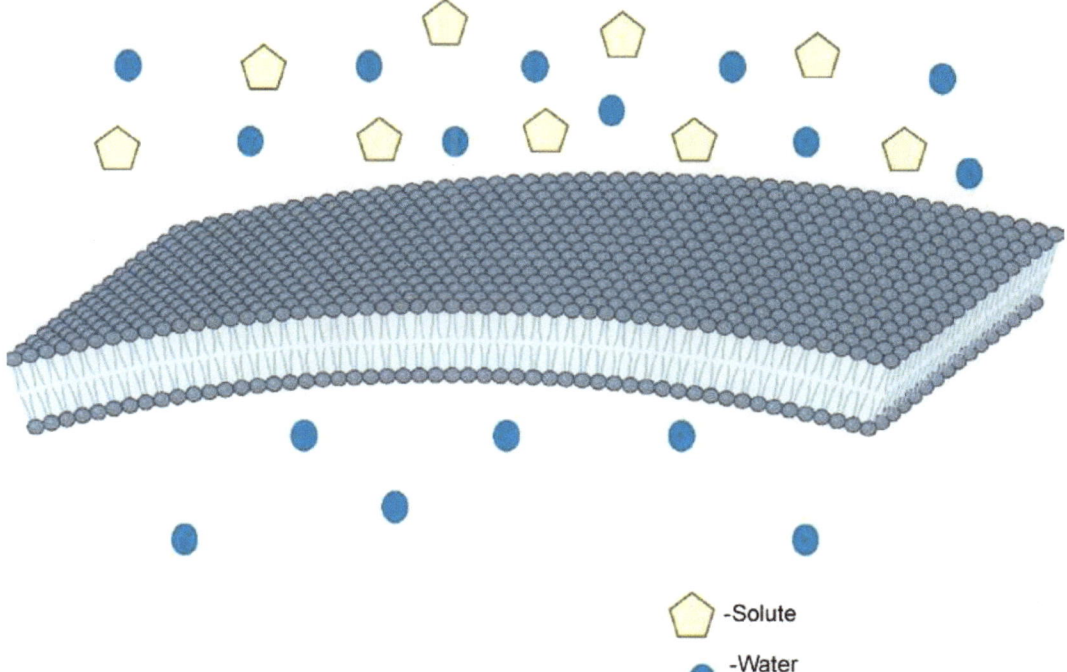

Fig. (3). Osmosis- the movement of water molecules through a semipermeable membrane.

Passive Diffusion

The solution reaches equilibrium across the membrane through passive diffusion, which doesn't require any extra energy source beyond that present in the electrochemical gradient of the solute. Passive diffusion can take two different forms: simple passive diffusion, in which the solute simply dissolves into and diffuses through the lipid layer, and facilitated passive diffusion, in which the solute only crosses the membrane at specific locations where diffusion is aided by carriers or facilitators that are specifically designed for the solute in question. Neither of the passive diffusions do not requires ATP for diffusion (Fig. **4**).

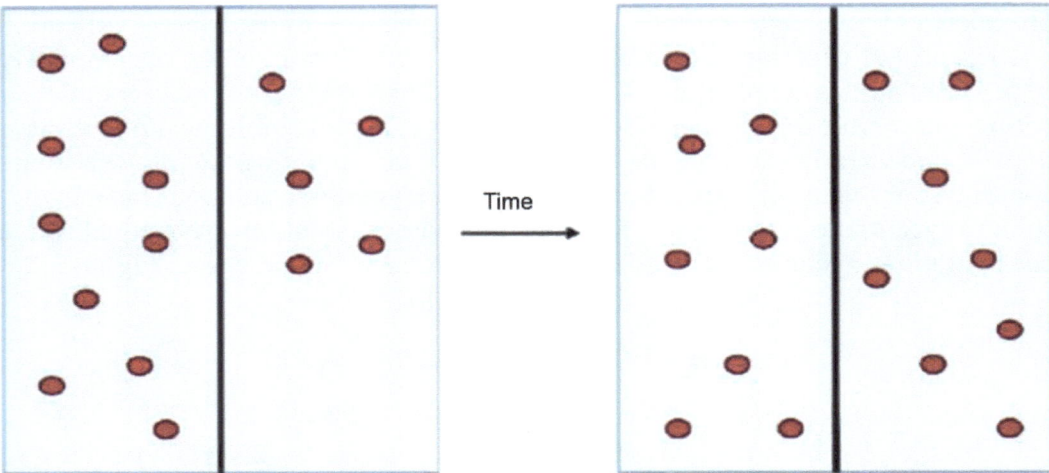

Fig. (4). Passive transport-movement of solutes without the requirement of energy.

Reasons for the high permeability of water

• Water is small compared to other solutes, so, dissolves better through the lipid bilayer.

• Water crosses through nonlamellar regions

• Water permeability is higher in places with regions of packing defect.

• There are certain integral proteins called aquaporins, which serve as water channels.

• Only water, tiny non-charged solutes, and gases are able to penetrate a membrane through simple passive diffusion. Large or charged solutes are practically prohibited from membranes and so cannot transverse a membrane by just passive diffusion alone.

Facilitated Diffusion

Similar to passive diffusion, facilitated diffusion (carrier-mediated) depends on the natural energy present in a solute gradient. The resulting solute distribution achieves equilibrium across the membrane without requiring additional energy for transport. Unlike passive diffusion, facilitated diffusion requires a transmembrane integral protein or carrier to assist in moving the solute across the membrane (Fig. 5). The main types of facilitators are carriers and gated channels. Facilitated diffusion of glucose across cell membranes, followed by enzymatic conversion, often deviates from typical Michaelis-Menten kinetics, showing a biphasic curve [13].

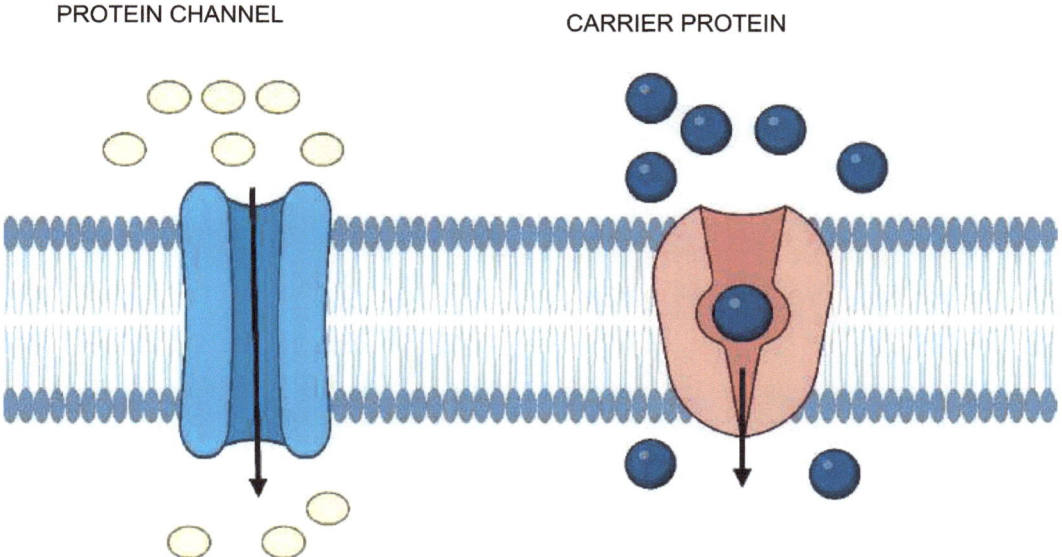

Fig. (5). Carrier proteins bind and transport molecules across the membrane *via* conformational changes, while protein channels provide a pore for passive diffusion of ions or small molecules.

Glucose Transporter

A prime example of a facilitated carrier is the glucose transporter (GLUT). Instead of being just over 100KJ/mol, glucose activation energy is merely 16KJ/mol. This significant difference is related to the existence of a diffusion carrier that facilitates glucose uptake. Nearly all cells have GLUTs; however, the cells lining the small intestine contain a higher density of these transporters. In a superfamily of transport facilitators, GLUTs are just one type. The membrane-spanning portion of GLUTs, which are essential membrane proteins, is made up of 12 helices. A typical membrane transport mechanism powers GLUTs. When glucose attaches to the membrane's outer surface site, a conformational shift related to membrane transport results. Glucose is released into the internal aqueous solution at the inner side of the membrane [14].

Potassium Channels

Potassium channels can be homotetramers or heterotetramers. The pore-loop structure in each subunit determines the channel's selectivity for potassium ions, allowing it to selectively permit potassium to pass through while excluding other ions. There are usually three states in K^+ channels: resting, activated, and inactivated. The channels will normally be closed at a resting state, open after stimulation, and then turn to a nonconductive state. The method of controlling a channel's opening and shutting is known as gating. There are two different types of gating mechanisms: the extracellular one contains the SF (Selectivity Filter), while the intracellular one is located where the inner helix bends. Although there is a connection between the two gates, the impact on various K^+ channels varies [15].

Calcium-activated Potassium Channel

It is open in response to increased intracellular calcium levels. When Ca^{2+} binds to specific sites on these channels, it triggers a conformational change that enables potassium ions to exit the cell. This efflux hyperpolarizes the cell membrane, decreasing excitability and affecting various physiological processes, including neurotransmitter release and muscle movement. It includes BK (big conductance) channels, which are sensitive to both calcium and voltage, and SK (small conductance) channels, which respond primarily to calcium. These channels play crucial roles in regulating neuronal activity and smooth muscle function [16].

Inwardly Rectifying Potassium Channel

Inwardly rectifying potassium channels facilitate the flow of K^+ ions into the cell more readily than out, helping to stabilize the resting membrane potential. They exhibit "inward rectification," meaning they preferentially conduct potassium ions inward at negative membrane potentials and are less effective at positive potentials due to internal blockades. This helps maintain a negative resting membrane potential and regulates cellular excitability. Key examples include KIR channels, which are important in neurons, cardiac cells, and various other tissues. By helping to counteract depolarization, these channels are essential for preserving cellular stability and regulating excitability [17].

Tandem Pore Domain Potassium Channel

Tandem pore domain potassium channels, commonly referred to as two-por--domain K^+ channels, have two pore-forming regions per subunit. These channels contribute to the "leak" current, which helps stabilize the membrane potential at rest and regulate cell excitability are characterized by their ability to open in reaction to various triggers, including Mechanical Stress, changes in temperature, and fluctuations in pH. Examples include TREK-1 and TRAAK channels. These channels play roles in sensory processes and maintaining background potassium conductance, which is crucial for regulating neuronal activity and responding to environmental changes [18].

Voltage-gated Potassium Channel

These are integral proteins that get activated or deactivated on alteration with the membrane potential. They are essential for the transmission of action potentials in both neurons and muscle cells. These channels have a voltage-sensing domain that detects changes in the electrical potential within the membrane. When depolarization occurs (a positive shift in membrane potential), the channel experiences a conformational change that opens the pore, permitting potassium ions to exit the cell. This efflux of potassium ions repolarizes the membrane, restoring the membrane potential and concluding the action potential. This process is essential for regulating neuronal firing rates and muscle contraction [15].

Sodium Channel

$Na+$ channels operate similarly to K^+ channels in several aspects. They both act as facilitated diffusion carriers, moving the cation along its electrochemical gradient. The ascending phase of action potentials is driven by Na^+ channels in excitable cells, including neurons, muscle cells, and certain glial cells. Poisonous neurotoxins represent substances that block Na^+ channels as well as nerve transmission. Na^+ channels come in two fundamental varieties: voltage-gated and ligand-gated. Na^+ is drawn to the selectivity filter at the entrance of the Na^+ channel. From there, the Na^+ ions go into a narrow, 3-5Å broad area of the channel. This is just big enough to let one Na^+ with one connected water flow through. The channel is Na^+ selective because the bigger K^+ cannot pass through [19].

Tetrodotoxin (TTX), commonly found in puffer fish, along with triggerfish, porcupine fish, and ocean sunfish, is a potent neurotoxin that blocks Na^+ channels without affecting K^+ channels. The golden poison frog is the most dangerous vertebrate in the world, alongside puffer fish. Golden poison frog is highly poisonous due to the presence of batrachotoxin, a potent neurotoxin that affects sodium channels. This toxin disrupts normal nerve function by blocking Na^+ channels, leading to paralysis or even death in predators [20].

Aquaporin

Aquaporins typically only permit the permeability of water and prevent the passage of any other solutes. Aqua-glyceroporins, a type of aquaporin, are capable of conducting several extremely tiny, uncharged solutes across the membrane, including glycerol, carbon dioxide, ammonia, and urea. But charged solutes are impermeable to all the aquaporins.

Active Transport

The transport of solutes against their concentration gradient requires a source of energy, frequently ATP, and causes a nonequilibrium distribution of solutes over the membrane. Active transport can be divided into Uniport and Cotransport, which have been discussed below:

Uniport is also called the primary membrane transport. The well-known example for uniport transport is Na^+, K^+, and H^+ gradients in thylakoid membranes. The active transport is inhibited by an enzyme called ouabain, which is a cardiac glycoside. Ouabain was initially found in an arrow poison used by Somali nomads for hunting. The term ouabain is derived from the Somali word Waabaayo, meaning "arrow poison". Ouabain is derived from the barks of specific plants in Africa. Na^+ and K^+ bind from inside and outside, respectively. Since ouabain binds to the external side of the cell membrane, it blocks K^+ transport. When transport of K^+ is blocked, it eventually inhibits dephosphorylation. By the breakdown of a single ATP molecule, two potassium ions are pumped in for every three sodium ions being pumped out. Sodium is required for phosphorylation [21].

A difference in ion concentration and electrical charge across the membrane is produced by the energy-dependent transport of an ion (such as H+). The transport of a solute in the same direction is referred to as symport, while the transport of a solute in the opposite direction is called antiport. Both kinds of co-transport have the same goal: to force one solute against another's gradient by harnessing the energy stored in an electrochemical gradient (Fig. **6**).

Group Translocation

A modified substance/transported solute emerging in a different form immediately after crossing the membrane is what is meant by the term group translocation. PTS (phosphotransferase system) is a very good example of group translocation. In order to absorb external sugars in bacteria, PTS, an energy-driven transport system, requires the energy derived from intracellular phosphoenolpyruvate (PEP). The system is driven by a high-energy phosphorylated compound called PEP. The plasma membrane and the cytoplasmic enzymes are essential to group translocation. PTS consists of many enzymes II that are specific to carbohydrates and two cytoplasmic energy-coupling proteins (Enzyme I and HPr) that catalyse simultaneous carbohydrate translocation and phosphorylation in bacteria. Catabolite suppression, inducer control, and chemotaxis are some phenomena that result from the signalling processes carried out by PTS and the related proteins (Fig. 7). These PTS characteristics provide bacteria with an integrated system that ensures the best possible utilisation of carbohydrates in challenging circumstances [22].

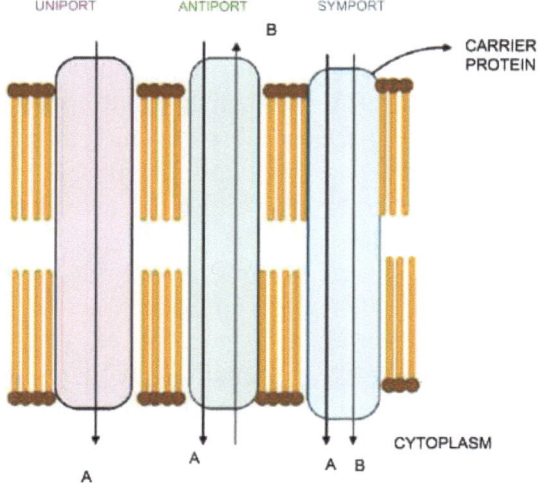

Fig. (6). Uniport-Single molecule transport in one direction; Symport- Two molecules transported in the same direction; Antiport-Two molecules exchanged in opposite directions.

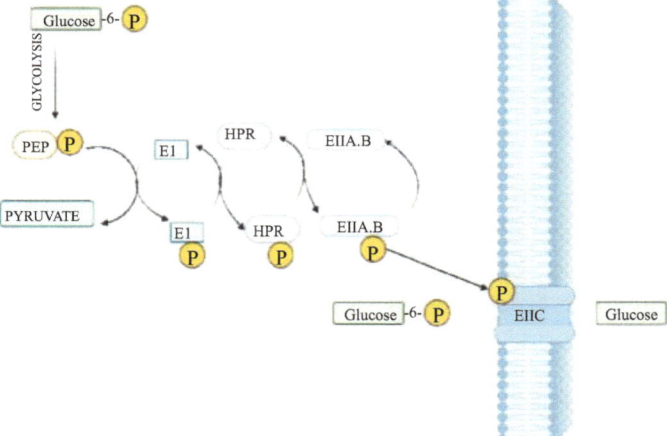

Fig. (7). PTS System-Facilitated transport mechanism that phosphorylates sugars during transport across the membrane.

ABNORMALITIES IN PLASMA MEMBRANE

In the phase of evolution, the eukaryotic cells have experienced damage in the plasma membrane due to several reasons, like osmotic stress, infections from pathogens, and mechanical stress. A defective plasma membrane is associated with various health conditions and diseases prevalent in today's world, including hypertension and sphingolipid-related disorders.

Hypertension

A lipid bilayer surrounds the proteins that make up the plasma membrane, which are intermittently embedded. These proteins perform crucial functions both as ion transport channels and as sensors for substances that control how cells operate. They also preserve the stability of the membrane. Thus, the membrane is essential for regulating the activity of vascular smooth muscle. In individuals with hypertension, the plasma membrane exhibits abnormal monovalent ion permeability and impaired calcium handling. Fewer calcium-binding sites are produced, and the ability of the membrane to stabilize itself is compromised. Erythrocytes, fat cells, and vascular smooth muscle cells have shown abnormalities in it. Defects in the ion transport system (in smooth muscle cells) are linked to the development of hypertension, as they contribute to elevated blood pressure.

Sphingolipids-related Disorders

Sphingolipids constitute a class of structural lipids found in cell membranes and are especially plentiful in nerve cells. Some sphingolipids, such as ceramide and sphingosine, behave as bioactive compounds that affect cellular signalling and proliferation in addition to having an impact on shape and fluidity. The hazardous buildup of undigested sphingolipids within lysosomes and other cellular membranes causes a family of hereditary disorders known as sphingolipidoses. Cells of the neural system are rich in sphingolipids. Excessive buildup of these substances during sickness has a substantial effect on the neurological system. The common aspect of many disorders is neurological malfunction, followed by neurological impairment; nevertheless. The fundamental processes that make neuron cells susceptible are yet to be fully understood. In many adult-onset neuropathies, malfunction of the plasma membrane has been linked to neuronal failure. In the case of sphingolipidoses, slow accumulation of unmetabolized lipids within cellular membranes results in localised abnormalities in lipid raft regions and dysregulation of numerous signalling pathways essential for the viability and function of neurons. In order to aid in cell survival, the membrane also takes part in important cellular processes like signalling and cell-to-cell communication. Highly polarized cellular membranes perform a variety of important functions within a neuron. Nerve impulses regulate myelination *via* glial signalling, while the synaptic membranes take in and discharge signalling molecules.

The soma and connected dendrites contain surface receptors that detect signals from the environment. Membrane integrity and function are essential for neuronal viability and development, and disruptions may result in neurodegeneration. Undigested sphingolipids would gradually build up in plasma membranes and create small-scale disturbances in the lipid raft domains, which could eventually lead to dysregulation of related signal pathways. Prominent disorders categorized as sphingolipidoses include Krabbe disease, characterized by the buildup of galactosyl sphingosine; Niemann-Pick disease, which involves the excess of sphingomyelin and cholesterol; Tay-Sachs disease, marked by the presence of GM2 gangliosides; and Fabry disease, associated with the storage of glycolipids [23].

Basis for Sphingolipidoses-neuronal Vulnerability

Different parts of a neuron may function differently and be more or less sensitive to membrane disruption during their activities. The type of lipid rafts present in each area and the extent to which their activity utilises signalling pathways that rely on intact lipid raft function would affect this differential sensitivity. Membrane disruption caused by lipid buildup in sphingolipidoses may affect synaptic function pathways more than those at the somatic membrane. The most likely damaged neural signalling pathways include neurotropic and IGF-receptor signalling, which both heavily rely on lipid raft stability [24].

Neurotrophic Signalling

For the survival, development, and differentiation of neurons, neurotrophic factors such as brain-derived neurotrophic factor (BDNF), nerve growth factor, neurotrophins 3 and 4, and glial cell line-derived neurotrophic factor are necessary. These factors are released from the target tissue and stimulate the Trk family of tyrosine kinase receptors, initiating subsequent signalling that includes the phosphatidylinositol 3-kinase, Ras/MAPK, and phospholipase C-γ pathways. Trk receptors are specifically located on synaptic membranes, and ligand interaction induces distinct signalling occurrences at presynaptic and postsynaptic sites that are essential for the viability of neurons. Trk activation is necessary for the development, shape, and synaptic plasticity of dendritic spines. Presynaptic stimulation regulates the secretion of neurotransmitters, the expansion of exons, and the maintenance of neurons. The ligand-activated Trk receptors and associated signalling molecules are taken up by endocytosis to create signalling vesicles after being activated at the presynaptic terminal. Dynein motors and microtubules regulate the backward movement of signalling vesicles from the synapse to the cell body. This type of long-distance survival signal transmission is essential for maintaining synaptic connections and cell-to-cell communication, which is mediated by neurotrophic action. Krabbe disease and Alzheimer's are two neurodegenerative illnesses with dying-back pathology that have been linked to impaired endosomal vesicle trafficking as a result of abnormalities in axonal transport, highlighting the significance of this pathway for neuronal survival.

Insulin Signalling

In vitro and in vivo neuronal growth and survival depend on insulin and insulin-like growth factor 1 (IGF-1), whose activities are mediated by the phosphatidylinositol 3-kinase /Akt and mitogen-activated protein kinase /extracellular signal-regulated kinase pathways, respectively. Insulin and IGF-1 reduce apoptosis while also promoting dendritic development and synaptic formation, and IGF-1 specifically contributes to axon guidance and synaptic plasticity IGF-1 modulates brain glucose metabolism in a manner akin to insulin *via* IGF-1 receptor-mediated Akt phosphorylation and GLUT4 expression, which are essential for maintaining energy levels to support swift neuronal development, particularly in the maturing brain. According to Huo, the IGF-1 receptor is mostly found in lipid rafts, and Romanelli found that disruption of lipid rafts caused by cholesterol deprivation causes an acute sensitivity in the downstream pathway activation. According to Hering, cholesterol deprivation also results in the reduction of dendritic synapses and spines, emphasizing the role that lipid rafts play in promoting dendritic and axonal development *via* IGF-1 receptor activity. The fact that IGF-1 Insulin signalling disruption is linked to neurodegenerative illness, including Parkinson's and Alzheimer's disease, highlights the importance of these pathways for neuronal health and the maintenance of brain function.

Impacts of Sphingolipid Accumulation on the use of Cellular Energy

Neurons, like other cells, require energy to operate, and both peripheral insulin action and IGF-1's insulin-like effects in the central nervous system help regulate cellular energy balance. GM3 accumulation in type 1 Gaucher patients correlates with peripheral insulin resistance, and cholesterol buildup in Niemann-Pick Type C impairs insulin signalling, as the functionality of IGF-1 and insulin receptors depends on the integrity of lipid rafts. Psychosine buildup inhibits In twitcher oligodendrocytes, IGF-1 signalling is impaired, reducing Akt and ERK1/2 phosphorylation.

Maintaining cellular energy homeostasis is crucial for peripheral neurons and central nervous system neurons due to their long axons and vesicle transport needs. In sphingolipid storage disease, defective energy metabolism from impaired insulin/IGF-1 pathways, combined with reduced synaptic activity and vesicular transport, contributes to neuronal vulnerability [25].

DIAGNOSTIC AND THERAPEUTIC ROLE OF PLASMA MEMBRANE

Membrane proteins and surface markers are invaluable in diagnostics. They act as unique identifiers for various cell types and states. For example, in cancer diagnostics, HER2/neu, a membrane protein overexpressed in some breast cancers, helps tailor treatments like trastuzumab (Herceptin). Similarly, CD4 and CD8 markers are crucial in identifying T cell subsets in HIV testing. Abnormalities in these markers can signal conditions like autoimmune diseases, where altered protein expression may trigger harmful immune responses. Moreover, detecting specific viral or bacterial proteins that interact with cell membranes can confirm infections, such as identifying the spike protein of SARS-CoV-2 in COVID-19. Therapeutically, membrane proteins are prime targets for drug development. Many treatments are designed to interact with these proteins to modify cellular responses. For instance, drugs like beta-blockers target beta-adrenergic receptors to manage heart disease. In gene therapy, correcting defective membrane proteins can restore normal function, as seen with cystic fibrosis, where gene therapy aims to fix the malfunctioning CFTR protein responsible for chloride transport [26].

RECENT TRENDS

CAR-T Therapy

The plasma membrane plays an essential role in the process of targeting and destroying tumor cells. This innovative immunotherapy entails genetically modifying a patient's T cells to produce synthetic receptors tailored to identify particular cancer antigens. The procedure starts with the extraction of T cells from the patient's blood *via* apheresis. The T cells are subsequently altered to incorporate a Chimeric Antigen Receptor (CAR) into their plasma membrane. The CAR is a fusion protein combining elements of an antibody and a T-cell receptor, specifically tailored to recognize antigens like CD19 found on the surface of certain cancer cells, such as those in B-cell malignancies. The plasma membrane's role here is vital as it incorporates these engineered CARs, equipping T cells with the ability to specifically target cancer cells.

Once the CAR-T cells are modified, they are expanded in the lab to produce a large quantity. Prior to reinfusion into the patient, the patient typically undergoes a conditioning regimen involving chemotherapy. This regimen reduces the number of normal immune cells and creates a more favorable environment for the CAR-T cells to thrive. When infused into the patient, the modified T cells use their CARs on the plasma membrane to bind specifically to cancer cells expressing the targeted antigen. This binding triggers a signalling cascade inside the T cells, activating them to release cytotoxic molecules that destroy the cancer cells. The plasma membrane is thus crucial for both the recognition of cancer cells and the subsequent activation of the immune response.

Additionally, the plasma membrane facilitates cell communication and migration. CAR-T cells must adhere to and migrate towards tumor sites, a process dependent on adhesion molecules and signalling cues present on the plasma membrane. This ensures effective targeting and accumulation in the tumor microenvironment.

CAR-T therapy can cause cytokine release syndrome, a severe response linked to plasma membrane signalling, requiring careful management during treatment. Overall, the plasma membrane is central to CAR-T therapy. It enables the CAR-T cells to specifically recognize and attack cancer cells, supports the signalling necessary for activation, and influences cell migration and side effect management. Its role is pivotal in the effectiveness and precision of this advanced cancer treatment.

Liposomal Drug Delivery Systems

Liposomal drug delivery systems are designed with liposomes—tiny, lipid-based vesicles that can encapsulate drugs. These liposomes can be engineered with surface ligands or antibodies that specifically attach to receptors on the target cell membrane. For instance, in cancer therapy, liposomes can be modified to display antibodies or ligands that target HER2 receptors, which are overexpressed on the surface of certain cancer cells. This targeted approach ensures that the therapeutic agents within the liposomes are delivered directly to the cancer cells, enhancing efficacy and minimizing damage to healthy tissues.

Once the targeted liposomes bind to the HER2 receptors on the cancer cell plasma membrane, they are internalized through endocytosis. The plasma membrane engulfs the liposomes, forming endocytic vesicles that enter the cell. The endocytic process can involve various pathways, such as clathrin-mediated endocytosis or caveolae-mediated endocytosis, based on the size and characteristics of the liposomes. Alternatively, some liposomes are designed to fuse directly with the cell membrane, delivering their contents into the cytoplasm without requiring endocytosis. This direct fusion is advantageous for delivering drugs or genetic material quickly and efficiently. The plasma membrane also serves an essential function in the regulated discharge of therapeutic agents contained within liposomes. These liposomes can be designed to expel their contents in response to specific cellular conditions like pH changes, which are detected by the plasma membrane. For example, pH-sensitive liposomes can release their drug payload once they reach the acidic environment inside the tumor cells, ensuring that the drug is activated and released specifically at the tumor site. The makeup of the plasma membrane, including lipids and proteins, influences how liposomes are internalized. The characteristics of the plasma membrane dictate the endocytic pathways utilized and affect the efficiency of drug delivery. To enhance circulation time and avoid detection by the immune system, liposomes may be wrapped in stealth materials, like polyethylene glycol. This coating assists the liposomes in evading immune cells, reducing phagocytosis, and allowing for more effective delivery to target cells [27].

CONCLUSION

The plasma membrane is an integral part of cellular life, a selective barrier that regulates the passage of substances and intercellular communication. Its complex structure, such as that represented by the fluid-mosaic model, represents evolutionary adaptations that have allowed it to fulfil all the functions. Mechanisms of transport across membranes allow homeostasis, and diseases caused by defects in sphingolipid metabolism underlie their role in health. Such diagnostic and therapeutic roles of the plasma membrane have become more and more appreciated by science, and trends in recent research hold the promise for deepening insight into its dynamic nature and pave the way to innovative treatments and interventions in cellular biology.

ACKNOWLEDGEMENTS

We would like to acknowledge Ms. Kruthika P, Ms. Raksa Arun, Ms. Srisri Satish Kartik, and Mr. Sasitharan T for their contribution in proofreading and editing the work.

REFERENCES

[1] Roualdes S, Rouessac V, Durand J. Plasma Membranes. Comprehensive Membrane Science and Engineering. Elsevier 2010; pp. 159-97.
[http://dx.doi.org/10.1016/B978-0-08-093250-7.00045-1]

[2] Nagle JF, Mathai JC, Zeidel ML, Tristram-Nagle S. Theory of passive permeability through lipid bilayers. J Gen Physiol 2008; 131(1): 77-85.
[http://dx.doi.org/10.1085/jgp.200709849] [PMID: 18166627]

[3] Frey F, Idema T. More than just a barrier: using physical models to couple membrane shape to cell function. Soft Matter 2021; 17(13): 3533-49.
[http://dx.doi.org/10.1039/D0SM01758B] [PMID: 33503097]

[4] van Deventer S, Arp AB, van Spriel AB. Dynamic Plasma Membrane Organization: A Complex Symphony. Trends Cell Biol 2021; 31(2): 119-29.
[http://dx.doi.org/10.1016/j.tcb.2020.11.004] [PMID: 33248874]

[5] Chifflet S, Hernández JA. The plasma membrane potential and the organization of the actin cytoskeleton of epithelial cells. Int J Cell Biol 2012; 2012: 1-13.
[http://dx.doi.org/10.1155/2012/121424] [PMID: 22315611]

[6] Davidson J. Selective permeability and the plasma membrane. Plant World 1916; 19(11): 331-49.
http://www.jstor.org/stable/43477486

[7] Brandl M, Eide Flaten G, Bauer-Brandl A. Passive Diffusion Across Membranes. Wiley Encyclopedia of Chemical Biology. Wiley 2008; pp. 1-10.
[http://dx.doi.org/10.1002/9780470048672.wecb432]

[8] ter Kuile BH, Cook M. The kinetics of facilitated diffusion followed by enzymatic conversion of the substrate. Biochim Biophys Acta Biomembr 1994; 1193(2): 235-9.
[http://dx.doi.org/10.1016/0005-2736(94)90158-9] [PMID: 8054344]

[9] Weigl J. Wasserstruktur und Permeation: Die Aktivierungsenergie und der molekulare Mechanismus der Wasserpermeation. Z Naturforsch B J Chem Sci 1967; 22(8): 885-90.
[http://dx.doi.org/10.1515/znb-1967-0818] [PMID: 4384767]

[10] Bauer WR, Nadler W. Stationary flow, first passage times, and macroscopic Fick's first diffusion law: Application to flow enhancement by particle trapping. J Chem Phys 2005; 122(24): 244904.
[http://dx.doi.org/10.1063/1.1940056] [PMID: 16035813]

[11] Kleinzeller A. Charles Ernest Overton's concept of a cell membrane.Curr Top Membr. Academic Press 1999; 48: pp. 1-22.
[http://dx.doi.org/10.1016/S0070-2161(08)61039-4]

[12] Kiil F. Kinetic model of osmosis through semipermeable and solute-permeable membranes. Acta Physiol Scand 2003; 177(2): 107-17.
[http://dx.doi.org/10.1046/j.1365-201X.2003.01062.x] [PMID: 12558549]

[13] Korla K, Mitra CK. Kinetic modelling of coupled transport across biological membranes. Indian J Biochem Biophys 2014 Apr; 51(2): 93-9.
[PMID: 24980012]

[14] Navale AM, Paranjape AN. Glucose transporters: physiological and pathological roles. Biophys Rev 2016; 8(1): 5-9.
[http://dx.doi.org/10.1007/s12551-015-0186-2] [PMID: 28510148]

[15] Kuang Q, Purhonen P, Hebert H. Structure of potassium channels. Cell Mol Life Sci 2015; 72(19): 3677-93.
[http://dx.doi.org/10.1007/s00018-015-1948-5] [PMID: 26070303]

[16] Minor DL Jr, Masseling SJ, Jan YN, Jan LY. Transmembrane structure of an inwardly rectifying potassium channel. Cell 1999; 96(6): 879-91.
[http://dx.doi.org/10.1016/S0092-8674(00)80597-8] [PMID: 10102275]

[17] Jan LY. Chapter 1 Studies of Voltage-Dependent and Inwardly Rectifying Potassium Channels. In 1999. p. 1–5.

[18] Frank R, Frances MA. Inwardly rectifying potassium channels,Current Opinion in Cell Biology, Volume 11, Issue 4, 1999, 503-508.
[https://doi.org/10.1016/S0955-0674(99)80073-8]

[19] Waxman SG, Cummins TR, Dib-Hajj S, Fjell J, Black JA. Sodium channels, excitability of primary sensory neurons, and the molecular basis of pain. Muscle Nerve 1999; 22(9): 1177-87.
[http://dx.doi.org/10.1002/(SICI)1097-4598(199909)22:9<1177::AID-MUS3>3.0.CO;2-P] [PMID: 10454712]

[20] Ritchie J. Tetrodotoxin and saxitoxin, and the sodium channels of excitable tissue. Trends Pharmacol Sci 1980; 1(2): 275-9.
[http://dx.doi.org/10.1016/0165-6147(80)90021-8]

[21] Kravtsova VV, Paramonova II, Vilchinskaya NA, *et al.* Chronic Ouabain Prevents Na,K-ATPase Dysfunction and Targets AMPK and IL-6 in Disused Rat Soleus Muscle. Int J Mol Sci 2021; 22(8): 3920.

[http://dx.doi.org/10.3390/ijms22083920] [PMID: 33920198]

[22] Kotrba P, Inui M, Yukawa H. Bacterial phosphotransferase system (PTS) in carbohydrate uptake and control of carbon metabolism. J Biosci Bioeng 2001; 92(6): 502-17.
[http://dx.doi.org/10.1016/S1389-1723(01)80308-X] [PMID: 16233138]

[23] Vanier MT. Disorders of Sphingolipid Metabolism. Inborn Metabolic Diseases. Berlin, Heidelberg: Springer Berlin Heidelberg 2006; pp. 479-94.
[http://dx.doi.org/10.1007/978-3-540-28785-8_38]

[24] Olsen ASB, Færgeman NJ. Sphingolipids: membrane microdomains in brain development, function and neurological diseases. Open Biol 2017; 7(5): 170069.
[http://dx.doi.org/10.1098/rsob.170069] [PMID: 28566300]

[25] Zaka M, Rafi MA, Rao HZ, Luzi P, Wenger DA. Insulin-like growth factor-1 provides protection against psychosine-induced apoptosis in cultured mouse oligodendrocyte progenitor cells using primarily the PI3K/Akt pathway. Mol Cell Neurosci 2005; 30(3): 398-407.
[http://dx.doi.org/10.1016/j.mcn.2005.08.004] [PMID: 16169744]

[26] Li F. Structure, Function, and Evolution of Coronavirus Spike Proteins. Annu Rev Virol 2016; 3(1): 237-61.
[http://dx.doi.org/10.1146/annurev-virology-110615-042301] [PMID: 27578435]

[27] Khafoor AA, Karim AS, Sajadi SM. Recent progress in synthesis of nano based liposomal drug delivery systems: A glance to their medicinal applications. Results in Surfaces and Interfaces 2023; 11: 100124.
[http://dx.doi.org/10.1016/j.rsurfi.2023.100124]

CHAPTER 10

Mitochondria and Chloroplasts: Evolutionary Engines of the Cell

K. Kumaran[1,†], **Keerthana Ramesh**[1,†], **Sanjana Dhayalan**[1], **Gargii Chatterjee**[1] and **K.N. Aruljothi**[1,*]

[1] *Department of Genetic Engineering, SRM Institute of Science and Technology, Kattankulathur, Chengalpattu, Tamil Nadu, India*

Abstract: Mitochondria and chloroplasts are essential organelles driving energy production and metabolism in eukaryotic cells. Mitochondria, known as the cell's "powerhouses," generate ATP through oxidative phosphorylation, utilizing nutrients to fuel cellular processes. According to the endosymbiotic theory, mitochondria originated from free-living bacteria, and their distinct mitochondrial DNA (mtDNA) is maternally inherited. Dysfunction in mitochondria—often due to accumulation of mtDNA mutations—is associated with neurodegenerative diseases, metabolic syndromes, and cancer, with recent studies revealing their roles in aging and disease. These insights are advancing therapies to improve mitochondrial health, and genome editing now holds the potential for correcting pathogenic mtDNA mutations. Chloroplasts in plant cells are essential for photosynthesis, synthesizing carbohydrates, amino acids, fatty acids, and membrane lipids. Chloroplasts are sensitive to temperature stress, and their malfunction can produce Reactive Oxygen Species (ROS), which may be fatal to plant cells. This sensitivity makes chloroplast health crucial for plant resilience, especially under climate stress. Future research in mitochondrial and chloroplast should concentrate on offering potential treatments for human diseases and developing stress-resistant crops. Genome editing technologies could address mitochondrial and chloroplast dysfunction, creating innovative therapies and sustainable agricultural practices to address health and environmental challenges.

Keywords: Cellular Respiration, Endosymbiotic Theory, Energy Production, Electron Transport Chain, Krebs Cycle, Mitochondria-Associated Membranes (MAMs), Mitochondrial DNA (mtDNA), Oxidative Phosphorylation, Reactive Oxygen Species (ROS), Uniparental Inheritance..

---

[*] **Corresponding author K.N. Aruljothi:** Department of Genetic Engineering, SRM Institute of Science and Technology, Kattankulathur, Chengalpattu, Tamil Nadu, India; E-mail: aruljotn@srmist.edu.in
[†] Equal contribution.

K.N. Aruljothi, Prakash Gangadaran, Krishnan Anand, Satish Ramalingam, K. Kumaran & Kruthika Prakash (Ed.)
All rights reserved-© 2026 Bentham Science Publishers

MITOCHONDRIA

Introduction

It was close to a billion years since cellular respiration had become a vital part of the cell when science discovered the pan portion of the cell during the middle of the nineteenth century. Based on light microscopy, the first diagrams of cellular structure showed the existence of granular structures known as mitochondria in the cell's cytoplasm. The first observation of the presence of mitochondria in muscle cells was made during the 1850s by the anatomist Kölliker (1856). However, in the 1890s, Altman devised a fantastic proposal: the granules, which Deichmann called "bioclasts," were symbiotes. This idea was expanded on by Mereschkowsky in 1905.

The same idea concerning chloroplasts was first proposed by C. Simons in 1883, before Schimper. The term "Mitochondrion" was first used to explain data by Benda in 1898. The name is derived from the Greek terms: "mitos" in thread, referring to an activity in the mitotic field, and "chondros" in grain, referring to an activity associated with spermatogenesis. As mitochondria are the primary site of energy conversion in the cell, or where "cellular respiration" takes place, they are also known as the organelles that produce blood inside the cell. However, the causes of such a phenomenon were not studied until the 1950s, before the genesis of mitochondria was considered. Very early in the 1950s, Mitchell (1952) and Ephrussi (1950) reported post-Mendelian genetic variation to control the forces determining the division of yeast mitochondria. A little while later, McLean *et al*. (1958) reported that mitochondria can synthesize polypeptides. Luck and Reich 1964), Nass, early in the nineteen sixties, since many groups. Their significant contributions regarding the mitochondria and their invention of the mitochondrion were related. However, the hypothesis on the endosymbiotic origin of mitochondria, which had existed, started to attract new and enthusiastic followers in the 1960s with the introduction of new biochemical methods. It would be superfluous to emphasize that the discovery of the DNA contained within the organelles and the non-classical inheritance of organelles were the two components that contributed to the resurgence of interest.

Once Margulis had rephrased the endosymbiotic theory into endosymbiosis (1970), it became recognized how mitochondria have evolved into the cell. It is accepted that the mitochondrion originated from a free-living bacterium. However, there is still an argument regarding how this endosymbiosis came about and which living things this process encompassed. There are about a hundred years between the first reports of mitochondria and the modern understanding of the structure, function, inheritance, and origin of mitochondria. Since 2003, the work of Tsang and Lemire has made it possible to identify the mitochondrial genomes of more than 250 species by examining the effects of mutations in several mitochondrial genes.

Origin and Evolution of Mitochondria

Mitochondria in eukaryotes are suspected of having an endosymbiotic origin. In the 1960s, Margulis Marshaled confirmed this hypothesis based on two factors, *i.e.*, the presence of DNA and a separate translational system distinct from one in the cytosol. Further elaborate studies of mitochondrial genomes and genes proclaim that mitochondria have a bacterial origin. However, another theory explained the mitochondrial origin, *i.e.*, the hydrogen hypothesis given by Martin and Muller in 1998. According to this, the origin of heterotrophic organelles and other eukaryotes was identical.

The symbiotic origin is based on two subjects: the "Archezoan Scenario" and the "Symbiogenesis Scenario." The former is based on endosymbiosis and the latter on the hydrogen hypothesis. The archean scenario explains the presence of a proto-mitochondrial endosymbiont in mitochondrial eukaryotes, and the symbiogenesis scenario explains the hold of an alpha proteobacterium in an archaebacterial cell [1, 2].

Mitochondrial Inheritance in Eukaryotes

The mitochondrion is usually inherited from a single parent (maternal inheritance). This is responsible for the offspring's homoplasmy (where all mtDNA are genetically identical)—the disruption of MT homoplasmy results in mitochondrial dysfunction and cancer in human beings. There are two methods for ensuring mtDNA homoplasmy: (i) Strict Maternal Inheritance (SMI) and (ii) the bottleneck mechanism. A strict maternal inheritance pattern is observed in the shell coiling of snails (*Limnaea stagnalis*). The three stages that control the bottleneck mechanism are:

(i) A bottleneck that occurs during oogenesis expresses mutations that allow purification before the resumption of replication during the oocyte's maturity, lowering the copy number of mtgenes.

(ii) Selective replication during oogenesis led to the establishment of a genetic bottleneck.

(iii) Another one forms during embryogenesis.

Recent reviews provide a glimpse into the status of maternal inheritance research (reviews are available in Jokinen and Battersby 2013 and Mishra and Chan 2014).

Autophagy of sperm in C. elegans prevents the transmission of mtDNA abrupt ending. Another evidence of autophagy is observed through the clearance of ubiquitinated sperm mitochondria. Thus, autophagy plays a critical role in ensuring maternal inheritance of mitochondria. Therefore, unipaternal inheritance of mitochondria is found through many pathways [3, 4]. Some unusual inheritance patterns in mtgenomes deviate from the strict Maternal Inheritance pattern. This includes the maternal inheritance of heteroplasmy, biparental, strictly paternal, and doubly unilateral patterns.

A condition with more than two mtgenomes in a single cell is heteroplasmy. It may be due to maternal inheritance, biparental transmission, or paternal leaking. The example of maternally inherited heteroplasmy is often found in the hosts of isopod crustaceans. This disease is characterized by a complex, heteroplasmic condition consisting of three nearly identical ~14 kb monomers, with two monomers forming a ~28 kb dimer and the other remaining as a monomer. Biparental inheritance and paternal leaking are rare conditions. It usually occurs in crosses between species. Biparental inheritance and paternal leaking are different, even though they may seem to pair. It occurs when two different mating types or both parents pass their mitochondrial genomes to zygotes as part of the normal reproductive process in a species. This is typified by the fact that both parental types of mitochondria persist throughout development, as in the inkcap mushroom. Schistosoma mansoni, the same parasite that causes blood schistosomiasis, has such a pattern of inheritance, too. Plants and marine algae, such as sequoia, bananas, cucumbers, and Chlamydomonas, follow strictly paternal inheritance. Bivalvia is recognized with uniparental inheritance in more than one instance. In the latter cases, two broadly divergent mitochondrial lineages are inherited in a sex-specific manner [5, 6].

Structure of Mitochondria

The sizes, shapes, and distributions of mitochondria are different in various species. These changes in morphology and distribution are commonly known as mitochondrial dynamics. However, the outer and inner membranes that encase every mitochondrion are generally divided by an area called the intermembrane space, as every other membranous organelle's mitochondrion membrane is also made up of phospholipids and proteins. The inner membrane leads to the matrix as membranous folding, cristae, and the inner boundary membrane. Connecting the inner membrane with cristae depends on Mitochondrial contact sites and the Cristae Organization System (MICOS). Sometimes, the void space inside the cristae, *i.e.*, the Intracoastal and inter-membrane spaces, are in association.

The inner membrane is composed of proteins involved in electron transport and oxidative phosphorylation. The outer mitochondrial membrane (OM), which is composed of a phospholipid, acts as a barricade and is responsible for cell signaling and transport of molecules like proteins and RNA in and out, respectively. The pore proteins form a channel in the outer membrane. The OM form interfaces with subcellular organelles to establish membrane contact sites. In addition, they also release mitochondrial-derived vesicles and mitochondrial-derived compartments, where the mitochondrial-derived vesicles interact with other organelles and mitochondrial-derived compartments to check the quality of proteins. The outer membrane consists of molecules that influence the fusion and fission processes, which enables proper function and restrains diseases.

An interesting fact about the mitochondrial outer membrane is that it is involved in the process of apoptosis (cell death, which occurs conventionally). The mitochondria comprise their own genome, the (mtDNA), and synthesize their proteins by translocating the RNAs to the cytosolic ribosomes. The mitochondrial genome is circular, with 16,569 base pairs. There are 37 genes in the mitochondrial genome, which code for 22 tRNAs, two rRNAs, and 13 proteins. The 13 proteins encoded by the mitochondrial genes instruct the cells on how to make the protein components for the oxidative phosphorylation system's enzyme complexes, which allow mitochondria to function as the cell's powerhouses (Fig. 1).

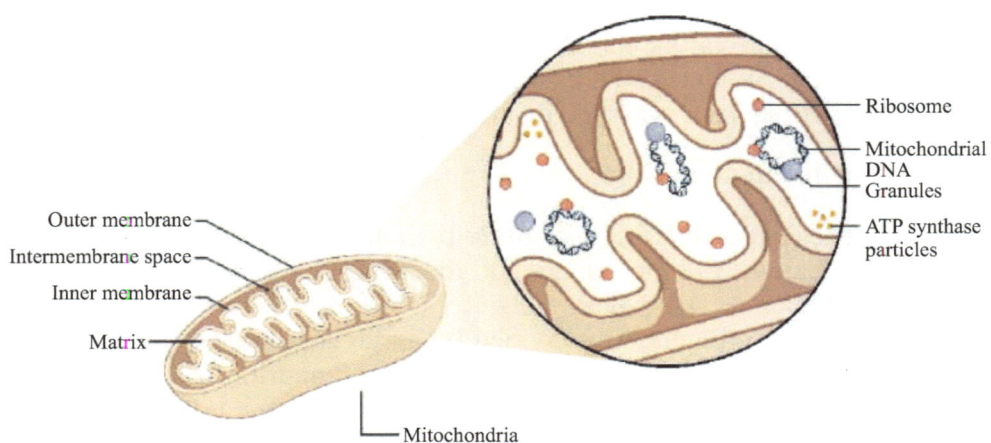

Fig. (1). Structure of mitochondrial matrix.

Since the mitochondrial genome is small, it cannot generate all the proteins required for proper function, so it mostly depends on imported nuclear gene products. Thus, the mitochondrial proteome combines nuclear and mitochondrial-encoded proteins. The mitochondrial proteome is characterized by its diversity and complexity. The exact number of proteins varies across different organisms, and their range is typically from hundreds to thousands [7 - 10].

Function of Mitochondria

The most essential functions of mitochondria are energy production and regulating ionic composition in a cell. This is done mainly in the inner membrane of mitochondria through electron transport and oxidative phosphorylation. Of the 13 genes linked to polypeptide synthesis, seven (*MT-ND1 to MT-ND6*) encode polypeptides linked to complex I of oxidative phosphorylation and contain subunits of NADH dehydrogenase. Some other functions are performed by mitochondria, like the Production of Reactive Oxygen Species (ROS), which is obtained as a by-product of oxidative phosphorylation. Complex 1 is the chief producer of ROS, but complex 2 is also involved in production under some physiological conditions. In Huntington's and Parkinson's diseases, dopamine degradation elevates ROS count. Proper balance is maintained at the ROS level, which controls another function, *i.e.*, mitochondrial biogenesis. Mitochondrial dynamics control the fission and fusion process. They are essential sites for calcium storage. They also support cell multiplication, differentiation, and apoptosis; these functions depend on mitochondrial proteins [10 - 12].

Electron Transport Chain and Oxidative Phosphorylation

After the glycolysis process yields pyruvate, the upcoming TCA or the Krebs cycle takes place in the mitochondrial matrix. The cycle mediates the breakdown of significant macromolecules by a series of eight enzymatic reactions to yield acetyl-CoA, CoA, NADH, and FADH, the primary electron donors in the electron transport chain in the inner mitochondrial membrane.

Oxidative phosphorylation is a series of steps involved in ATP production. Here, the electrons move through various protein complexes, and oxygen is absorbed, creating an electrochemical gradient. The electron shuttle occurs through four complexes in a final electron acceptor, *i.e.*, the oxygen that forms water. In complex I, NADH $(H)_2$ oxidizes into NAD+, releasing electrons and a pair of protons. The electrons released pass through the FMN (Flavin mononucleotide) and Fe-S (Iron-containing sulphur) in complex I, and the protons are pumped out to the intermembrane space.

The electrons from Fe-S enter the Q cycle, where ubiquinone (COQ) gets reduced to (UQH2) ubiquinol. And then electron passage continues to complex III. At the same time, the pair of protons is pumped to the intermembrane space. The electrons are delivered to cytochrome b and then pass through Fe-S and then to cytochrome c1, where the electrons are transferred to complex IV by mobile charge carrier cytochrome c. In complex IV, the electrons delivered by cytochrome c move on from CuA1 to cytochrome a and to cytochrome a3, then to CuA2, from where it is delivered to the final electron acceptor, *i.e.*, the oxygen atom to form water, and two protons are given out, to the intermembrane space.

The second cyclic process of the electron transport chain continues with the involvement of complexes II, III, and IV. In complex II, FAD(H)2 oxidizes to FAD+ and gives out an electron, passing through the FAD (flavin adenine dinucleotide). To, Fe-S, two protons are pumped to the intermembrane space, and the electrons enter the Q cycle again. The exact process takes place where UQ is converted into UQH2, from which electrons are delivered to complex III, and another pair of protons is pumped out of complex III. The electrons then continue the same passage through cytochrome c and complex IV to the final electron acceptor, forming water and generating another pair of protons [13, 14]. At the end of the electron transport chain are ten protons, two of which are from complex I. Four from complex III and four from complex IV. It is accumulated in the intermembrane space, due to which an electrochemical gradient is established. To equalize the potential following chemiosmotic theory, the proton transport from the intermembrane space occurs through complex V through the membranous (F1) and transmembrane (F0). Each pair of protons transferred generates ATP from ADP (Fig. **2**).

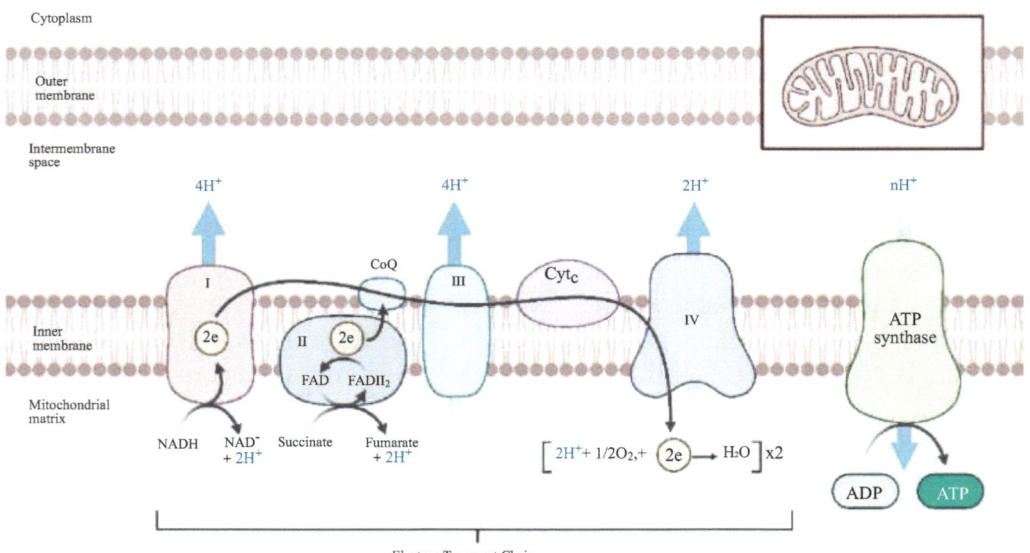

Fig. (2). Electron transport chain.

Mitochondria-associated Membranes

Mitochondria and endoplasmic reticulum are associated with forming Mitochondria-Associated Membranes (MAMs). The Nlrp3 complex is an inflammasome complex associated with MAMs. The activation of the Nlrp3 complex is done in two steps: (i) the First step is the interaction of Toll-Like endosomal Receptors (TLR) with their ligands that enables the transcription of Nlrp3 and pro-IL-1β, (ii) an activation step involving K+ efflux, lysosomal rupture, and ROS production [15, 16].

Mitochondrial DNA Mutations

Mitochondrial DNA (mtDNA) is always passed down from mothers to their children. This means that people who are related through their mothers, like siblings, cousins, and grandparents, have the same mtDNA. However, mtDNA changes a lot, so there are significant differences between the mtDNA of people who aren't related. However, people who are related through their mothers, like a grandmother and her grandchild, have very similar mtDNA. This makes it easy to compare and match their mtDNA. Closely related people have very similar mtDNA, while people who aren't related have very different mtDNA. Scientists can more easily collect and study mtDNA from deceased relatives because there are many more copies of mtDNA in cells than nuclear DNA.

Each cell in eukaryotes possesses numerous mitochondria. The mitochondrial genome is more susceptible to mutations than the nuclear genome because a different population of heterogeneous mitochondria is observed within the same cell. At the genome level, differences exist within the same genome, making them heteroplasmic (Fig. 3). During cell division, the mitochondria are split between the two daughter cells; this process is entirely random and is not controlled correctly and segregated as in the nuclear genome case, resulting in similar mitochondrial copies that are not identical in the daughter cells. A gene called POLG (DNA polymerase gamma), a nuclear gene, helps make a protein that copies DNA in mitochondria.

This protein has two parts: one that checks for mistakes in DNA copying and fixes them, the exonuclease domain, and the other that copies the DNA, which is the catalytic domain. A new study suggests that there might be an uneven supply of nucleotides for DNA in mitochondria. This could lead to more mistakes in the DNA and make the copying protein less accurate. Mitochondrial diseases are passed down from mothers to their children, not fathers. While each cell has many mitochondria, each can have its own DNA, and some might have errors (mutations). When discussing these mutations, it's essential to consider the whole group of mitochondria in a cell, not just one.

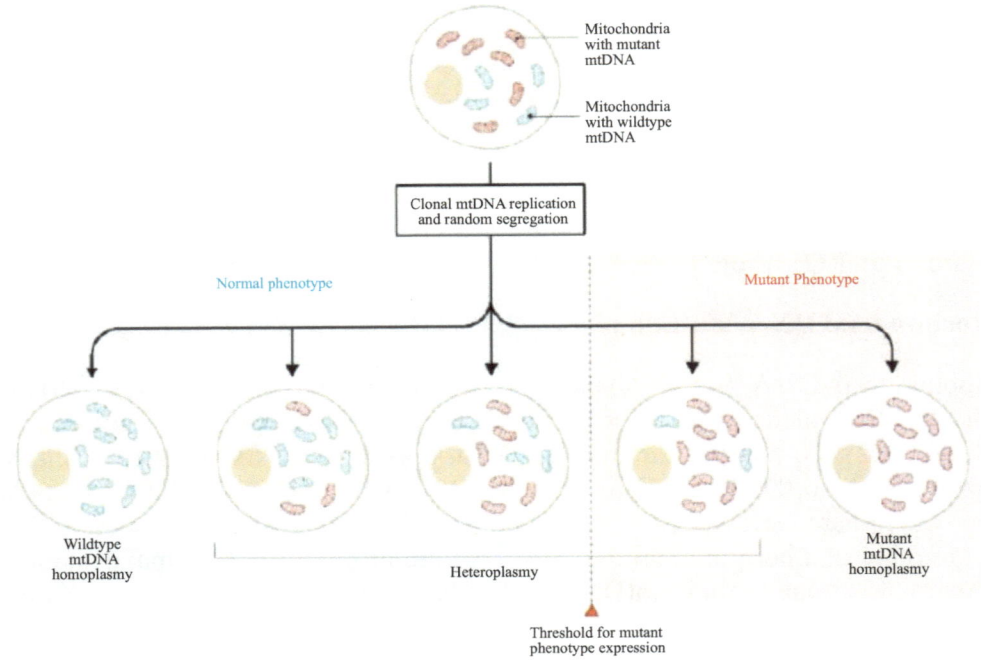

Fig. (3). mtDNA heteroplasmy.

Mothers can have a lot of mitochondria with the same error, which means their children will also have it. But sometimes, even though children get the same mutations from their mother, the way it affects them can be different because each child gets a mix of mitochondria with and without the mutation [17, 18]. Many other clinical problems can be linked to mitochondrial diseases. These diseases often affect tissues that need a lot of energy, like the brain, eyes, muscles, and heart. This is because mitochondria are like the powerhouses of our cells. Some examples include Leigh syndrome, a severe brain disorder; progressive external ophthalmoplegia, which causes eye muscle weakness; rhabdomyolysis, a muscle breakdown; and muscle problems after exercise. Apart from rare clinical conditions, mitochondrial mutations are also linked to some common diseases like Diabetes, Alzheimer's, and Parkinson's disease, to name a few [3, 19, 20].

Role of Mitochondria in Cancer and Aging

Mitochondria, the energy centers of our cells, produce Reactive Oxygen Species (ROS). These can cause damage to the DNA in mitochondria, leading to problems with how cells produce energy. This damage can then cause more damage, creating a cycle of problems. ROS can also cause cells to die, make it harder for cells to copy their DNA, and reduce the energy a cell produces. All of these things contribute to the aging of tissues. Following recent research about the role of mitochondria in cancer, it was found that a single cell with a growth advantage from the somatic mitochondrial alterations multiplied faster than the surrounding cells. Moreover, they proposed that the mutation may have given the mutant mitochondrial genome a replicative advantage based on the homoplasmic character of the mitochondrial DNA mutation. Subsequent research in the years that followed also found links between somatic mitochondrial mutations and other types of cancer, such as solid tumors and leukemias. It is still unclear whether mitochondrial alterations and cancer are related.

Recent Trends in the Biology of Mitochondria

A popular genome editing method is TALEN(Transcription Activator-Like Effector Nuclease), composed of artificially produced nucleases by fusing the FoK1 type-II nuclease domain with the TALE DNA binding domain from Xanthomonas. For example, MTS and UTR sequences from ATP5B and SOD2 are brought into the mitochondria. In mitochondrial diseases ranging from Leigh's syndrome, MELAS, and LHON to others, it has been reported that the TALENs targeted and mitochondrially cleaved the damaged or mutant mtDNA. Although TALEN is considered relatively more straightforward to manufacture than ZFN, it is known to present more target effects than ZFN. One of the most significant advantages of CRISPR/Cas systems in genome editing technologies is their multiplicity and ease of use. This can be supported by retargeting the Cas in CRISPR/Cas systems by changing 20 nucleotides in the guide RNA. In addition to the mitochondrial genome editing, the other two genome editing methods, ZFNs and TALENs, necessitate recording proteins with long DNA segments of about 500–1500 base pairs for each new target site. In 2015, Jo and colleagues became the first to successfully apply CRISPR/Cas9, a groundbreaking advancement in genome editing science. Targeting two mitochondrial proteins, cox1 and cox3, part of oxidative phosphorylation, was shown to modify mtDNA and impair normal mitochondrial function. Instead of the traditional Cas9, this one targets the nuclear genome by utilizing nuclear localization signals. The process of mitochondrial oxidative phosphorylation (OXPHOS) produces adenosine triphosphate (ATP), which is the energy needed to sustain cell viability. Still, on the other hand, damaged mitochondria release cytochrome c, Reactive Oxygen Species (ROS), and other signals, which activate caspase family proteins on binding with particular proteins, leading to apoptosis.

Events arising from mitochondrial injury include morphological abnormalities, changes in membrane potential and permeability to Ca^{2+}, reductions in membrane phosphate esters, and impairment of coupling in oxidative phosphorylation. These changes affect normal cell functioning and contribute to various diseases. For instance, pathological changes from the injury of the mitochondria lead to conditions such as mitochondrial myopathy, cerebral myopathy, and Leber's hereditary optic neuropathy, among others.

Besides, deranged mitochondrial metabolism and structural impairments are thought to play a critical role in the causation and course of many disorders. For example, there is evidence that mitochondrial failure has some etiopathogenesis relationship with Parkinson's Disease, an endemic neurodegenerative disorder. Abnormal structure and function of the mitochondria are associated with wide-ranging disorders, including mental illnesses, tumors, aging, cardiovascular disease, diabetes, *etc.* In some cases, treatments that target the mitochondria also produce beneficial results. The commonality among these disorders is that mitochondrial dysfunction mainly manifests as an impaired production capacity because of reduced oxidative phosphorylation, increased ROS, and aberrant apoptotic signals, even though these diseases arise in diverse tissue areas and have varied symptoms. To date, the oldest identified mitochondrial-derived peptide is human, which is encoded in mitochondrial DNA 16S rRNA.

Many age-related disorders, including AD, cancer, fibrosis, heart disease, and age-related macular degeneration models, have been linked to higher humanin levels, and treatment with humanin is protective against many diseases. Other known anti-aging pathways, such as IGF-I, also modulate Humanin. Humanin levels are decreased across an organism's life span in several species, including humans, mice, and monkeys. Another aging modulator known to be produced from the mitochondrion is MOTS-c. A small open reading frame inside the mitochondrial 12S rRNA gene encodes MOTS-c. This peptide was first called an "exercise mimetic" because it improved insulin sensitivity and blood glucose control in age-dependent, insulin-resistant individuals.

Neurodegenerative disorders NDs consist of a family of heterogeneous diseases, including amyotrophic lateral sclerosis, Parkinson's disease, Alzheimer's disease, and Huntington's disease. Since mitochondrial balance is essential for oxidative balance and metabolic activity in neurons, NDs are associated with mitochondrial dysfunction, and mitochondria have been considered a candidate for therapeutic intervention in these diseases. There are two barriers to drug penetration: the BBB and cell/mitochondrial membranes. The layer may create a limitation to drug penetration. Several techniques have been developed in the field of neuron mitochondrial targeting. Of them, treatments based on nanotechnology stand out in the expected results. Its lipophilic surface, proper charge, and tunable size make nanoparticles (NPs) optimal as the perfect theranostic system that can penetrate the blood-brain barrier and target the mitochondria of neurons.

The conventional description of mitochondrial dysfunction is a hallmark of aging and the underlying common characteristic of most chronic diseases, like cancer, diabetes, or neurodegenerative disorders, which is the decline in efficiency of ATP generation. Recently, reversing the shape and function of mitochondria has been a domain of intense investigation. For example, genome editing technologies can be applied for the correction of anomalies in the mitochondrial DNA as a therapeutic approach for Angio cardiopathy, while such products of medicines with diagnostic and therapeutic purposes, which target mitochondria, such as Gboxin, an inhibitor of oxidative phosphorylation, target suppression of diabetes or glioma. Despite extensive efforts to cure disorders associated with mitochondria, most cases tend to cause irreparable damage to the structure and functionality of the mitochondria, discouraging most treatments. Conversely, a newly invented therapy known as MRT, short for mitochondrial replacement therapy, where a functioning mitochondrion is inserted into the cells and expressed fundamentally, might be critical in curing mitochondria-related disorders. Combined treatments, such as phototherapy and small-molecule anticancer medications, have been developed in response to the synergistic treatment.

Some therapies targeting Mitochondria—like using lipophilic cations as a transport medium—can specifically transport antioxidants to mitochondria. This approach can be used with a wide range of bioactive compounds, especially the hydrophobic ones that are difficult for cells and mitochondria to take in. Triphenylphosphonium (TPP) has been used most to prepare MTAs in recent decades. Exploratory TPP-associated MTAs include SkQ1, MitoE, Mito-TEMPO, and MitoQ. Animal models and human clinical studies have been extensively investigated for MitoQ and SkQ1. One significant limitation of using TPP-linked antioxidants in treating disease is their toxicity to mitochondria. During the transfer of the TPP-linked antioxidants, TPPs adhere increasingly to the inner mitochondrial membrane. This process may damage the integrity of the mitochondrial membrane and, thereby, respiration and ATP production, but administration of lower levels of MitoQ had no toxic effect. MitoQ given orally at 20 mg/day increased brachial artery flow-mediated dilatation, decreased aortic stiffness, and lowered plasma low-density lipoprotein in a clinical trial including twenty healthy older persons with endothelial dysfunction aged 60 to 79. Oral MitoQ supplementation at 40 or 80 mg/day lowered blood alanine transaminase in patients infected with the hepatitis C virus, suggesting a reduction in hepatic necroinflammation.

Reports on the antioxidative effects of SS-31 in kidney glomerular mitochondria have also emerged. It has been demonstrated that treatment with SS-31 averts pathological changes in the indices of chronic renal disease models. More recently, nanopolyplexes have been developed to deliver SS-31 targeting AKI, suggesting that combined treatment with SS-31 and nanopolyplexes benefits the oxidatively stressed and inflamed kidney. Administration of SS-31 in a traumatic brain injury model was also observed to cause an increase in the degradation of O2 in the cytoplasm and mitochondria through the regulation of the expression of the NADPH oxidase component NOX4. Increasing disorders associated with mitochondria have focused on drugs targeting the mitochondria, bringing extensive potential and an urgent need for research and development of drugs targeting the mitochondria. Agents that target mitochondria are specific to various illnesses and cellular settings. It is necessary to minimize the toxicity of the targeted drugs to treat mitochondrial disease. However, these agents also mediate cytotoxicity, which may help kill harmful cells like cancerous cells. Moreover, a lot of work has been done in the creation of mitochondria-targeting drugs that can be applied for the imaging and monitoring of mitochondria to make further study on the physiological concepts of the mitochondrion, as well as the creation of new methods to control the mitochondrion for therapeutic purposes [21 - 28].

CHLOROPLAST

Introduction

Chloroplast was the reason life could evolve to a complex state. The sun is the supreme key behind all the types of energy on earth; this tiny organelle could directly absorb it and convert it into glucose. This led to early plants colonizing and ruling the surface of the planet. This meant the energy could be released and used whenever and wherever required. Such is the greatness behind the process of photosynthesis. This stored energy is used by all other complex living beings on the food web. Chloroplasts can be built into an entire massive rainforest from thin air and light from the sun.

Origin and Evolution of Chloroplasts

Chloroplasts were once believed to be a self-sustaining microbial living organism. It was a eubacteria-like cell that, by accident, was consumed by an ancient Eukaryotic-type cell. Our eukaryotic cell has found it helpful to keep this ancient small cell with an extraordinary ability to produce energy in the form of ATP and other substances. This is the theory of endosymbiosis. A compelling scenario was put forward by Lynn Margulis in her 1967 article "On the Origin of Mitosing Cells" in the Journal of Theoretical Biology.

Structure of Chloroplast

They are relatively large organelles and measure in size 5-10 micrometers. Like its other counterpart, mitochondria, it also possesses a double-layered membrane known as a chloroplast envelope. The dense fluid inside the inner membrane space is called the stroma, the site where CO2 to carbohydrate conversion occurs. A third membrane-bound structure consists of flattened sac-like membrane structures called thylakoids, where light energy is converted into chemical energy. Light-harvesting complexes are all found in this thylakoid system, including the electron transport chains used in photosynthesis and pigments like chlorophyll and carotenoids. A stack of 10 to 20 thylakoids is collectively called a granum. Inside the chloroplast, different grana are connected by the stroma lamella. There's also the presence of circular DNA of 120-200 kb in length and ribosomes for producing certain of its required proteins (Fig. **4**).

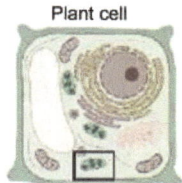

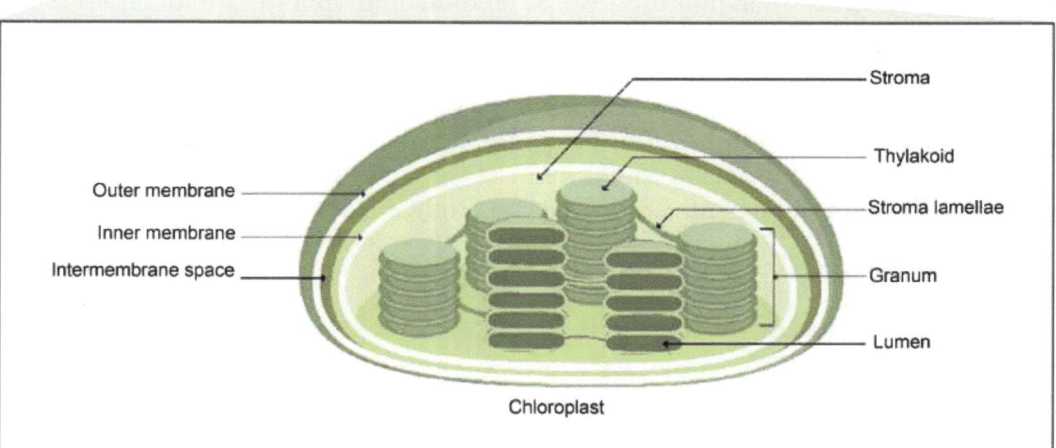

Fig. (4). Structure of the chloroplast.

DNA of chloroplasts throughout evolution, the chloroplast gene has found its way to the nuclear genome, making it responsible for encoding its required proteins. These genomes are conserved mainly in structural aspects among most land plants, with studies finding over 81% similarity in the genome with ancient algal species. The DNA within chloroplasts holds up to 120 genes, which include various genes, including *rRNA* genes, *tRNA* genes, at least three subunits of prokaryotic RNA polymerases, and some other protein-coding genes like thylakoid proteins and ribosomal proteins. Like mitochondria, the outer membrane is permeable to all kinds of small molecules with the help of protein structures on the outer membrane. The inner membrane is impermeable to all metabolites and molecules and requires unique shuttle pathways to facilitate the exchange across the inner membrane. Data on the ultrastructure of the thylakoid membrane about the protein organization is not sufficiently available due to the electron microscopy of the stroma. It appears unstructured and packed with dispersed particles of the enzyme Rubisco. The stroma has a protein concentration of c. 0.1 mM [29] translates to 400 mg protein ml−1 or c. 50% of total soluble protein in leaves [30].

Dynamic Thylakoid Architecture

The intensity of the exposure often varies across different parts of the plant, leading to a dynamic thylakoid membrane that adapts to the dynamic changes in the environment. The relocation of chloroplasts occurs dynamically in the cell to attenuate the amount of solar radiation absorbed by the cell by adjusting the incidence angle in the cell. The chloroplast relocation is regarded as a cell-autonomous intracellular light response [31]. The Chloroplast relocation is regulated by blue light photoreceptors, phototropin (photo 1 and photo 2) in higher plants or by red light-absorbing neochromes [32]. Chloroplast is connected to a unique actin network called cp-actin. The phototropin then remodels the actin network to relocate the chloroplast within the cell [32]. Thylakoid membranes, the intricate structures within chloroplasts, are highly dynamic and responsive to environmental cues, especially light intensity and color. These membranes house the machinery for photosynthesis, converting light energy into chemical energy.

Structural Plasticity

- Whole Membrane Level: The entire thylakoid membrane can change shape and size.
- Supramolecular Level: The arrangement of protein complexes within the membrane can also be altered.

Light Intensity and Thylakoid Structure

Medium Light

Thylakoid lumens swell, creating more space for the movement of proteins like plastocyanin. This swelling facilitates electron transport between photosystem II

(PSII) and photosystem I (PSI). This mechanism helps balance the flow of electrons to PSI and CO2 fixation.

High Light

Photoprotection mechanisms kick in to prevent damage caused by excess light energy. Thylakoid membranes likely undergo structural changes to dissipate excess energy or protect key components.

This could involve alterations in protein-protein interactions, energy transfer pathways, or the formation of protective complexes.

Significance

The ability of thylakoid membranes to adapt to changing light conditions is essential for the survival and efficiency of plants. By fine-tuning their structure and function, thylakoids can optimize photosynthesis and protect themselves from harmful effects. Understanding these dynamic processes can provide insights into plant stress responses, crop improvement, and the development of sustainable bioenergy technologies [33 - 36].

Light and Dark Reactions

The primary function of chloroplasts is to utilize light energy and convert it into chemical energy, also known globally as photosynthesis.

Photosynthesis is split into two parts: the light cycle and the dark cycle. The light cycle is a photochemical cycle that includes the process of light absorption, hydrolysis, and the release of oxygen. The light reaction produces photo-induced electrons that release free energy that is trapped in the phosphate bonds of ATP. Photophosphorylation is divided into cyclic and non-cyclic pathways. The cyclic pathway produces only ATP, with no change in any reactants' net oxidative and reductive state. The photo-induced electrons are transported from water to ferredoxin, which leads to the formation of ATP coupled with the evolution of oxygen molecules in the non-cyclic pathway. Plants have two photosystems: Photosystem I and Photosystem II. Photosystem I absorbs a longer wavelength of light (about 700nM), and Photosystem II absorbs a shorter wavelength of light (about 650nM). The incidence of light on the chloroplast excites the chlorophyll α to a higher energy level in the grana of the chloroplast, which leads to the synthesis of ATP and NADP. Plants utilize cyclic pathways for immediate emergency requirements and use photosystem 1. The photo-induced electrons are transferred back to the p700 instead of moving into the NADP from the electron acceptor. This cyclic reaction leads to the formation of ATPs in a cyclic manner.

In a non-cyclic pathway, the photoelectrons from Photosystem II don't return to them but are transferred to Photosystem 1. The movement is, hence, noncyclic or unidirectional. The electrons from Photosystem NADP then accept me for long-term storage. Based on the complexity of the plant, they perform either the C3 cycle or the C4 cycle for the dark reaction of photosynthesis. The cycle type is based on 0whether the first produced is a three-carbon or four-carbon molecule. CO_2 and H_2O give end products glucose and other sugar metabolites during the biosynthetic phase. The first chemically induced stable product from light reactions is PGA (Phosphoglyceric Acid). The Discovery of PGA led Calvin and his associates to formulate the carbon cycle (reductive pentose phosphate cycle), which was part of the dark reactions. One mole of CO_2 to the level of hexose phosphate requires 3 moles of ATP and 2 moles of NADP. The carbon cycle starts with the Phosphorylation of ribulose monophosphate to give ribulose diphosphate. The next step is assimilating CO_2 with the ribulose diphosphate to give 2-phosphoglycerate. The two phosphoglycerates react with another 2 moles of ATP to give two di-phosphoglycerates, which, on further oxidation by 2 moles of NADPH, give triose phosphates. The triose phosphate could again be converted into ribulose monophosphate to be used again for CO_2 assimilation. The triose phosphates are then converted to hexose phosphate and stored in starch [37].

The C4 pathway is seen in plants of tropical desert regions. A 4-carbon compound called Oxaloacetic acid (OAA) is the first product of carbon fixation in the C4 cycle. Anatomical and biochemical changes were required to evolve the C4 cycle from the C3 cycle. The pathway involves the CO_2 pump to have a spatial separation from the site of Rubisco, and it has evolved into Kranz anatomy over time [38 - 43].

Recent Trends in Malfunctions in Chloroplasts

Autophagy and Chloroplast Degradation

A recent study has highlighted the significance of autophagy in the selective destruction of chloroplasts, referred to as chlorophagy. This breakdown is essential during leaf senescence, carbon deficiency, or intense light stress and is mediated by both macroautophagy and microautophagy pathways. These activities facilitate the sequestration of damaged chloroplast components into vacuoles for breakdown, preventing the buildup of harmful by-products that might jeopardize the plant's health.

Sensitivity to Thermal Stress

Chloroplasts are very vulnerable to thermal stress, potentially causing considerable harm. This encompasses possible cell death and a reduction in photosynthetic efficiency, which may result from the weakening of Photosystem II (PSII) and the inactivation of essential proteins, such as Rubisco activase. Moreover, chloroplasts activate defensive mechanisms, such as synthesizing protein chaperones, to preserve cellular integrity under heat stress.

Production of Reactive Oxygen Species (ROS)

Chloroplasts generate deleterious Reactive Oxygen Species (ROS) in response to unfavorable circumstances, such as dryness or elevated temperatures, which may intensify damage. A recent study reveals that plants have developed ways to degrade these ROS-producing chloroplasts before they cause significant cellular harm. This process entails the targeted elimination of damaged chloroplasts marked by ubiquitin, highlighting the quality control mechanism inherent in plant cells.

Impact on Mitochondrial Function

Moreover, chloroplast malfunction may affect mitochondrial functioning. Studies indicate that mitochondrial DNA levels and respiration rates are modified in variegated leaves exhibiting chloroplast deficiencies. This suggests that mitochondria may augment their functioning as a compensating reaction to diminished chloroplast activity. As leaves age and enter senescence, chloroplasts undergo a degradation process that plays a key role in nutrient recycling. This process is essential for the plant to manage oxidative stress and re-allocate nutrients to other parts of the plant. Environmental disturbances may expedite chloroplast breakdown during leaf senescence since regulating redox homeostasis and Reactive Oxygen Species (ROS) generation significantly influences this process. Recent research trends indicate that chloroplast dysfunctions due to environmental stressors trigger intricate responses, including autophagy, regulation of reactive oxygen species, and inter-organellar communication with mitochondria. It is crucial to understand these mechanisms to develop methods that enhance plant resistance to stress.

CONCLUSION

The mitochondria and chloroplasts serve as cellular powerhouses, each with distinct roles rooted in their evolutionary origins as endosymbiotic bacteria. Mitochondria, with their dual-membrane structure, generate cellular energy through oxidative phosphorylation, and mutations in mitochondrial DNA often leading to diseases in energy-demanding tissues. On the other hand, the chloroplast captures solar energy for photosynthesis, converting CO_2 into carbohydrates, thus sustaining the food web. Advances in genome editing, drug discovery, and environmental adaptation research are expanding the ability to recognize and potentially treat mitochondrial dysfunction and enhance chloroplast resilience to stress. Emerging studies on autophagy, inter-organelle communication, and the regulation of reactive oxygen species underscore the need for deeper exploration to link mitochondrial and chloroplast function to aging, stress response, and disease progression. Future research must focus on these dynamic interactions to develop targeted therapies for mitochondrial diseases and innovations in crop bioengineering for improved resilience and productivity.

ACKNOWLEDGEMENTS

We want to acknowledge Ms. Kruthika P, Ms. Raksa Arun, Ms. Srisri SatishKartik, and Mr. Sasitharan T for contributing to proofreading and editing the work.

REFERENCES

[1] Zachar I, Boza G. Endosymbiosis before eukaryotes: mitochondrial establishment in protoeukaryotes. Cell Mol Life Sci 2020; 77(18): 3503-23.
[http://dx.doi.org/10.1007/s00018-020-03462-6] [PMID: 32008087]

[2] Gray MW, Burger G, Lang BF. Mitochondrial Evolution. Science (1979). 1999 Mar 5;283(5407):1476–81.

[3] Taylor RW, Turnbull DM. Mitochondrial DNA mutations in human disease. Nat Rev Genet 2005; 6(5): 389-402.
[http://dx.doi.org/10.1038/nrg1606] [PMID: 15861210]

[4] Wallace DC, Chalkia D. Mitochondrial DNA genetics and the heteroplasmy conundrum in evolution and disease. Cold Spring Harb Perspect Biol 2013; 5(11): a021220-0.
[http://dx.doi.org/10.1101/cshperspect.a021220] [PMID: 24186072]

[5] Ladoukakis ED, Zouros E. Evolution and inheritance of animal mitochondrial DNA: rules and exceptions. J Biol Res (Thessalon) 2017; 24(1): 2.
[http://dx.doi.org/10.1186/s40709-017-0060-4] [PMID: 28164041]

[6] Moritz C, Dowling TE, Brown WM. Evolution of animal mitochondrial DNA: Relevance for population biology and systematics. Annu Rev Ecol Syst 1987; 18(1): 269-92.
[http://dx.doi.org/10.1146/annurev.es.18.110187.001413]

[7] Giacomello M, Pyakurel A, Glytsou C, Scorrano L. The cell biology of mitochondrial membrane dynamics. Nat Rev Mol Cell Biol 2020; 21(4): 204-24.
[http://dx.doi.org/10.1038/s41580-020-0210-7] [PMID: 32071438]

[8] Palmer CS, Osellame LD, Stojanovski D, Ryan MT. The regulation of mitochondrial morphology: Intricate mechanisms and dynamic machinery. Cell Signal 2011; 23(10): 1534-45.
[http://dx.doi.org/10.1016/j.cellsig.2011.05.021] [PMID: 21683788]

[9] Ingman M, Kaessmann H, Pääbo S, Gyllensten U. Mitochondrial genome variation and the origin of modern humans. Nature 2000; 408(6813): 708-13.
[http://dx.doi.org/10.1038/35047064] [PMID: 11130070]

[10] Annesley SJ, Fisher PR. Mitochondria in Health and Disease. Cells 2019; 8(7): 680.
[http://dx.doi.org/10.3390/cells8070680] [PMID: 31284394]

[11] Herst PM, Rowe MR, Carson GM, Berridge MV. Functional Mitochondria in Health and Disease. Front Endocrinol (Lausanne) 2017; 8: 296.
[http://dx.doi.org/10.3389/fendo.2017.00296] [PMID: 29163365]

[12] Sharma P, Sampath H. Mitochondrial DNA Integrity: Role in Health and Disease. Cells 2019; 8(2): 100.
[http://dx.doi.org/10.3390/cells8020100] [PMID: 30700008]

[13] Mitchell P. Coupling of phosphorylation to electron and hydrogen transfer by a chemi-osmotic type of mechanism. Nature 1961; 191(4784): 144-8.
[http://dx.doi.org/10.1038/191144a0] [PMID: 13771349]

[14] Zhao RZ, Jiang S, Zhang L, Yu ZB. Mitochondrial electron transport chain, ROS generation and uncoupling (Review). Int J Mol Med 2019; 44(1): 3-15.
[http://dx.doi.org/10.3892/ijmm.2019.4188] [PMID: 31115493]

[15] Missiroli S, Patergnani S, Caroccia N, et al. Mitochondria-associated membranes (MAMs) and inflammation. Cell Death Dis 2018; 9(3): 329.
[http://dx.doi.org/10.1038/s41419-017-0027-2] [PMID: 29491386]

[16] Giorgi C, Missiroli S, Patergnani S, Duszynski J, Wieckowski MR, Pinton P. Mitochondria-associated membranes: composition, molecular mechanisms, and physiopathological implications. Antioxid Redox Signal 2015; 22(12): 995-1019.
[http://dx.doi.org/10.1089/ars.2014.6223] [PMID: 25557408]

[17] Polyak K, Li Y, Zhu H, et al. Somatic mutations of the mitochondrial genome in human colorectal tumours. Nat Genet 1998; 20(3): 291-3.
[http://dx.doi.org/10.1038/3108] [PMID: 9806551]

[18] Prezant TR, Agapian JV, Bohlman MC, et al. Mitochondrial ribosomal RNA mutation associated with both antibiotic–induced and non–syndromic deafness. Nat Genet 1993; 4(3): 289-94.
[http://dx.doi.org/10.1038/ng0793-289] [PMID: 7689389]

[19] Schubert C. Profile: Mary-Claire King. Nat Med 2003; 9(6): 633-3.
[http://dx.doi.org/10.1038/nm0603-633] [PMID: 12778148]

[20] Song S, Pursell ZF, Copeland WC, Longley MJ, Kunkel TA, Mathews CK. DNA precursor asymmetries in mammalian tissue mitochondria and possible contribution to mutagenesis through reduced replication fidelity. Proc Natl Acad Sci USA 2005; 102(14): 4990-5.
[http://dx.doi.org/10.1073/pnas.0500253102] [PMID: 15784738]

[21] Chan DC. Mitochondria: dynamic organelles in disease, aging, and development. Cell 2006; 125(7): 1241-52.
[http://dx.doi.org/10.1016/j.cell.2006.06.010] [PMID: 16814712]

[22] Rajasekaran A, Venkatasubramanian G, Berk M, Debnath M. Mitochondrial dysfunction in schizophrenia: Pathways, mechanisms and implications. Neurosci Biobehav Rev 2015; 48: 10-21.
[http://dx.doi.org/10.1016/j.neubiorev.2014.11.005] [PMID: 25446950]

[23] Baldassarro VA, Kręzel W, Fernández M, Schuhbaur B, Giardino L, Calzà L. The role of nuclear receptors in the differentiation of oligodendrocyte precursor cells derived from fetal and adult neural stem cells. Stem Cell Res (Amst) 2019; 37: 101443.
[http://dx.doi.org/10.1016/j.scr.2019.101443] [PMID: 31022610]

[24] Winklhofer KF, Haass C. Mitochondrial dysfunction in Parkinson's disease. Biochim Biophys Acta Mol Basis Dis 2010; 1802(1): 29-44.
[http://dx.doi.org/10.1016/j.bbadis.2009.08.013]

[25] Chen Z, Li G, Liu J. Autonomic dysfunction in Parkinson's disease: Implications for pathophysiology, diagnosis, and treatment. Neurobiol Dis 2020; 134: 104700.
[http://dx.doi.org/10.1016/j.nbd.2019.104700] [PMID: 31809788]

[26] Taanman JW. The mitochondrial genome: structure, transcription, translation and replication. Biochim Biophys Acta Bioenerg 1999; 1410(2): 103-23.
[http://dx.doi.org/10.1016/S0005-2728(98)00161-3] [PMID: 10076021]

[27] Prakash R, Kannan A. Mitochondrial DNA modification by CRISPR/Cas system: Challenges and future direction. Prog Mol Biol Transl Sci 2021: 178: 193-211.

[28] Kim SJ, Miller B, Kumagai H, Silverstein AR, Flores M, Yen K. Mitochondrial-derived peptides in aging and age-related diseases. Geroscience 2021; 43(3): 1113-21.
[http://dx.doi.org/10.1007/s11357-020-00262-5] [PMID: 32910336]

[29] Oguchi R, Hikosaka K, Hirose T. Does the photosynthetic light-acclimation need change in leaf anatomy? Plant Cell Environ 2003; 26(4): 505-12.
[http://dx.doi.org/10.1046/j.1365-3040.2003.00981.x]

[30] Robinson JM. Does O_2 photoreduction occur within chloroplasts *in vivo*? Physiol Plant 1988; 72(3): 666-80.

[31] Kirchhoff H, Hall C, Wood M, *et al.* Dynamic control of protein diffusion within the granal thylakoid lumen. Proc Natl Acad Sci USA 2011; 108(50): 20248-53.
[http://dx.doi.org/10.1073/pnas.1104141109] [PMID: 22128333]

[32] Kong SG, Wada M. Molecular basis of chloroplast photorelocation movement. J Plant Res 2016; 129(2): 159-66.
[http://dx.doi.org/10.1007/s10265-016-0788-1] [PMID: 26794773]

[33] Kirchhoff H. Chloroplast ultrastructure in plants. New Phytologist 2019; 223(2): 565-74.
[http://dx.doi.org/10.1111/nph.15730]

[34] Li L, Yuan H. Chromoplast biogenesis and carotenoid accumulation. Arch Biochem Biophys 2013; 539(2): 102-9.
[http://dx.doi.org/10.1016/j.abb.2013.07.002] [PMID: 23851381]

[35] Takahashi S, Badger MR. Photoprotection in plants: a new light on photosystem II damage. Trends Plant Sci 2011; 16(1): 53-60.
[http://dx.doi.org/10.1016/j.tplants.2010.10.001] [PMID: 21050798]

[36] Ruban AV. Evolution under the sun: optimizing light harvesting in photosynthesis. J Exp Bot 2015; 66(1): 7-23.
[http://dx.doi.org/10.1093/jxb/eru400] [PMID: 25336689]

[37] Wada K, Arnon DI. Three Forms of Cytochrome b55g and Their Relation to the Photosynthetic Activity of Chloroplasts (electron carriers/redox potential/light-induced electron transport/spinach) [Internet]. Vol. 68. 1971. Available from: https://www.pnas.org

[38] Furbank RT. Walking the C4 pathway: Past, present, and future. Vol. 67, Journal of Experimental Botany. Oxford University Press; 2016. p. 4057–66.

[39] Kutschera U, Ray PM. Forever young: stem cell and plant regeneration one century after Haberlandt 1921. Vol. 259, Protoplasma. Springer; 2022. p. 3–18.

[40] Cazzonelli CI. Carotenoids in nature: insights from plants and beyond. Funct Plant Biol 2011; 38(11): 833-47.
[http://dx.doi.org/10.1071/FP11192] [PMID: 32480941]

[41] Shumskaya M, Wurtzel ET. The carotenoid biosynthetic pathway: Thinking in all dimensions. Plant Sci 2013; 208: 58-63.
[http://dx.doi.org/10.1016/j.plantsci.2013.03.012] [PMID: 23683930]

[42] Vallabhaneni R, Bradbury LMT, Wurtzel ET. The carotenoid dioxygenase gene family in maize, sorghum, and rice. Arch Biochem Biophys 2010; 504(1): 104-11.
[http://dx.doi.org/10.1016/j.abb.2010.07.019] [PMID: 20670614]

[43] Giuliano G, Tavazza R, Diretto G, Beyer P, Taylor MA. Metabolic engineering of carotenoid biosynthesis in plants. Trends Biotechnol 2008; 26(3): 139-45.
[http://dx.doi.org/10.1016/j.tibtech.2007.12.003] [PMID: 18222560]

CHAPTER 11

Cytoskeleton: The Cell's Backbone and Highway

Danyal Reyaz[1], Sparshika Mishra[1], Ranjini Sengupta[1], T. Sasitharan[1], S. Gnanavel[2] and K.N. Aruljothi[1,*]

[1] *Department of Genetic Engineering, SRM Institute of Science and Technology, Kattankulathur, Chengalpattu, Tamil Nadu, India*

[2] *Department of Biomedical Engineering, SRM Institute of Science and Technology, Kattankulathur, Chengalpattu, Tamil Nadu, India*

Abstract: The cytoskeleton is an essential framework in eukaryotic cells, providing structural support, maintaining shape, and facilitating intracellular transport and movement. This chapter explores the key components of the cytoskeleton: microfilaments, intermediate filaments, and microtubules, discussing their structures, assembly mechanisms, and roles in cellular processes. Microfilaments, composed of actin, contribute to cellular shape, movement, and muscle contraction through mechanisms like treadmilling and interaction with myosin. Intermediate filaments, including keratin and vimentin, provide mechanical strength and support cellular integrity under stress. Microtubules, composed of tubulin, are involved in mitosis, intracellular transport, and the maintenance of cell polarity. Additionally, this chapter delves into the role of motor proteins like kinesin and dynein in facilitating molecular transport along microtubules and how cytoskeletal dynamics are crucial for both healthy cellular function and pathological conditions like cancer and neurodegenerative diseases.

Keywords: Actin, Cancer, Cell Motility, Dynein, Intermediate filaments, Keratin, Kinesin, Locomotion, Microfilaments, Microtubules, Muscle, Neurodegenerative Diseases.

INTRODUCTION

In addition to the cytoplasm, plasma membrane, and organelles, eukaryotic cells have something like a hidden superhero called the cytoskeleton. This internal framework comprises tiny fiber proteins, which we call cytoskeletal fibers. The cytoskeletal fibers play a critical role in maintaining cellular integrity by preserving cell shape and providing mechanical support. The plasma membrane functions as a selective barrier that protects the intracellular environment. Within

---

[*] **Corresponding author K.N. Aruljothi:** Department of Genetic Engineering, SRM Institute of Science and Technology, Kattankulathur, Chengalpattu, Tamil Nadu, India; E-mail: aruljotn@srmist.edu.in

K.N. Aruljothi, Prakash Gangadaran, Krishnan Anand, Satish Ramalingam, K. Kumaran & Kruthika Prakash (Ed.)
All rights reserved-© 2026 Bentham Science Publishers

the cytoplasm, distinct organelles carry out specialized biochemical processes: mitochondria facilitate ATP production through cellular respiration, the Golgi apparatus is responsible for protein modification and trafficking, and the endoplasmic reticulum is involved in protein synthesis and intracellular storage. Now, let us talk about the cytoskeleton. Think of it as the house's skeleton. Just like our bones give our bodies structure, these cytoskeletal fibers give the cell its shape and help it stay strong. They're like the support beams in our house. But the cytoskeleton does more than that. It's like the cell's personal trainer. It helps the cell move around, divide into two when it needs to make a copy of itself, and even helps things move around inside the cell. So, in simple terms, the cytoskeleton is like the cell's own hero, working behind the scenes to keep everything in order, maintain shape, and ensure the cell can do everything it needs to do [1].

Types of Cytoskeletal Fibers

The cytoskeleton is a complex network of fibers that provides structural support, shape, and organization to cells. It plays a critical role in various cellular processes, including movement, division, and intracellular transport. This dynamic system is made up of different types of protein-based filaments, each contributing to specific cellular functions (Table 1).

Table 1. Types of protein present in the cytoskeleton, along with the protein size.

Type	Protein	Site	Size (kDa)
I	Acidic Keratins	Epithelial Cells	40-60
II	Neutral or Basic Keratins	Epithelial cells	50-70
III	Vimentin Desmin Glial fibrillary acidic proteins Peripherin	Fibroblasts, white blood cells, and other cell types Muscle cells Glial cells Peripheral neurons	54 53 51 51
IV	Neurofilament proteins NF-L NF-M NF-H α-Nexin Nestin	Neurons Stem cells	67 150 200 66 200
V	Nuclear lamins	Nuclear lamina of all cell types	60-75

The three types are:

- Microfilaments
- Intermediate filaments
- Microtubules

MICROFILAMENTS

Structures that were a part of the cytoskeleton that supported the flexibility and increased the strength of the cell were found in 1974 by B. Paleviz *et al*. They are microfilaments, as we now refer to them.

Structure

Microfilaments are little rod-shaped structures approximately 6 nm in diameter, found inside the cytoskeleton of eukaryotic cells. Microfilaments are constituted by the aggregation of actin monomers. It comprises two strands of actin protein subunits coiled in a helical formation, hence referred to as actin filaments. The actin subunits that constitute microfilaments are globular actin (G-actin), which, upon polymerization, are referred to as filamentous actin (F-actin). The actin subunits assemble directionally to produce microfilaments. Subunits possess a "top" and a "bottom," with the top of one subunit consistently interacting with the bottom of another. The terminal subunit of a filament is referred to as the negative (-) end, while the opposing end, which experiences greater polymerization, is designated as the plus (+) end. Microfilaments possess polarity, although this pertains solely to their directionality and is unrelated to electrical charge [2] (Fig. 1).

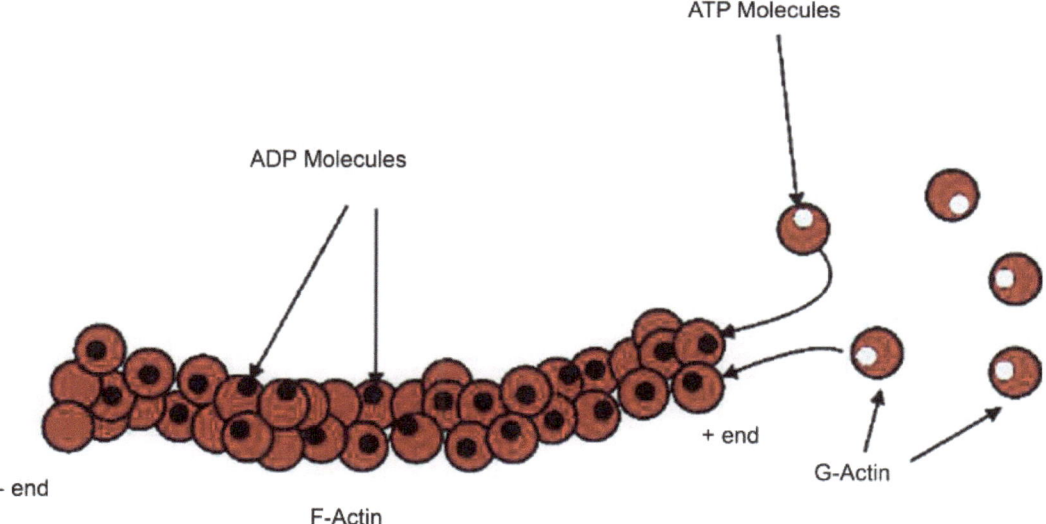

Fig. (1). Structure and assembly of microfilaments.

Organisation

The formation of a small aggregate consisting of three actin monomers is referred to as the nucleation phase of actin polymerisation. Each actin monomer (G actin) possesses binding sites that facilitate head-to-tail contacts with two other actin monomers, resulting in the polymerisation of actin monomers into filaments (F actin). The reversible addition of monomers to both termini of an actin filament facilitates its growth; however, only one terminus—the plus end—elongates five to ten times more rapidly than the minus end. Additionally, ATP binds to actin monomers and is hydrolysed to ADP during filament assembly.

Actin polymerisation is reversible; thus, filaments can depolymerise through the dissociation of subunits. The incorporation rate of monomers correlates with

their concentration, establishing a threshold concentration at which the rates of association and dissociation are equivalent. A discernible balance exists between actin monomers and filaments.

Treadmilling

Actin filaments have differential growth rates at their termini, with the plus end elongating five to ten times more rapidly than the minus end due to the addition of monomers. The essential concentration of monomers necessary for polymerisation at both termini varies due to the slower dissociation rate of ATP-actin compared to ADP-actin. Treadmilling, a phenomenon illustrating the dynamic nature of actin filaments, may arise from this difference. The concentration of free actin monomers must be positioned at the midpoint between the critical concentrations required for polymerisation at the plus and minus ends of the actin filaments to maintain an overall steady state in the system. Under these circumstances, there is a net depletion of monomers from the minus end, counterbalanced by a net accumulation at the plus end. Treadmilling necessitates ATP, as ATP-actin polymerises at the plus end of filaments while ADP-actin disassociates from the minus end.

Organization of Actin Filaments

Individual actin filaments can be assembled into two types of structures, called actin bundles and actin networks.

In bundles, actin filaments are crosslinked into closely packed parallel arrays.

Actin filaments that are closely spaced and parallelly oriented make up the first type of bundle, which supports plasma membrane protrusions like microvilli. All of the filaments in these bundles are similarly polarised, with their plus ends next to the plasma membrane. Fimbrin is an example of a bundling protein involved in the development of these structures.

The second kind of actin bundles, which include the actin bundles of the contractile ring that splits cells in half after mitosis, are made up of filaments that are more loosely spaced and have the ability to contract. These bundles, also known as contractile bundles, have a looser shape that is a result of the α-actinin protein's ability to crosslink proteins [3].

In networks, the filaments generate three-dimensional mesh networks with the characteristics of semisolid gels by loosely crosslinking in orthogonal arrays.

Functions

Protrusions of the Cell Surface

Actin-based cell surface projections include microvilli. They are particularly prevalent on the surfaces of cells engaged in absorption, such as the epithelial cells lining the intestine, and are finger-like extensions of the plasma membrane.

The comprehensive structural investigation of intestinal microvilli, which comprise closely packed parallel bundles of 20 to 30 actin filaments, has been made possible by their abundance and simplicity of isolation. Fimbrin contributes to the crosslinking of the filaments in these bundles.

Muscle Contraction

Muscle cells are highly specialized for a single task—contraction. In both skeletal and cardiac muscle, the contractile elements of the cytoskeleton are present in highly organized arrays, giving rise to characteristic patterns of cross-striations.

The sliding filament model, first put forth in 1954 by Hugh Huxley and Jean Hanson as well as Andrew Huxley and Ralph Niedergerke, serves as the foundation for understanding muscular contraction [4].

The Z discs become closer together as a result of each sarcomere shortening during muscular contraction. The I band and the H zone nearly entirely vanish, while the A band's breadth remains unchanged. Actin filaments moving into the A band and H zone as a result of myosin filaments slipping past one another cause these alterations. Muscle contraction thus arises from an interaction between the actin and myosin filaments that creates their movement relative to one another. Myosin's association with actin filaments, which enables it to operate as a motor to promote filament sliding, provides the chemical basis for this relationship. Several hundred myosin molecules are arranged in a parallel staggered pattern within the thick filaments of muscle through interactions between their tails. Actin is bound by the myosin globular heads, creating cross-bridges between the thick and thin filaments. At the sarcomere's M line, the orientation of myosin molecules in the thick filament reverses. The relative orientation of myosin and actin filaments is the same on both halves of the sarcomere because the polarity of actin filaments, which are connected to Z discs at their barbed ends, reverses like this at the M line. The myosin heads bind and hydrolyse ATP in addition to actin, which supplies the energy needed to drive filament sliding binding. Myosin and actin are separated by ATP binding. The location of the myosin head group is then changed by ATP hydrolysis as a result of a conformational shift. The myosin head is then bound to a new location on the actin filament, and Pi is released after that. Actin filament sliding is triggered by the myosin head returning to its initial configuration and the release of ADP [5, 6].

Step-by-step sequence of skeletal muscle contraction and relaxation (Fig. 2):

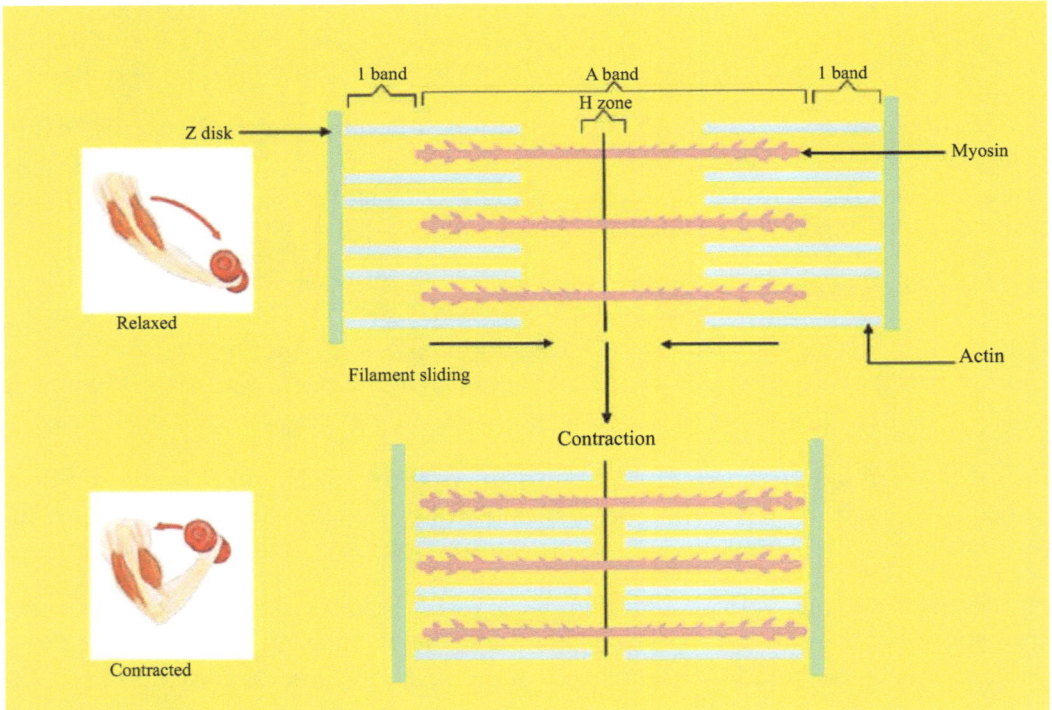

Fig. (2). Sliding filament mechanism of muscle contraction.

- Signal from Motor Neuron
- Nerve Impulse Arrives at Neuromuscular Junction
- Release of Acetylcholine (ACh) into the Synaptic Cleft

- Binding of ACh to Muscle Cell Receptors
- Generation of Action Potential in Muscle Cell
- Propagation of Action Potential Along the Sarcolemma
- T-Tubule Activation
- Release of Calcium Ions (Ca^{2+}) from Sarcoplasmic Reticulum
- Calcium Ions Bind to Troponin
- Conformational Change in Troponin-Tropomyosin Complex
- Exposure of Myosin Binding Sites on Actin Filaments
- Cross-Bridge Formation (Myosin Heads Bind to Actin Filaments)
- Power Stroke (Myosin Pulls Actin Filament Toward Sarcomere Center)
- Cross-Bridge Detachment
- ATP Hydrolysis to Reset Myosin Head
- Shortening of Sarcomere (Z-discs move closer)
- Contraction of the I-band (Actin filaments slide over Myosin filaments)
- H-zone Narrows (Myosin filaments partially overlap with each other)
- Muscle Contraction Continues (Repeated Cross-Bridge Cycling)
- Nerve Impulse Ceases
- Calcium Ions Pumped Back into Sarcoplasmic Reticulum
- Troponin-Tropomyosin Complex Blocks Myosin Binding Sites
- Muscle Relaxation

Contractile Assemblies of Actin and Myosin in Non-muscle Cells

Stress fibres and circumferential belts are two instances of contractile assemblies seen in non-muscle cells.

The contraction of stress fibres causes tension across the cell, allowing the cell to draw on a substratum (*e.g.*, the extracellular matrix) to which it is tethered.

During embryonic development, when epithelial cell sheets fold into shapes like tubes, the contraction of adhesion belts is crucial because it changes the morphology of epithelial cell sheets.

Cytokinesis

In yeast and mammalian cells, a contractile ring made of myosin II and actin filaments forms immediately beneath the plasma membrane towards the conclusion of mitosis.

Its contraction gradually pulls the plasma membrane inward, causing the cell's centre to narrow and split in half.

It is interesting to note that the contractile ring's thickness does not change as it contracts, suggesting that actin filaments break down as the contraction progresses.

Cell division is followed by full dispersion of the ring.

As a bacterial protein with an actin-like activity that also participates in cell division, this function of actin seems to be an evolutionary one.

Locomotion

Locomotion is one of the various ways microfilaments contribute to the cell. The process usually starts with the plasma membrane or the cell cortex becoming specialised in order for cells to establish an initial polarity.

To create the cell's leading edge, protrusions are expanded. These include

- Pseudopodia
- Lamellipodia

When the actin proteins polymerize and form chains in the cell body, pseudopodia may develop. Actin polymerization creates a protrusive force that propels cell protrusion. It appears that the force pushing the cell membrane in the desired direction is produced by actin chains. The remainder of the cytoplasm moves forward as a projection forms, moving the cell along with it. Amoeboid movement is the name for this type of movement.

Lamellipodia are thin sheet-like protrusions generated by a branched actin network pushing out against the plasma membrane. Filopodia are thin tubular protrusions that can elongate up to 30 µm at the front of lamellipodia but can also form independently [7].

INTERMEDIATE FILAMENTS

Intermediate filaments are a long, rope-like group of proteins that are fibrous and made up of subunits of Keratin and its derivatives. Intermediate filaments are not directly associated with locomotory functions of the cell, but provide resistance against shear and mechanical stresses, as well as aid in providing a scaffold for processes of the cell to take place [8].

Intermediate Filaments in the Cell

The discovery and understanding of Intermediate proteins took place with advancements in technology and tools. William. T. Astbury was the first person to publish an article in 1932 regarding the structure of Keratin, with the applications of X-ray diffraction and crystal structure theory [9]. The molecular structure of keratin was deduced by alpha-keratin diffraction patterns of light discovered by Linus Pauling. George Rogers and Peter Steinert discovered the polymerized structure of Intermediate filaments. Individual subunits were visualized upon the advent of Electron microscopy (Fig. 3).

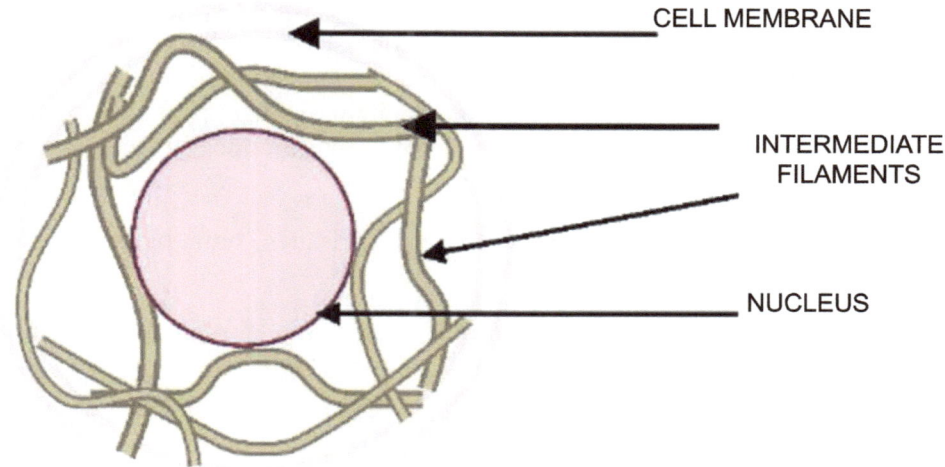

Fig. (3). Intermediate filaments distribution.

The term "Intermediate filaments" was coined by Keith. R.Porter. It was named so because it has an intermediate diameter concerning the other two cytoskeletal elements (actins, microtubules) [10].

Composition

Intermediate filaments have dimensions of width 10-12 nm, which is between actin filaments (7nm) and microtubules (25nm). Intermediate filaments are made up of alpha keratin and its derivatives [11]. Intermediate filaments are highly stable owing to Keratin filaments in their structure. They also sustain for longer periods (eg- nails, hair) [12]. The alpha keratins are made up of two subunits, forming a dimer. The subunit alpha-helices twist around each other to form a stable coiled, helical structure. Globular domains with binding sites cap these regions [13]. These assemble to give a rope-like structure of intermediate filaments.

Intermediate filaments are polymers of various proteins expressed differently in various cells [14].

Types of Intermediate Filaments

Seventy different proteins are associated with the framework of intermediate filaments in different cells. These can be classified into 5 types [15].

TYPE I: Acidic Keratins

Acidic Keratins are predominantly found in the outermost epithelium, such as hair, nails, and other epidermal cells. These consist of smaller and less stable keratins than Type II (basic keratins). This group of Type I intermediate filaments consists of 28 genes. It is a huge contributor to mechanical support and resists the deformation of outer epithelial cells. They have a huge concentration of amino acids [16].

Location: They are found in outermost epidermal tissues, hair, nails, and some mucous membranes.

Diseases: Mutations in Type-1 genes result in genetic disorders like-keratinopathies like Keratinopathic ichthyosis and blistering effects due to mutations in KRT9 and KRT10 causing Epidermolytic ichthyosis [17].

TYPE II: Basic Keratins

Basic Keratins are found in the outermost epithelial tissues of skin, hair, nails, and even in the linings of respiratory and digestive tracts. These can also be seen in other animals in the form of feathers, hooves, *etc* [58]. These are more stable than Type I. These can be classified into two types: alpha keratins and beta keratins. Alpha Keratins are present in mammals, whereas Beta Keratins are present in reptiles, birds, and other lower animals. Basic Keratin has protective roles in various cells and also provides mechanical strength [18].

Location: They are found in the outermost layer of skin, hair, nails, respiratory tract, digestive tract, feathers, hooves, and other parts [19].

Diseases: Genetic mutations of basic keratins can cause blistering effects like-Epidermolysis bullosa or diseases that affect the nails and skin like Pachyonychia congenita [20].

TYPE III: Vimentin, Desmin, Glial Fibrillary Acidic Proteins, Peripherin

Vimentin

Vimentins are mainly found in connective tissues in the body. It is mainly of mesenchymal origin and forms cells that are found in embryonic connective tissues, blood vessels, and the immune system. They provide mechanical support to cells by maintaining their shape and integrity [21].

In humans, vimentin is encoded by a single gene located on chromosome 10. Vimentins are used as markers for Immunohistochemistry procedures, which provide a piece of detailed information regarding the spatial arrangement of proteins or antigens that are localized at a particular section of the tissues.

Diseases: Mutations in the vimentin expression may lead to breast, lung, pancreatic, and prostate cancer, due to epithelial-mesenchymal transition (EMT) that increases the cancer progression. The agglomeration of vimentin can be a reason for neurodegenerative diseases like inclusion body myositis and Alexander disease, which affects muscle contraction and shoots directly up to the central nervous system. Vimentins are also involved in cardiovascular diseases, such as atherosclerosis. The amount of vimentin expression is observed to have increased in these regions [22].

Desmin

These intermediate filaments are found in muscle cells, which include both skeletal and cardiac muscle cells. Desmins provide a network for the functioning of actin and myosin. They play a key role in how the myofibrils align and organize themselves within muscle cells.

In smooth cardiac muscles, desmins provide structural integrity and function in the mechanical coupling of muscle cells. Desmin is encodedby a single gene located on chromosome 2. It is also used as a marker in a biopsy to identify muscle cells [23].

Diseases-Mutations in the gene that encodes Desmin cause "Desmin-related myopathies" that cause muscle degeneracy and stunted growth with weakness. This includes Myofibrillar myopathy, which disintegrates myofibrils and causes abnormal accumulation of proteins. This causes muscle atrophy and cardiovascular issues [24].

Glial Fibrillary Acidic Proteins (GFAP)

Glial fibrillary acidic proteins are composed of monomers that then add up to become dimers and finally form filaments after their assembly. They are located in astrocytes that function in supporting neurons.

These intermediate filaments support the glial cells and provide mechanical stability. During any injury to the Central Nervous System, the GFAP expression stimulates the hypertrophy or enlargement of astrocytes and thus helps them to function in place of neurons. GFAP assists astrocytes in protecting neurons from toxicity. GFAP proteins also signal cerebral blood cells to dilate or constrict to allow blood to flow.

The GFAP gene is located on chromosome 17 in humans. They are used as markers in pathology to detect the distribution of astrocytes. GFAP expressions are monitored during the testing of neurological diseases like Alzheimer's Disease and Multiple sclerosis.

Diseases: Mutations to GFAP can lead to neurodegenerative diseases like Alexander's Disease, with the characterization of excess GFAP in the astrocytes, which may lead to stunted growth, seizures, and intellectual disorder [25].

Peripherin

Peripherins form filamentous structures in the peripheral axons. They are mainly found in the Peripheral Nervous System, in the axons of sensory and motor neurons. Peripherins provide protection and mechanical support to nerve axons. They also assist in the transport of components along neurofibrils along axons. Peripherin encoding gene PRPH is located on chromosome 12 [26].

Diseases: Mutations in the gene may cause the accumulation of Peripherin leading to Amyotrophic Lateral Sclerosis, which leads to neuronal degeneracy [27].

TYPE IV: Neurofilament Proteins, NF-L, NF-M, NF-H, A-Internexin, Nestin

Neurofilament Proteins, NF-L, NF-M, NF-H:

Neurofilaments are intermediate filament protein families that provide neurons with stability and support. There are three types of neurofilaments: neurofilament-light, neurofilament-medium, and neurofilament-heavy. All of them provide axonal stability and assist in related processes [28].

Diseases: Mutations or abnormal expression of these proteins can cause neurodegenerative diseases like Alzheimer's Disease, which is characterized by increased levels of NF-L in cerebrospinal fluid. Another disease associated with mutations of the gene that encodes NF-L protein is Hereditary Spastic Paraplegia, which causes degeneration of motor neurons [29].

A-Internexin

a-Internexin or Internexin neuronal intermediate filament alpha (INA) are intermediate filaments found in the CNS and integrate neurons. It works with neurofibrils to transport cellular components [30].

Diseases- Diseases like Charcot-Marie-Tooth disease are caused due to mutations in INA, which lead to neuronal degeneracy [31].

Nestin

Nestins are intermediate filament proteins that are found in neural stem cells and are expressed in the embryonic stage, in regions where the development of neurons is actively taking place. They are also found in the brains of adults. Nestin is found in certain muscle stem cells and cancer cells. These are also used as markers and are used to study neural stem cells and progenitor cells. The gene that encodes Nestin is located on chromosome 1 in humans [32].

Diseases: An increased number of Nestin expression is seen in Glioblastoma, which is a type of brain cancer. Its mutation is observed in pancreatic cancer and cancer-related to sarcoma.

Nestin expression abnormalities can harm muscle regeneration. Studying nestin expression in neurodegenerative cells or injured neurons helps us understand the mechanisms underlying these conditions [33].

TYPE V: Nuclear Lamins

Nuclear lamins are found in the nucleus, especially in the nuclear envelope, inner nuclear membrane, and nuclear lamina. Nuclear lamins provide structural support to the nucleus of the cell, helping to maintain its size and shape. Nuclear lamins play a major role in the morphological properties of the nucleus. They contribute to the stiffness or elasticity of the nucleus. Moreover, it has an indirect role in the nuclear processes, including gene expression. These intermediate proteins also interact with signaling and transcription factors, thus making an influence in cell signaling pathways [34].

Diseases: mutations in nuclear lamin proteins can lead to the formation of diseases that are collectively called laminopathies. Diseases such as Progeria and muscular dystrophy can occur due to mutations in these intermediate proteins [35].

Structure And Assembly of Intermediate Filaments

Though intermediate filaments show a vast contrast in terms of their localizations and their functions, they are morphologically much the same. All intermediate filaments contain an alpha-helical rod domain in the center. These are made up of 310 amino acids. Only nuclear lamins are made up of 350 amino acids.

The alpha-helical rod is bordered by several carboxyl and amino terminals. The alpha-helical rod domain is divided into (i) the N-terminal head domain, and (ii) the C-terminal tail domain (Fig. **4**).

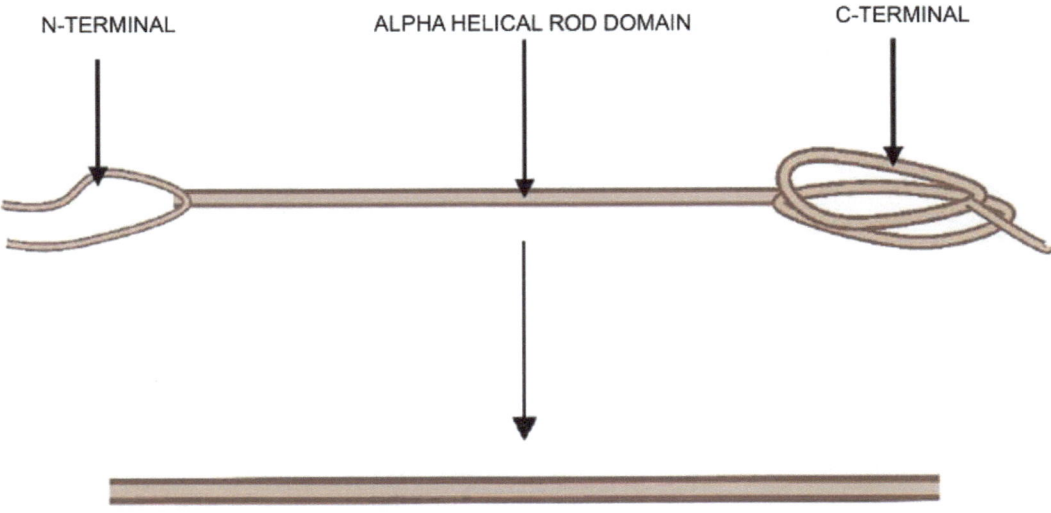

Fig. (4). Assembly of intermediate filaments.

Two polypeptides have a central rod supercoil to form a dimer. Dimers then join in a staggered anti-parallel fashion to form Tetramers. The next step is the formation of the Protofilament. The end-to-end association of tetramers leads to its formation [36].

Approximately 8 protofilaments coil in a rope-like structure to form a Filament.

Intermediate Filament: Organization And Intracellular Localization

Intermediate filaments cover a widespread area within the nucleus, ranging from the plasma membrane to the nucleus.

These not only act alone but also associate with other cytoskeletal elements like microtubules and actin. Vimentin and keratin intermediate filaments anchor the nucleus in the cell [37].

Intermediate filaments thus act as a scaffold providing essential assistance in cellular processes and integrity to the cell in all [38].

Intermediate Filaments and Cell Signalling

Intermediate filaments like Vimentin, Glial fibrillary acidic proteins, and keratins are associated with cell signalling processes. Several protein kinases, such as RAF, Protein Kinase C (PKC), c-Bcr, Bcr–Abl, and phosphatidylinositol 3-kinase, bind to 14-3-3 proteins, which are regulated by Intermediate filaments to carry out important cellular metabolism processes like cell apoptosis. It induces the mitogen-activated protein kinase (MAPK) kinase signal transduction pathway during periods of cell stress. 14-3-3 proteins are molecular chaperones that change their conformation [39]. Intermediate filaments offer scaffolds for these cell-signalling proteins and thus integrate these processes. Intermediate filaments also activate mechanosensitive pathways, such as the focal Adhesion Pathway (FAK). Intermediate filaments provide strength and facilitate mechanotransduction, where mechanical stimuli are converted into biochemical signals [40]. Thus, intermediate filaments play an integral part in the integration of the cell.

MICROTUBULES

Microtubules, intermediate filaments, and actin are all crucial parts of the cytoskeleton [41]. They appear as straight hollow rods with a diameter of 25nm. Microtubules mediate numerous dynamic actions, such as cell migration and chromosome movement during anaphase in mitosis. These actions occur because of the constant construction and disassembly of microtubules. Tubulin dimers, which are 55 kDa polypeptides known as alpha tubulin and beta tubulin, make up microtubules. When these polypeptides are present in longitudinal connection, they form a protofilament. Each microtubule is made up of 13 linear protofilaments organized in a radial pattern around a hollow core. Thus, microtubules are polar entities with two distinct ends called the plus and minus ends, just like actin filaments. The direction of movement along microtubules is determined in large part by this polarity.

The direction of myosin migration is determined by filaments. Microtubules are extremely dynamic structures capable of fast cycling. Both alpha and beta-tubulin bind GTP. GTP bound to beta-tubulin hydrolyzes GDP as soon as polymerization occurs. GTP hydrolysis reduces the tubulin dimer binding affinity for each other. This can cause GDP-bound tubulin to dissociate from the terminals of microtubules rapidly. As a result, the minus ends of microtubules within cells must be safeguarded against degradation.

Positive-end tracking proteins called MAPs attach to the plus ends of microtubules and may monitor microtubules or encourage their attachment to the plasma membrane [41].

Microtubules Assembly

The protofilament number of microtubules seen in cells is uniform. Theodor Boveri reported in 1888 that the majority of animal cells' microtubules emerge from the centrosome and are located close to the nucleus, near the cells' interphase center [41]. The centrosomes of the majority of animals have two centrioles that are orthogonal to one another and are surrounded by pericentriolar amorphous substances [42]. The centrosome's essential protein, γ-tubulin, is joined by eight other proteins to form a ring. The gamma-tubulin ring is a structure seen in many proteins. The centrioles are cylindrical in shape, built with nine triplets comparable to the cilia and flagella basal bodies (Fig. 5) [43].

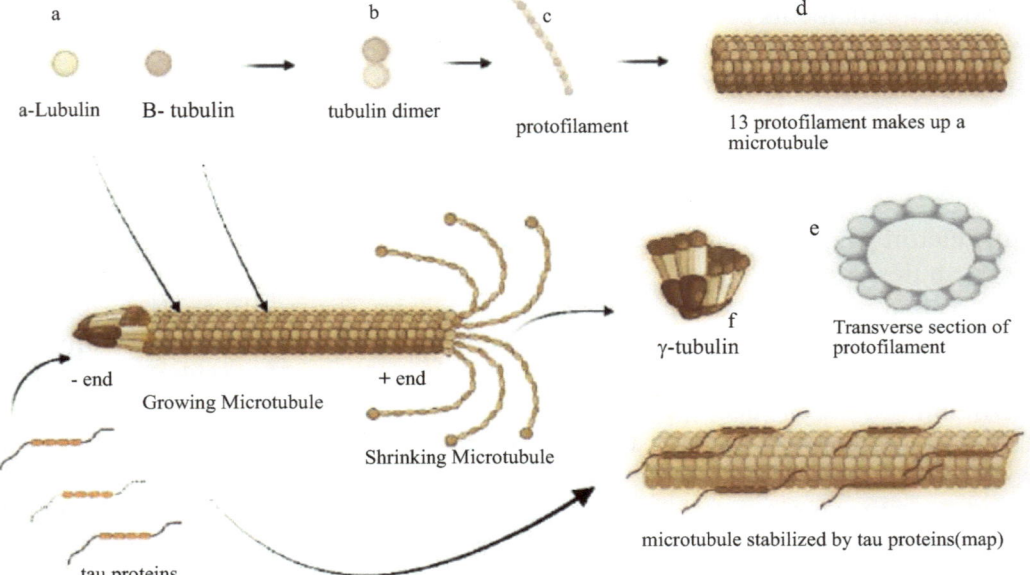

Fig. (5). Microtubules assembly.

Role of MAPs in the Organization of Microtubules in Cells

MAPs or microtubule-associated proteins consist of the following proteins:

- XMAP215
- CLASP
- Katanin
- Kinesins 8 and Kinesin 13
- Gamma-tubulin ring complex

Tau protein serves as a microtubule-associated protein found in the brain and axons. The tau sequence revealed three to four tandem repeats, which have been shown to comprise the microtubule binding site. In contrast to full-length tau or tau fragments that retain both the amino terminus and the repeat domain, microtubule bundle formation happens when tubulin joins together in the presence of an amino- and carboxyl-terminally shortened tau protein [44] (Table 2).

Table 2. Proteins present in the family of MAPs and their functions.

Proteins	Functions
Gamma tubulin ring complex	Microtubule initiation
XMAP215	Plus-end polymerization
Kinesins 8 and 13	Plus-end depolymerization
Katanin	Severing
Tau, MAP1, MAP2	Lengthwise stabilization
CLASP	Rescue

The MAP 1 family is an important class of microtubule-binding proteins in which both MAP1A and MAP1B serve to connect with actin and signalling proteins and stabilize microtubules [45]. MAP kinase and glycogen synthase kinase-3 β pathways play an important role in regulating the function of this class of protein [45].

Microtubule-related proteins that attach to the plus ends of the microtubule control the microtubule dynamics. Polymerases promote development by boosting the incorporation of tubulin, which is bound to GTP. Depolymerases, on the other hand, detach GTP-tubulin from its plus end, causing microtubule shrinking. CLASP proteins stabilize microtubules by preventing disassembly and promoting growth [46]. In nerve cells, some cells contain tau protein, whereas others contain MAP2 such as dendrites contain MAP2, while axons contain tau [41]. Axon microtubules are positioned with their plus sides facing the axon tip, whereas dendritic microtubules are arranged with their plus ends pointing both toward and away from the cell body [41].

Microtubule Motors

Numerous cell movements, such as the movement and positioning of organelles and membrane vesicles within cells, the separation of chromosomes during mitosis, and the beating of cilia and flagella, are regulated by microtubules. Movement along microtubules is brought about by motor proteins, which employ ATP hydrolysis-derived energy to create force and motion [47]. Kinesins and dyneins are two major protein families involved in movements caused by microtubules. These proteins usually move in opposite directions along the microtubule, as proteins from the Kinesin family move towards the plus end and dynein family proteins move towards the minus end. Ian Gibbons first identified dynein in 1965 [48].

The structure of dynein consists of two or three heavy chains, several light chains, and intermediate chains. The heavy chains' globular head domains are the motor domains. Kinesin type 1 is composed of two heavy chains and two light chains. The heavy chains' globular head domains, which bind microtubules, constitute the molecule's motor domains [49]. The proteins responsible for functions like chromosomal movement and organelle transport are found in cytoplasm at very low quantities, making it difficult to identify additional microtubule-based motors [50].

Cargo Transport by Microtubules

- One of the primary roles of microtubules is to transport organelles, membrane vesicles, and macromolecules through the cytoplasm of cells in eukaryotic organisms [51].
- Such cytoplasmic organelle trafficking is particularly evident in nerve cell axons, which can be up to one meter long, as was previously described [52].
- Proteins, membrane vesicles, organelles (including mitochondria), and dendrites must be transferred from the cell body to the axon since ribosomes are only found in the cell body [52, 53].
- Kinesin, for instance, transports secretory vesicles carrying neurotransmitters from the Golgi apparatus to the terminal branches of the axon [54].
- Similar to how macromolecules, membrane vesicles, and organelles are transported by microtubules in other cell types (Fig. **6**).
- Because microtubules are typically oriented with their plus end reaching toward the cell periphery and their minus end fixed in the centrosome, it is believed that the kinesin and dynein families transfer cargo traveling through the cytoplasm in opposing directions [51, 54].

Diseases Associated with Microtubules

- The microtubule-associated protein tau builds up as clumps of linked helical filaments in neurons experiencing neurodegeneration due to Alzheimer's disease [55].
- Variations in the density, stability, and post-translational modifications of the microtubule network define pathological cardiac remodeling. In various types of cardiac disorders, altered microtubules can directly impairhe ability of cardiomyocytes to contract [56].

- Axons and dendrites are formed and maintained throughout a neuron's life, and they are prone to degeneration and disorder in several types of neurodegenerative diseases. The tools or techniques used to identify these were genotyping, phenotyping, and sequencing studies [57].
- Tubulin-binding agents (TBA), or microtubule-targeted agents, are among the most often utilized chemotherapeutic medicines in cancer therapy [58]. The intracellular target of tubulin-binding agents (TBAs) is the tubulin/microtubule system [45].
- A few studies have found unique mutations in c-actin in TBA-resistant cancer cell lines, resulting in the loss of wild-type protein. Research is now being conducted to elucidate the molecular basis of the interaction between microtubules and c-actin in resistance to tubulin-binding agents [41, 45].
- Many substances are in clinical use and development as antimitotic medicines by changing microtubule functions. Drugs that attach to one of three identified domains (taxane, vinca alkaloid, or colchicine site) influence microtubule dynamics [59, 60].
- Various vinca-site binding agents modify the configuration of tubulin polymers. Vinca alkaloids attach to a location next to the hydrolyzable GTP on the dimer, inducing structural alterations in the monomers. It reduces the rate of GTP hydrolysis and hence inhibits microtubule polymerization. To change the heterodimer's configuration, colchicine adheres to the free heterodimer at the intradimer junction [59].
- In contrast to Vinca alkaloids, colchicine binds to a location adjacent to the nonhydrolyzable GTP, increasing the rate of GTP hydrolysis and thereby reducing microtubule polymerization while increasing polymerization [52, 59].
- The binding of Taxol to the interior edge of β-tubulin promotes the development and integrity of microtubules without altering the shape of the dimer [52, 59].

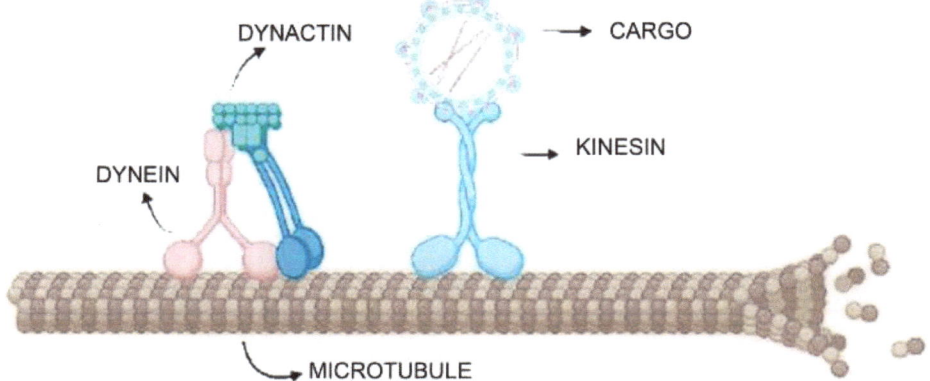

Fig. (6). Cargo transport on microtubule with the help of motor proteins.

CONCLUSION

The cytoskeleton serves as the backbone of cellular architecture, integrating structure with function in eukaryotic cells. This chapter has explored the three main components—microfilaments, intermediate filaments, and microtubules—highlighting their distinct roles in maintaining cell shape, enabling movement, and facilitating intracellular transport. The dynamic nature of cytoskeletal elements, including the involvement of motor proteins (Fig. **6**), underscores their importance in both cellular health and disease states such as cancer and neurodegenerative disorders. Understanding these intricate processes paves the way for further exploration into therapeutic targets and biomedical applications related to cytoskeletal functions.

ACKNOWLEDGEMENTS

We would like to acknowledge Ms. Kruthika P, Mr. Kumaran K, Ms. Raksa Arun, Ms. Srisri SatishKartik, Ms. Sanjana Dhayalan, and Mr. Sasitharan T for their contribution in proofreading and editing works.

REFERENCES

[1] Behnke O. Microtubules and microfilaments. Triangle 1974; 13(1): 7-15.
    [PMID: 4598515]
[2] Bretscher A. Microfilament structure and function in the cortical cytoskeleton. Annu Rev Cell Biol 1991; 7(1): 337-74.
    [http://dx.doi.org/10.1146/annurev.cb.07.110191.002005] [PMID: 1839710]
[3] Cooper GM. The Cell: A Molecular Approach. 2nd edition. Sunderland (MA): Sinauer Associates; 2000. Structure and Organization of Actin Filaments. Available from: https://www.ncbi.nlm.nih.gov/books/NBK9908/
[4] Sweeney HL, Hammers DW. Muscle Contraction. Cold Spring Harb Perspect Biol 2018; 10(2): a023200.
    [http://dx.doi.org/10.1101/cshperspect.a023200] [PMID: 29419405]
[5] Spooner BS, Yamada KM, Wessells NK. Microfilaments and cell locomotion. J Cell Biol 1971; 49(3): 595. Available from: /pmc/articles/PMC2108496/?report=abstract
[6] Huxley AF, Niedergerke R. Structural changes in muscle during contraction; interference microscopy of living muscle fibres. Nature 1954; 173(4412): 971-3.
    [http://dx.doi.org/10.1038/173971a0] [PMID: 13165697]
[7] Bosgraaf L, Van Haastert PJM. The Ordered Extension of Pseudopodia by Amoeboid Cells in the Absence of External Cues. PLoS One 2009; 4(4): 5253. Available from: /pmc/articles/PMC2668753/
[8] Cooper GM. The Cell: A Molecular Approach. 2nd edition. Sunderland (MA): Sinauer Associates; 2000. Intermediate Filaments. Available from: https://www.ncbi.nlm.nih.gov/books/NBK9834/
[9] Oshima RG. Intermediate Filaments: A Historical Perspective. Exp Cell Res 2007; 313(10): 1981. Available from: /pmc/articles/PMC1950476/
[10] Bragulla HH, Homberger DG. Structure and functions of keratin proteins in simple, stratified, keratinized and cornified epithelia. J Anat 2009; 214(4): 516. Available from: /pmc/articles/PMC2736122/

[11] Xiao S, McLean J, Robertson J. Neuronal intermediate filaments and ALS: A new look at an old question. Biochim Biophys Acta Mol Basis Dis 2006; 1762(11-12): 1001-12.
[http://dx.doi.org/10.1016/j.bbadis.2006.09.003] [PMID: 17045786]

[12] Lane EB. Keratin Intermediate Filaments and Diseases of the Skin. In: Madame Curie Bioscience Database [Internet]. Austin (TX): Landes Bioscience; 2000-2013. Available from: https://www.ncbi.nlm.nih.gov/books/NBK6247/

[13] Yu B, Kong D, Cheng C, et al. Assembly and recognition of keratins: A structural perspective, Seminars in Cell & Developmental Biology, 128, 2022, 80-89, https://doi.org/10.1016/j.semcdb.2021.09.018.

[14] Goldman RD, Grin B, Mendez MG, Kuczmarski ER. Intermediate Filaments: Versatile Building Blocks of Cell Structure. Curr Opin Cell Biol 2008; 20(1): 28. Available from: /pmc/articles/PMC3243490/

[15] Omary MB, Coulombe PA, McLean WHI. Intermediate filament proteins and their associated diseases. N Engl J Med 2004; 351(20): 2087-100.
[http://dx.doi.org/10.1056/NEJMra040319] [PMID: 15537907]

[16] Langbein L, Rogers MA, Winter H, Praetzel S, Schweizer J. The catalog of human hair keratins. II. Expression of the six type II members in the hair follicle and the combined catalog of human type I and II keratins. J Biol Chem 2001; 276(37): 35123-32.
[http://dx.doi.org/10.1074/jbc.M103305200] [PMID: 11445569]

[17] Smith FJD, Kreuser-Genis IM, Jury CS, Wilson NJ, Terron-Kwiatowski A, Zamiri M. Novel and Recurrent Mutations in Keratin 1 cause Epidermolytic Ichthyosis (EI) and Palmoplantar Keratoderma. Clin Exp Dermatol 2019; 44(5): 528. Available from: /pmc/articles/PMC7116359/

[18] Feroz S, Muhammad N, Ratnayake J, Dias G. Keratin - Based materials for biomedical applications. Bioact Mater 2020; 5(3): 496-509.
[http://dx.doi.org/10.1016/j.bioactmat.2020.04.007] [PMID: 32322760]

[19] Coulombe PA, Kopan R, Fuchs E. Expression of keratin K14 in the epidermis and hair follicle: insights into complex programs of differentiation. J Cell Biol 1989; 109(5): 2295-312.
[http://dx.doi.org/10.1083/jcb.109.5.2295] [PMID: 2478566]

[20] Chamcheu JC, Siddiqui IA, Syed DN, Adhami VM, Liovic M, Mukhtar H. Keratin Gene Mutations in Disorders of Human Skin and its Appendages. Arch Biochem Biophys 2011; 508(2): 123. Available from: /pmc/articles/PMC3142884/
[http://dx.doi.org/10.1016/j.abb.2010.12.019]

[21] Ferrari S, Cannizzaro LA, Battini R, Huebner K, Baserga R. The gene encoding human vimentin is located on the short arm of chromosome 10. Am J Hum Genet 1987; 41(4): 616. Available from: /pmc/articles/PMC1684315/?report=abstract

[22] Usman S, Waseem NH, Nguyen TKN, et al. Vimentin Is at the Heart of Epithelial Mesenchymal Transition (EMT) Mediated Metastasis. Cancers (Basel) 2021; 13(19). Available from: /pmc/articles/PMC8507690/

[23] Agnetti G, Herrmann H, Cohen S. New roles for desmin in the maintenance of muscle homeostasis. FEBS J 2022; 289(10): 2755-70.
[http://dx.doi.org/10.1111/febs.15864] [PMID: 33825342]

[24] Goldfarb LG, Olivé M, Vicart P, Goebel HH. Intermediate filament diseases: desminopathy. Adv Exp Med Biol 2008; 642: 131-64.
[http://dx.doi.org/10.1007/978-0-387-84847-1_11] [PMID: 19181099]

[25] Yang Z, Wang KKW. Glial Fibrillary acidic protein: From intermediate filament assembly and gliosis to neurobiomarker. Trends Neurosci 2015; 38(6): 364. Available from: /pmc/articles/PMC4559283/

[26] Keddie S, Smyth D, Keh RYS, *et al.* Peripherin is a biomarker of axonal damage in peripheral nervous system disease. Brain 2023; 146(11): 4562-73.
[http://dx.doi.org/10.1093/brain/awad234] [PMID: 37435933]

[27] Cluskey S, Ramsden DB. Mechanisms of neurodegeneration in amyotrophic lateral sclerosis. Molecular Pathology 2001; 54(6): 386. Available from: /pmc/articles/PMC1187128/

[28] Yuan A, Rao M V., Veeranna, Nixon RA. Neurofilaments and Neurofilament Proteins in Health and Disease. Cold Spring Harb Perspect Biol 2017; 9(4). Available from: /pmc/articles/PMC5378049/

[29] Liu Q, Xie F, Alvarado-Diaz A, *et al.* Neurofilamentopathy in Neurodegenerative Diseases. Open Neurol J 2011; 5(1): 58. Available from: /pmc/articles/PMC3170930/
[http://dx.doi.org/10.2174/1874205X01105010058]

[30] Yuan A, Rao M V., Sasaki T, *et al.* α-Internexin Is Structurally and Functionally Associated with the Neurofilament Triplet Proteins in the Mature CNS. J Neurosci 2006; 26(39): 10006. Available from: /pmc/articles/PMC6674481/

[31] Yoshioka Y, Taniguchi JB, Homma H, *et al.* AAV-mediated editing of PMP22 rescues Charcot-Marie-Tooth disease type 1A features in patient-derived iPS Schwann cells. Communications Medicine 2023; 3(1): 1-15. Available from: https://www.nature.com/articles/s43856-023-00400-y

[32] Singh KN, Ramadas MN, Veeran V, Naidu MR, Dhanaraj TS, Chandrasekaran K. Expression Pattern of the Cancer Stem Cell Marker "Nestin" in Leukoplakia and Oral Squamous Cell Carcinoma. Rambam Maimonides Med J 2019 10(4):24. Available from: /pmc/articles/PMC6824828/
[http://dx.doi.org/10.5041/RMMJ.10378]

[33] Szymańska-Chabowska A, Świątkowski F, Jankowska-Polańska B, Mazur G, Chabowski M. Nestin Expression as a Diagnostic and Prognostic Marker in Colorectal Cancer and Other Tumors. Clin Med Insights Oncol 2021; 15. Available from: /pmc/articles/PMC8377314/
[http://dx.doi.org/10.1177/11795549211038256]

[34] Dechat T, Pfleghaar K, Sengupta K, *et al.* Nuclear lamins: major factors in the structural organization and function of the nucleus and chromatin. Genes Dev 2008; 22(7): 832. Available from: /pmc/articles/PMC2732390/
[http://dx.doi.org/10.1101/gad.1652708]

[35] Kang S mi, Yoon MH, Park BJ. Laminopathies; Mutations on single gene and various human genetic diseases. BMB Rep 2018; 51(7): 327. Available from: /pmc/articles/PMC6089866/

[36] Sanghvi-Shah R, Weber GF. Intermediate filaments at the junction of mechanotransduction, migration, and development. Front Cell Dev Biol 2017; 5(SEP): 81. Available from: www.frontiersin.org
[http://dx.doi.org/10.3389/fcell.2017.00081] [PMID: 28959689]

[37] Windoffer R, Beil M, Magin TM, Leube RE. Cytoskeleton in motion: the dynamics of keratin intermediate filaments in epithelia. J Cell Biol 2011; 194(5): 669. Available from: /pmc/articles/PMC3171125/

[38] Jones JCR, Kam CY, Harmon RM, Woychek A V., Hopkinson SB, Green KJ. Intermediate Filaments and the Plasma Membrane. Cold Spring Harb Perspect Biol 2017; 9(1). Available from: /pmc/articles/PMC5204322/
[http://dx.doi.org/10.1101/cshperspect.a025866]

[39] Liao J, Omary MB. 14-3-3 proteins associate with phosphorylated simple epithelial keratins during cell cycle progression and act as a solubility cofactor. J Cell Biol 1996; 133(2): 345-57.
[http://dx.doi.org/10.1083/jcb.133.2.345] [PMID: 8609167]

[40] Infante E, Etienne-Manneville S. Intermediate filaments: Integration of cell mechanical properties during migration. Front Cell Dev Biol 2022; 10: 951816.
[http://dx.doi.org/10.3389/fcell.2022.951816] [PMID: 35990612]

[41] Janmey PA. The cytoskeleton and cell signaling: component localization and mechanical coupling. Physiol Rev 1998; 78(3): 763-81.
[http://dx.doi.org/10.1152/physrev.1998.78.3.763] [PMID: 9674694]

[42] Kirschner M, Mitchison T. Beyond self-assembly: From microtubules to morphogenesis. Cell 1986; 45(3): 329-42.
[http://dx.doi.org/10.1016/0092-8674(86)90318-1] [PMID: 3516413]

[43] Wittmann T, Hyman A, Desai A. The spindle: a dynamic assembly of microtubules and motors. Nature Cell Biology 2001; 3(1): E28–34. Available from: https://www.nature.com/articles/ncb0101_e28
[http://dx.doi.org/10.1038/35050669]

[44] Aranson IS, Tsimring LS. Theory of self-assembly of microtubules and motors. Phys Rev E Stat Nonlin Soft Matter Phys 2006; 74(3): 031915.
[http://dx.doi.org/10.1103/PhysRevE.74.031915] [PMID: 17025675]

[45] Hohmann T, Dehghani F. The Cytoskeleton—A Complex Interacting Meshwork. Cells 2019; 8(4). Available from: /pmc/articles/PMC6523135/

[46] Amos LA, Schlieper D. Microtubules and Maps. Adv Protein Chem 2005; 71: 257-98.
[http://dx.doi.org/10.1016/S0065-3233(04)71007-4] [PMID: 16230114]

[47] Field JJ, Kanakkanthara A, Miller JH. Microtubule-targeting agents are clinically successful due to both mitotic and interphase impairment of microtubule function. Bioorg Med Chem 2014; 22(18): 5050-9.
[http://dx.doi.org/10.1016/j.bmc.2014.02.035] [PMID: 24650703]

[48] Sedbrook JC. MAPs in plant cells: delineating microtubule growth dynamics and organization. Curr Opin Plant Biol 2004; 7(6): 632-40.
[http://dx.doi.org/10.1016/j.pbi.2004.09.017] [PMID: 15491911]

[49] Balabanian L, Chaudhary AR, Hendricks AG. Traffic control inside the cell: microtubule-based regulation of cargo transport. Biochem (Lond) 2018; 40(2): 14-7. Available from: /biochemist/article/40/2/14/447/Traffic-control-inside-the-cell-microtubule-based
[http://dx.doi.org/10.1042/BIO04002014]

[50] Whitley BN, Engelhart EA, Hoppins S. Mitochondrial dynamics and their potential as a therapeutic target. Mitochondrion 2019; 49: 269. Available from: /pmc/articles/PMC6885535/
[http://dx.doi.org/10.1016/j.mito.2019.06.002]

[51] Yildiz A. Sorting out microtubule-based transport. Nature Reviews Molecular Cell Biology 2020 22:2 [Internet]. 2020 Available from: https://www.nature.com/articles/s41580-020-00320-y

[52] Wadsworth P. Cytoskeleton 2020 paper of the year. Cytoskeleton (Hoboken) 2021; 78(2): 21-2. Available from: https://onlinelibrary.wiley.com/doi/full/10.1002/cm.21653
[http://dx.doi.org/10.1002/cm.21653] [PMID: 33634565]

[53] Fourriere L, Jimenez AJ, Perez F, Boncompain G. The role of microtubules in secretory protein transport. J Cell Sci 2020; 133(2): jcs237016.
[http://dx.doi.org/10.1242/jcs.237016]

[54] Loncar A, Rincon SA, Lera Ramirez M, Paoletti A, Tran PT. Kinesin-14 family proteins and microtubule dynamics define *S. pombe* mitotic and meiotic spindle assembly, and elongation. J Cell Sci 2020; 133(11): jcs240234.
[http://dx.doi.org/10.1242/jcs.240234] [PMID: 32327557]

[55] Caviston JP, Holzbaur ELF. Microtubule motors at the intersection of trafficking and transport. Trends Cell Biol 2006; 16(10): 530-7.
[http://dx.doi.org/10.1016/j.tcb.2006.08.002] [PMID: 16938456]

[56]   Goldman C. A Hopping Mechanism for Cargo Transport by Molecular Motors on Crowded Microtubules. J Stat Phys 2010; 140(6): 1167-81.
[http://dx.doi.org/10.1007/s10955-010-0037-2]

[57]   Marchisella F, Coffey ET, Hollos P. Microtubule and microtubule associated protein anomalies in psychiatric disease. Cytoskeleton (Hoboken) 2016; 73(10): 596-611.
[http://dx.doi.org/10.1002/cm.21300] [PMID: 27112918]

[58]   Matamoros AJ, Baas PW. Microtubules in health and degenerative disease of the nervous system. Brain Res Bull 2016; 126(Pt 3): 217-25.
[http://dx.doi.org/10.1016/j.brainresbull.2016.06.016] [PMID: 27365230]

[59]   Pasquier E, Kavallaris M. Microtubules: A dynamic target in cancer therapy. IUBMB Life 2008; 60(3): 165-70.
[http://dx.doi.org/10.1002/iub.25] [PMID: 18380008]

[60]   Benitez-King G, Ramírez-Rodríguez G, Ortíz L, Meza I. The neuronal cytoskeleton as a potential therapeutical target in neurodegenerative diseases and schizophrenia. Curr Drug Targets CNS Neurol Disord 2004; 3(6): 515-33.
[http://dx.doi.org/10.2174/1568007043336761] [PMID: 15581421]

CHAPTER 12

Signal Transduction Pathways Orchestrate Cellular Communication: A Narrative Review

Kruthika Prakash[1], Raksa Arun[1], Srisri Satishkartik[1], Sayantani Chattopadhyay[1], Mashira Rahman[1], Prakash Gangadaran[2,3,4,*] and K.N. Aruljothi[1]

[1] *Department of Genetic Engineering, SRM Institute of Science and Technology, Kattankulathur, Chengalpattu, Tamil Nadu, India*

[2] *Department of Nuclear Medicine, School of Medicine, Kyungpook National University, Daegu, Republic of Korea*

[3] *Cardiovascular Research Institute, Kyungpook National University, Daegu, Republic of Korea*

[4] *BK21 FOUR KNU Convergence Educational Program of Biomedical Sciences for Creative Future Talents, School of Medicine, Kyungpook National University, Daegu, Republic of Korea*

Abstract: Cell signaling and signal transduction coordinate cellular communication and the execution of diverse biological functions. This review synthesizes how cells sense and integrate external and internal cues through core pathways, with emphasis on G Protein–Coupled Receptor (GPCR) and Mitogen-Activated Protein Kinase (MAPK) cascades that govern growth, differentiation, and apoptosis. We outline the roles of second messengers, protein kinases, and transcription factors in propagating and tuning signals, and describe regulatory mechanisms that maintain fidelity and specificity. We also summarize how pathway dysregulation underlies disease, including cancer, and discuss the therapeutic implications of targeting signaling networks.

Keywords: Cell signaling, Signal transduction, GPCR, MAPK, Second messengers, Protein kinases, Transcription factors, Apoptosis.

INTRODUCTION

Cell signaling enables cells to communicate with each other and with their environment to sustain physiological function. This communication network is essential for organismal growth, adaptation, and survival in dynamic contexts [1]. Signal transduction converts extracellular and intracellular cues into specific

---

* Corresponding author Prakash Gangadaran: Department of Nuclear Medicine, School of Medicine, Kyungpook National University, Daegu, Republic of Korea; E-mail: prakashg@knu.ac.kr

K.N. Aruljothi, Prakash Gangadaran, Krishnan Anand, Satish Ramalingam, K. Kumaran & Kruthika Prakash (Ed.)
All rights reserved-© 2026 Bentham Science Publishers

cellular responses, ensuring appropriate reactions to stimuli associated with immune defense, growth, differentiation, and other core processes [2].

Signals include hormones, growth factors, neurotransmitters, and environmental inputs such as temperature and light [3]. Cells operate within integrated networks that support organismal health. Cell signaling coordinates growth and development, maintains homeostasis, mediates immune responses, and ensures proper nervous system function [4].

Multicellular organisms develop from a single fertilized egg through cell division, differentiation, and controlled tissue formation, all tightly regulated by signaling pathways [5]. During embryogenesis, signaling molecules guide cell positioning and organogenesis by activating gene programs for specialization [6]. To preserve a stable internal environment, cells continuously sense changes in temperature, nutrient levels, and pH and adjust their activity to restore equilibrium [7]. Hormones such as insulin and glucagon exemplify signaling molecules that regulate metabolism and energy balance [8].

The immune system relies on signaling to detect and eliminate pathogens. Immune cells such as T cells and B cells use signaling to recognize invaders, activate other immune cells, and coordinate an effective defense through cell–cell communication [9]. Neurons communicate *via* electrical and chemical signals to transmit information across extensive networks; neurotransmitters enable rapid, coordinated responses underlying movement, sensation, and cognition [10].

TYPES OF CELL SIGNALING

Cell signaling employs distinct modes of communication defined by distance and delivery. The four principal forms are autocrine, paracrine, endocrine, and juxtacrine signaling.

Autocrine Signaling

In autocrine signaling, a cell releases signaling molecules that bind receptors on its own surface. This feedback modulates the cell's activity and is especially important in development and immune responses [11]. Cancer cells often exploit autocrine loops to drive their growth and survival.

Paracrine Signaling

In paracrine signaling, cells secrete molecules that act on nearby cells, mediating local communication [12]. During tissue development and repair, growth factors stimulate adjacent cells to proliferate and divide; paracrine cues similarly coordinate wound healing [13].

Endocrine Signaling

Endocrine signaling involves hormone release into the bloodstream, allowing signals to reach distant targets. This long-range communication coordinates metabolism, development, and reproduction across organ systems; endocrine glands such as the pituitary, thyroid, and pancreas orchestrate these responses [14].

Juxtacrine Signaling

In juxtacrine signaling, cells communicate through direct physical contact. Membrane-bound ligands on one cell engage receptors on an adjacent cell, triggering local responses. During embryonic development and tissue differentiation, where precise cell-to-cell interactions build ordered structures, juxtacrine signaling is essential [15].

Role of Signal Transduction in Cellular Communication

After a signal reaches its target cell, it must be converted into a specific intracellular response—a process termed signal transduction. Signal transduction comprises sequential molecular events that amplify and relay information from the cell surface to the intracellular machinery that produces the appropriate response. Major components include:

Receptors

Receptors reside on the cell surface or within the cell and bind specific signaling molecules; they are the first points of contact for external cues [16]. Two principal classes of cell-surface receptors mediate signal transduction: G Protein-Coupled Receptors (GPCRs) and Receptor Tyrosine Kinases (RTKs) [17]. Ligand binding induces a conformational change that initiates the intracellular signaling cascade [18].

Signaling Molecules

Signaling molecules—such as hormones, growth factors, and neurotransmitters—are released by a cell and bind receptors on target cells to initiate transduction [19]. Insulin, for example, binds its receptor to regulate glucose uptake and metabolism [20].

Signal Transduction Proteins

Once receptors are activated, intracellular signaling proteins propagate and amplify the signal, often through cascades in which one protein activates the next

[21]. Common examples include protein kinases, which phosphorylate substrates to modulate activity, and second messengers such as calcium ions or cAMP (cyclic AMP), which diffuse within the cell to activate downstream components. These mechanisms transmit the signal throughout the cell [22].

Effector Proteins

Effector proteins execute the terminal cellular response, including changes in gene expression, metabolic reprogramming, or cell-cycle progression. For instance, in response to a growth factor, effectors can induce transcription of genes required for proliferation [23].

The architecture of signaling networks can be linear or branched, depending on signal complexity and the required response. A single signal often activates multiple pathways, enabling integration and coordination of outputs to external stimuli.

CELL-CELL SIGNALING

Cell-cell signaling underpins processes such as development, tissue repair, and immune responses. A signaling cell releases molecules that locate target cells and bind specific receptors. Through biochemical pathways involving protein kinases, phosphatases, and second messengers like cAMP and calcium ions (Ca^{2+}), this interaction triggers signal transduction, converting an extracellular cue into an intracellular response [24]. Resulting outcomes include altered gene expression, enzyme activity, proliferation, differentiation, or programmed cell death (apoptosis) [25]. From local to long-distance communication, paracrine, autocrine, endocrine, and synaptic modes have distinct mechanisms and ranges of action.

Modes of Cell-cell Communication

Cell-cell signaling operates *via* two primary modes: direct cell-cell contact and chemical signaling (Fig. **1**).

Direct Cell-cell Contact

Direct cell-cell communication occurs through physical interactions mediated by specialized surface structures or ligands. This mode is efficient in densely packed tissues, enabling rapid, localized information transfer.

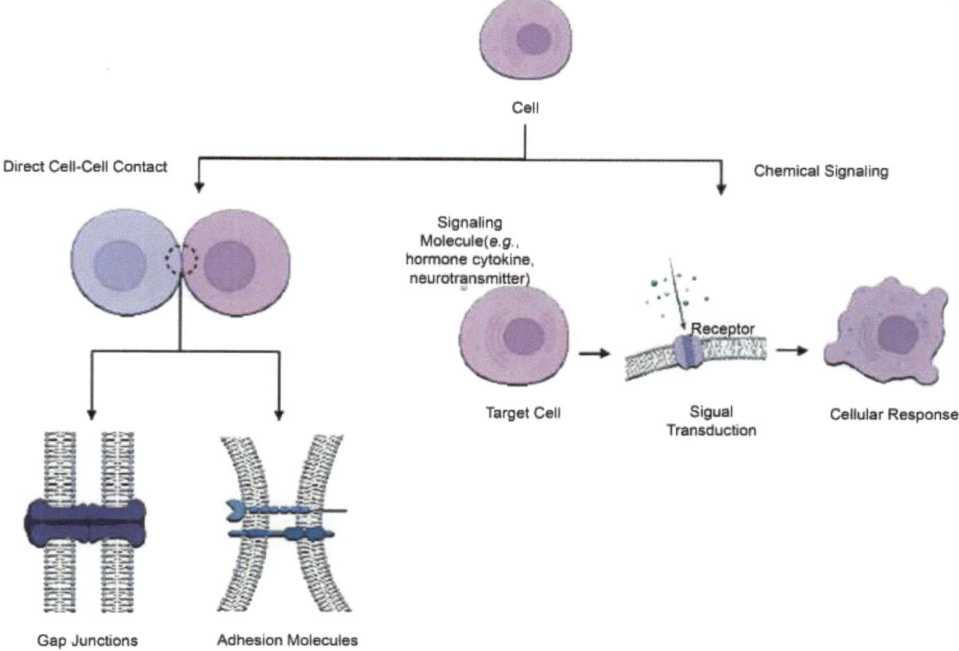

Fig. (1). Mechanisms of cell communication: direct contact and chemical signaling.

Gap Junctions

Gap junctions are protein channels that connect neighboring cells, creating direct cytoplasmic continuity. These channels permit ions, small metabolites, and selected signaling molecules to pass freely, allowing cells to exchange information and coordinate activity. In cardiac tissue, gap junctions synchronize cardiomyocyte contraction *via* direct ion fluxes that conduct electrical impulses across the myocardium [26].

Adhesion Molecules

Cell-surface adhesion molecules, including cadherins and integrins, mediate physical attachment between cells. Beyond maintaining tissue architecture, they transduce mechanical and biochemical cues from the microenvironment [27]. Cell-cell adhesion can activate intracellular pathways that regulate growth, migration, and differentiation [28].

Chemical Signaling

Cells also communicate *via* diffusible signaling molecules that act on nearby or distant targets. After binding receptors on target cells, these cues initiate intracellular cascades that produce a biological response. Based on distance, chemical communication is categorized as paracrine (nearby cells), endocrine (distant cells), or autocrine (the signaling cell itself).

Hormones

Endocrine glands secrete hormones that travel through the bloodstream to distant targets, coordinating processes such as development, metabolism, and reproduction. Hormones fall into three major classes: steroid hormones, peptide/protein hormones, and amine hormones. Steroid hormones (*e.g.*, testosterone and estrogen) are lipid-soluble and cross membranes to bind intracellular receptors, thereby modulating gene expression [29]. Water-soluble peptide and protein hormones such as insulin and growth hormone bind cell-surface receptors to activate intracellular signaling pathways and elicit diverse responses [30]. Amine hormones, derived from amino acids (*e.g.*, thyroid hormones and epinephrine), can act *via* membrane or intracellular receptors depending on the molecule. Insulin, for example, signals muscle and liver cells to increase glucose uptake, thereby controlling blood glucose levels [31].

Cytokines

Essential for regulating immune responses and inflammation, cytokines are small proteins produced by multiple cell types, including immune cells. Key cytokines include IL-2, which promotes T cell development and proliferation, and IL-1, a central mediator of inflammation [32]. Interferons regulate immunity and inflammation and act as antiviral agents [33]. Tumor Necrosis Factor (TNF) modulates inflammation and apoptosis and can influence tumorigenesis [34]. Chemokines are essential for immune surveillance and wound healing, recruiting immune cells to sites of infection or inflammation [35]. As messengers, cytokines direct macrophages and T cells to mobilize and eliminate pathogens, thereby coordinating immune responses [36].

Neurotransmitters

Released at synapses, neurotransmitters enable neurons to communicate with target cells such as muscles and glands. These molecules underlie synaptic transmission and support motor control, mood, and cognition [37]. Monoamines derived from amino acids—dopamine, norepinephrine, epinephrine, and serotonin—govern mood, motivation, reward, and arousal [38]. Amino acid transmitters include glutamate (excitatory) and GABA (inhibitory) [39]. Choline derivatives such as acetylcholine regulate muscle contraction, memory, and learning [40]. Together, these signaling molecules coordinate complex neural, endocrine, and immune responses.

Types of Signaling Molecules

Signaling molecules that mediate cell–cell communication include proteins, hormones, and small metabolites. Their diverse structures confer distinct functions in biological systems.

Proteins and Peptides

Many signaling molecules—including growth factors, hormones, and cytokines—are proteins or peptides. Because they are generally hydrophilic, they signal *via* cell-surface receptors rather than diffusing across membranes [41]. For example, Epidermal Growth Factor (EGF) binds its receptor to activate intracellular pathways that drive cell growth and division [42].

Steroids

Steroid hormones, such as testosterone and estrogen, are cholesterol-derived, lipid-soluble molecules that diffuse across membranes to bind intracellular receptors [43]. The hormone–receptor complex translocates to the nucleus to regulate gene expression. Steroid signaling is essential in metabolism, inflammation, and reproductive development [44].

Small Molecules

Some signaling molecules are small metabolites or gases (*e.g.*, nitric oxide, carbon monoxide) that readily cross membranes to initiate signaling. In the cardiovascular system, nitric oxide diffuses into smooth muscle to trigger cascades that relax vessels and lower blood pressure [45].

Importance in Tissue Homeostasis

Effective cell–cell communication is fundamental to tissue homeostasis—the maintenance of a stable, functional internal environment. It enables cells to coordinate growth and repair, adapt to changing conditions, and combat infection.

Examples of how cell–cell communication preserves tissue homeostasis include:

Immune Response

The immune system relies on cell–cell signaling to detect and clear pathogens. Immune cells, including macrophages and T cells, use cytokines and other mediators to coordinate their responses [46]. Upon pathogen recognition, cytokines recruit additional immune cells to the infection site, amplifying the response and accelerating pathogen clearance [47]. Antigen-presenting cells also engage T cells through direct contact and soluble cues to initiate pathogen-specific responses [48].

Growth Regulation

Within tissues, cell–cell signaling controls growth, proliferation, and differentiation. Cells secrete growth factors, such as transforming growth factor beta (TGF-β), to regulate neighboring cells and maintain tissue architecture [49]. Dysregulated signaling drives unchecked proliferation and tumorigenesis, underscoring the need for tight control to preserve homeostasis [50].

Tissue Repair

Following injury, cell–cell signaling orchestrates repair. During wound healing, growth factors and cytokines promote the migration and proliferation of fibroblasts and other cells at the damage site. These pathways are subsequently attenuated to terminate responses once repair is complete [51].

PATHWAYS OF INTRACELLULAR SIGNAL TRANSDUCTION

Signal transduction is the process by which cells sense and respond to external cues. It involves first messengers (ligands), receptors, transducers, and second messengers that relay information from the cell surface to intracellular effectors. The process also involves allosteric regulation and post-translational modifications, including methylation, acetylation, glycosylation, ubiquitination, and phosphorylation. GTP-binding proteins mediate nucleotide exchange and regulate protein phosphorylation, while kinases and phosphatases control phosphate transfer through defined catalytic mechanisms [52, 53].

Signal transduction begins when a ligand binds a specific receptor at the plasma membrane. Binding induces conformational changes that activate enzymatic functions or expose docking sites for intracellular signaling proteins. The receptor–ligand interaction initiates cascades propagated by second messengers such as cAMP, Ca^{2+}, and diacylglycerol (DAG), which amplify and disseminate signals throughout the cell [52 - 54].

Signal transduction pathways amplify extracellular cues. The binding of a single ligand to its receptor can initiate a cascade in which each activated component triggers many downstream effectors, thereby magnifying the intracellular signal and the resulting cellular response. Outputs include changes in gene expression, metabolism, and behaviors, such as growth, differentiation, and apoptosis. These reactions preserve cellular homeostasis and support diverse physiological functions [55].

Tight regulation ensures appropriate responses and prevents overactivation. Mechanisms include receptor desensitization or internalization, second-messenger degradation, and inactivation of downstream signaling proteins, enabling cells to remain responsive to new cues [56].

The mitogen-activated protein kinase (MAPK) pathway exemplifies signal transduction. It converts extracellular inputs, including growth factors, into cellular programs such as proliferation, differentiation, and survival [54]. The pathway begins when a signaling molecule, such as EGF, binds its receptor, EGFR, at the cell surface, inducing receptor dimerization and autophosphorylation. This event activates downstream effectors. A central module involves the small GTPase Ras, which, upon GTP loading, initiates a kinase cascade: Ras activates Raf, which activates MEK, which in turn activates ERK (extracellular signal–regulated kinase). Activated ERK translocates to the nucleus and phosphorylates transcription factors to control gene expression, promoting proliferation and survival. Dysregulation of MAPK signaling is common in cancer, underscoring its critical role in normal cellular function [57, 58].

Phases of Signal Transduction

Signal transduction proceeds through five coordinated stages—reception, transduction, response, amplification, and termination—each converting an external cue into a defined cellular action.

Reception

In the reception stage, a ligand binds a specific receptor at the cell surface or within the cell, initiating signaling. For example, insulin binding to its receptor triggers pathways that regulate glucose uptake and maintain systemic glucose levels.

Transduction

Ligand–receptor engagement activates intracellular signaling modules that relay and transform the signal. In the GPCR pathway, ligand binding activates a G protein, which stimulates second messengers such as cAMP to propagate and amplify the signal.

Response

The transmitted signal elicits specific actions, including altered gene expression, enzyme activity, or physiology. In the MAPK pathway, when kinases such as ERK are activated, they enter the nucleus and stimulate gene expression programs that drive cell growth and differentiation.

Amplification

Amplification allows a single ligand–receptor event to produce a large response through branching cascades. For instance, epinephrine binding can activate multiple protein kinases, leading to extensive glycogen breakdown and glucose release in liver cells. This greatly magnifies the effect of the original ligand.

Termination

Termination mechanisms halt signaling once an adequate response is achieved. Processes include ligand dissociation and degradation, receptor deactivation, and reversal of modifications by enzymes such as phosphatases. For instance, cAMP is degraded by phosphodiesterases, and calcium ions are pumped out of the cytoplasm to restore resting levels. This regulation allows cells to reset and be ready for new stimuli [59 - 61].

In summary, signal transduction is an organized, multistage process that enables cells to detect, process, and respond to external stimuli through reception, transduction, response, amplification, and termination. Proper regulation of these pathways, exemplified by the MAPK pathway, maintains cellular homeostasis; dysregulation contributes to diseases such as cancer. Understanding these phases is fundamental to advancing research and therapeutic interventions that restore or modulate cellular communication in disease contexts (Fig. 2).

CELL SURFACE RECEPTORS

Cell surface receptors are membrane proteins that bind specific extracellular molecules, including hormones, neurotransmitters, and nutrients. Binding initiates cellular responses, including proliferation, immune activation, apoptosis, differentiation, changes in gene expression, and metabolic adjustments (*e.g.*, increased glucose uptake or altered lipid metabolism) [62]. These receptors transmit extracellular signals to intracellular machinery and mediate stimulus-specific responses. Receptors can be studied using direct radioligand-binding approaches (*e.g.*, saturation and competition binding assays). These approaches quantify ligand–receptor interactions and receptor properties [63]. There are four main types of cell surface receptors: GPCRs, receptor tyrosine kinases (RTKs), ligand-gated ion channels, and cytokine receptors.

Types of Cell Surface Receptors

Cell surface receptors are specialized membrane proteins that detect extracellular signals and initiate intracellular responses. The main classes include:

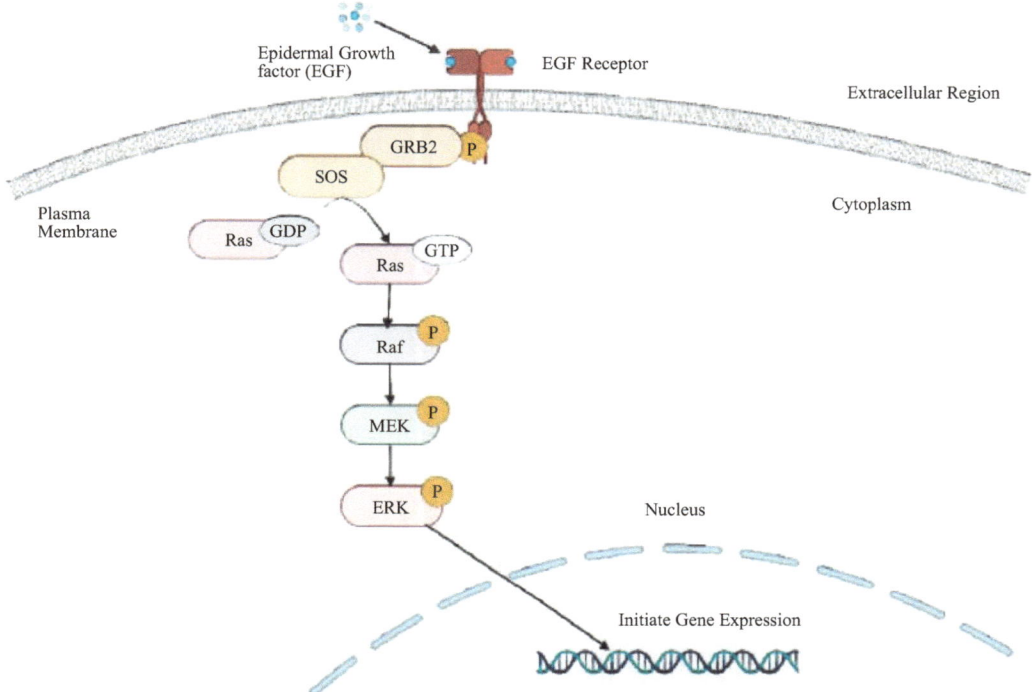

Fig. (2). Signal transduction through the MAPK pathway. Epidermal growth factor (EGF) binds the epidermal growth factor receptor (EGFR), initiating a cascade of downstream events.

GPCRs

GPCRs constitute a large, diverse superfamily of cell-surface receptors. GPCRs transduce signals *via* membrane-associated heterotrimeric guanine nucleotide–binding proteins (G proteins). The human genome encodes more than 1,000 GPCRs, representing approximately 5–10% of genes [64]. Upon agonist binding, GPCRs engage heterotrimeric G proteins and catalyze guanosine diphosphate–guanosine triphosphate (GDP–GTP) exchange to activate the G protein [65]. GPCRs regulate fundamental processes, including cell growth, differentiation, motility, and death. They also interact with growth factors and microtubule networks to promote neurite outgrowth through Gsα internalization and microtubule dynamics [66]. GPCR signaling activates Rho GTPases—molecular switches that cycle between GDP- and GTP-bound states—which are central to cell motility [67]. GPCRs share a canonical architecture of a single polypeptide with seven transmembrane helices, known as the seven-transmembrane (7TM) domain [68]. This configuration permits ligand binding and coupling to intracellular signaling pathways. Activation of phospholipase C by GPCRs results in the production of inositol trisphosphate

(IP3), which releases Ca^{2+} from the endoplasmic reticulum, and DAG, which recruits and activates signaling molecules, thereby propagating the signal. GPCRs can also undergo allosteric modulation, whereby binding of an allosteric modulator stabilizes alternative receptor conformations and tunes signaling output [69].

RTKs

RTKs are membrane-bound receptors that play central roles in cellular communication, particularly in the central nervous system [70]. RTKs control cell growth, differentiation, survival, migration, and metabolism by transducing signals through ligand-induced dimerization or oligomerization in the plasma membrane [71]. Binding of specific ligands (*e.g.*, growth factors, insulin, nerve growth factor) promotes receptor dimerization and autophosphorylation of tyrosine residues, initiating downstream signaling that supports cell growth. One key pathway is the MAPK cascade; upon RTK activation, components such as Ras, Raf, and MEK are phosphorylated and activated, ultimately activating MAPK and regulating genes required for proliferation [72]. RTKs feature a single-pass transmembrane helix that links an extracellular domain to an intracellular kinase domain. Ligand binding to the extracellular domain drives receptor dimerization and kinase activation [73].

Ligand-gated Ion Channels

Ligand-gated ion channels are membrane proteins that mediate fast synaptic signaling in neurons. These channels open in response to extracellular neurotransmitters or intracellular ligands such as Ca^{2+} or cyclic nucleotides [74]. Typically composed of three to five subunits, each contributing to the ion-conducting pore, these proteins operate cooperatively to achieve precise gating. Their architecture enables tight control of ion flux across the membrane, which underpins neuronal signal transmission and other physiological processes [75].

Cytokine Receptors

Cytokine receptors are cell-surface proteins that bind cytokines—signaling molecules that regulate diverse mammalian physiological processes. Ligand binding induces receptor dimerization or oligomerization, triggering intracellular signaling cascades [76]. Many cytokine receptors are homodimeric class I receptors, including erythropoietin (EPOR), thrombopoietin (TPOR), granulocyte colony-stimulating factor 3 (CSF3R), growth hormone (GHR), and prolactin receptors (PRLR). These single-pass transmembrane glycoproteins regulate cell growth, proliferation, and differentiation. The active signaling complex consists of a receptor homodimer with one or two ligands bound to the extracellular domains [77]. More than 40 cytokine receptor subunits in mammals can activate the JAK (Janus kinase)/STAT (signal transducer and activator of transcription) pathway. Binding of cytokines to their receptors brings associated JAKs into proximity, leading to activation; JAKs then phosphorylate tyrosine residues on the receptor, creating docking sites for STAT proteins. STATs are recruited and phosphorylated by JAKs, dimerize, and translocate to the nucleus, where they regulate genes involved in immune responses and cell growth (Fig. 3) [78].

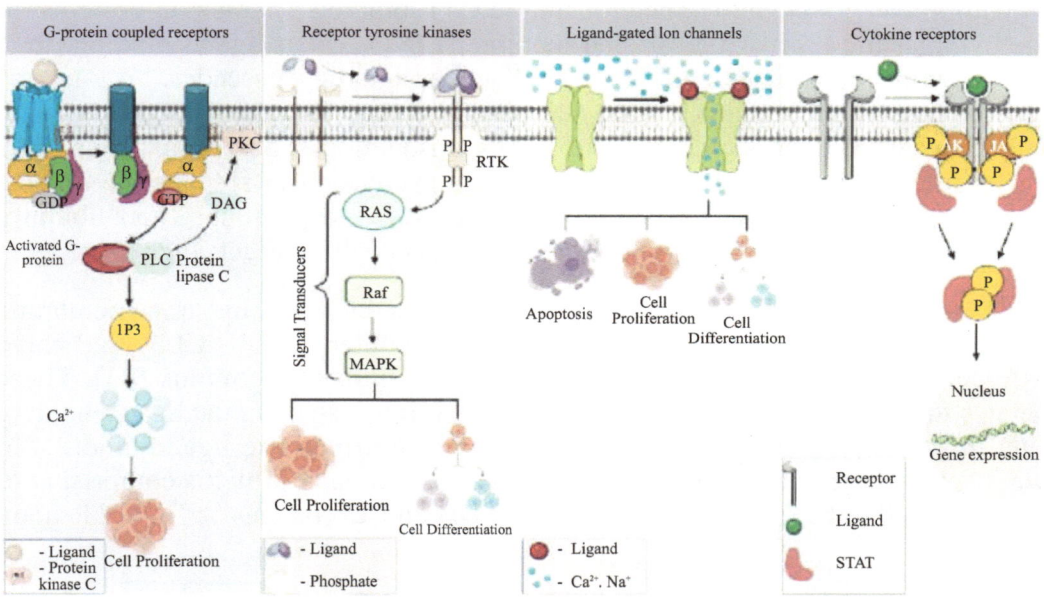

Fig. (3). Four major types of cell-surface receptors: G protein–coupled receptors (GPCRs), receptor tyrosine kinases (RTKs), ligand-gated ion channels, and cytokine receptors.

Receptor–ligand Interaction

Receptor–ligand binding involves direct interactions and diffusion dynamics. Signal transduction begins when a ligand binds its cognate receptor, triggering intracellular events that culminate in defined cellular responses. Ligand binding induces a receptor conformational change that enables interactions with intracellular proteins or signaling molecules. Activated receptors often generate second messengers, such as cyclic AMP (cAMP) or calcium ions, which amplify the signal within the cell. These signaling events elicit specific cellular responses essential for cell viability [79]. Diffusion dynamics primarily follow two mechanisms: the Eley–Rideal (ER) and Langmuir–Hinshelwood (LH) pathways. In ER, ligands diffuse in three dimensions and interact directly with surface receptors, forming receptor–ligand complexes that initiate signaling. In LH, ligands first diffuse in 3D to the cell surface, adsorb, and then undergo two-dimensional (2D) diffusion to reach receptors. Both mechanisms can operate simultaneously, with coupled 2D and 3D dynamics enhancing receptor–ligand encounter rates and signaling [80].

GPCR PATHWAYS

GPCRs constitute one of the largest protein superfamilies in the human genome and across eukaryotes. They are also termed seven-transmembrane (7TM) receptors. GPCRs signal *via* heterotrimeric GTP-binding proteins (G proteins) at the cytoplasmic face of the plasma membrane [81]. Upon ligand binding, they detect extracellular cues and trigger cell-wide signaling cascades. Among the approximately 800 human GPCRs, many have sensory functions, including taste, olfaction, pheromone signaling, and phototransduction. They regulate immune responses, secretion, metabolism, and neurotransmission in response to diverse stimuli such as neurotransmitters, ions, hormones, and photons. Their ubiquity and roles in physiology and disease make them prime drug targets [82, 83].

GPCRs share an extracellular N terminus; seven hydrophobic transmembrane helices (TM1–TM7) connected by three intracellular (ICL1–ICL3) and three extracellular (ECL1–ECL3) loops; and an intracellular C terminus [84]. These helices form a barrel-like fold, with extracellular loops shaping the ligand-binding site. In some receptors, the N-terminal tail also contributes to ligand binding. In the inactive state, GPCRs associate with heterotrimeric G proteins comprising α, β, and γ subunits. The α subunit binds GDP and exchanges it for GTP upon activation [85].

GPCR ligands include odorants and photons; neurotransmitters such as γ-aminobutyric acid (GABA) and biogenic amines (*e.g.*, serotonin, dopamine); peptide hormones (*e.g.*, oxytocin, follicle-stimulating hormone [FSH], gonadotropin-releasing hormone [GnRH]); and modulators such as opioid peptides and endocannabinoids [86].

Mechanism of GPCR Activation

Extracellular agonist binding induces conformational rearrangements that activate the associated heterotrimeric G protein. Twisting and reorientation of the transmembrane helices widen the intracellular cavity, promoting GDP–GTP exchange on the α subunit and dissociation of Gα-GTP from the βγ dimer. In this process, the GPCR functions as a guanine nucleotide exchange factor (GEF). During the active period, Gα-GTP and the membrane-retained βγ dimer engage effectors that produce second messengers [87].

Downregulation of signaling occurs with receptor desensitization and removal of agonists [88]. In addition, intrinsic GTP hydrolysis on the α subunit terminates activity and is accelerated by regulators of G protein signaling (RGS), which are GTPase-activating proteins (GAPs) [89]. GAPs lower the activation energy for GTP hydrolysis on active Gα, favoring reformation of the inactive GDP-bound α subunit and reassociation with the βγ dimer [90].

Although GPCR signaling is often considered solely G protein–dependent, β-arrestins and GPCR-regulating kinases (GRKs) also mediate receptor desensitization. These transducers promote sequestration and endocytosis preceding ligand removal; downstream effectors frequently include members of the MAPK family [91]. Dynamin, a large GTPase, regulates the rate of GPCR internalization. GRK-mediated GPCR phosphorylation and β-arrestin binding sterically block G protein coupling, triggering endocytosis. After β-arrestin dissociates and the receptor is dephosphorylated, ligands are removed, terminating signaling and returning the GPCR and G protein subunits to an inactive state [92].

Signaling by GPCRs

Ligand binding stabilizes active receptor conformations and recruits transducers such as β-arrestins or G proteins. Depending on the transducer, signaling proceeds through G protein–dependent or G protein–independent pathways (Fig. **4**).

Multiple closely related G protein isoforms exist, and any Gα subtype can be activated. Functional selectivity (agonist-directed trafficking) denotes ligand-dependent preference for a particular Gα subtype. Limited availability of a preferred Gα can redirect coupling to an alternative isoform [93].

cAMP Pathways

cAMP is a second messenger generated from ATP that drives cAMP-dependent signaling. In 1971, Earl Sutherland received the Nobel Prize "for his discoveries concerning the mechanisms of the action of hormones," notably the cAMP-dependent mechanism of epinephrine action [94]. cAMP transduces extracellular hormone signals that cannot cross the plasma membrane. Ligand-activated GPCRs promote GDP–GTP exchange on Gα, which dissociates from the heterotrimer [95]. Activated Gα then engages adenylyl cyclase, which converts ATP to cAMP.

cAMP activates protein kinase A (PKA) to regulate gene expression, metabolism, secretion, and membrane permeability. PKA is a tetramer composed of two regulatory and two catalytic subunits. cAMP binding to the regulatory subunits releases the active catalytic subunits [96], which can enter the nucleus to modulate transcription. Active PKA phosphorylates targets that increase vascular tone *via* muscle contraction, promote glycogenolysis, and stimulate renin production in the kidney [97].

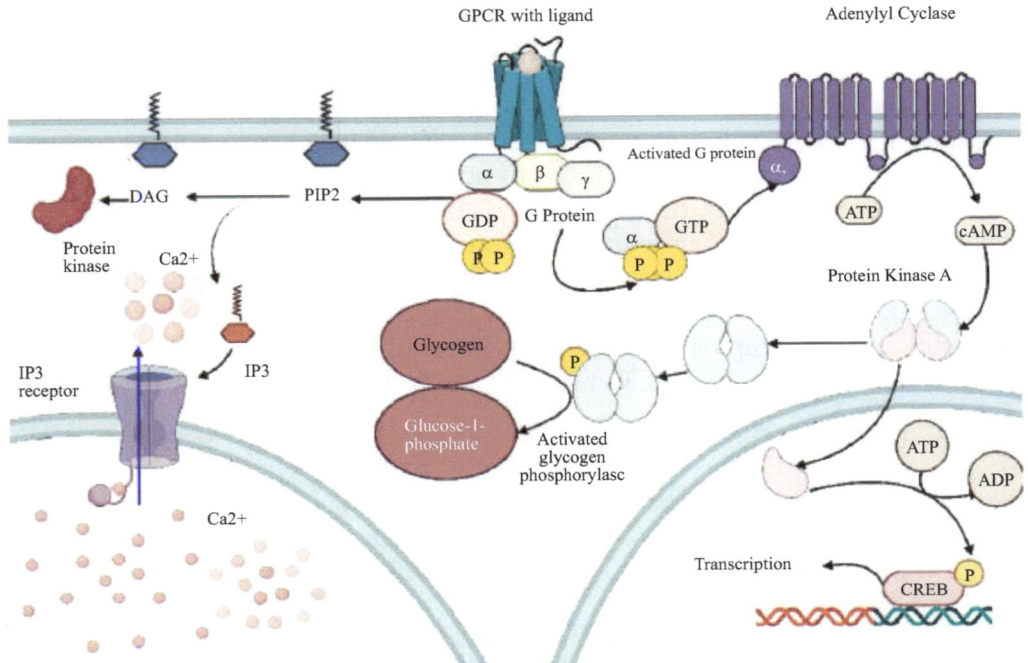

Fig. (4). Signaling pathways mediated by G protein–coupled receptors: activation of adenylyl cyclase and phospholipase C.

Gα self-terminates signaling by hydrolyzing GTP to GDP, switching off adenylyl cyclase. Downstream, dephosphorylation of PKA targets also attenuates responses. Receptors are further downregulated by endocytic internalization and lysosomal degradation [98].

Regulation *via* cAMP is essential for control of heart rate, memory formation, and renal water reabsorption. Dysregulated cAMP signaling has been linked to cancer progression [99].

IP3/DAG Pathway

IP3 is a second messenger generated by hydrolysis of phosphatidylinositol 4,5-bisphosphate (PIP2), a plasma-membrane phospholipid. Together with DAG, IP3 activates Ca^{2+}-regulated intracellular pathways [100]. IP3 diffuses through the cytoplasm to the endoplasmic reticulum (ER), where it binds IP3 receptors—ligand-gated Ca^{2+} channels—to release Ca^{2+}. The resulting rise in cytosolic Ca^{2+} regulates processes such as cell proliferation, muscle contraction, and hemostasis [101].

Upon ligand activation, Gα dissociates from the heterotrimeric G protein. Gα activates PLCβ, which cleaves membrane PIP2 into IP3 and DAG. IP3 then diffuses to the ER or, in muscle, the sarcoplasmic reticulum (SR). There it binds IP3 receptors, gated ER Ca^{2+} channels [102]. Cytosolic Ca^{2+} further amplifies release *via* calcium-induced calcium release in cardiomyocytes or by opening plasma-membrane Ca^{2+} channels in response to elevated intracellular Ca^{2+} [103, 104].

Physiological and Pathological Roles

In the body, GPCRs are essential for sensory perception. Rhodopsin converts photon energy into electrical signals that the visual system interprets as vision. Activation of taste GPCRs by bitter, umami, or sweet ligands stimulates the G protein gustducin, triggering an IP3/DAG signaling cascade. Neurotransmitters that regulate mood and behavior—including dopamine, serotonin, GABA, and histamine—signal through numerous brain GPCRs [105]. GPCR pathways also govern parasympathetic and sympathetic outputs, controlling autonomic functions such as blood pressure and heart rate.

More than 30 diseases have been linked to pathogenic GPCR variants. For example, retinitis pigmentosa (RP) arises from mutations in the gene encoding rhodopsin, whereas cholera reflects toxin-mediated dysregulation of G protein signaling.

Retinitis pigmentosa is an eye disease associated with retinal damage. Patients develop retinal dystrophy with progressive visual field loss, blurred vision, and poor low-light vision. Early-onset night blindness is common, and lost vision cannot currently be restored. This disorder is caused by inactivating mutations in the rhodopsin gene, *RHO*, which encodes the GPCR of the same name [106]. Rhodopsin is required for low-light vision because it is found in the rods of the retina [107]. Upon photon absorption, rhodopsin initiates a biochemical cascade that is transduced into electrical signals interpreted as vision by the brain.

Whereas RP is genetic, pathogens can also disrupt GPCR pathways, as in cholera. In cholera, the bacteria secrete a toxin that penetrates the intestinal lining and modifies the G protein. The α subunit is altered so that GTP hydrolysis to GDP is prevented, locking the protein in a constitutively active state. This persistent signaling keeps ion channels open, driving excessive chloride and water efflux into the intestinal lumen. The resulting diarrhea and dehydration can be fatal without treatment [108].

MAPK PATHWAYS

MAPK pathways are highly conserved signaling modules that regulate cell growth, proliferation, differentiation, motility, survival, stress responses, and apoptosis by transducing extracellular stimuli [109]. Each cascade comprises three kinases: a MAP kinase kinase kinase (MAP3K), a MAP kinase kinase (MAP2K), and a MAP kinase (MAPK). Sequential phosphorylation transmits and amplifies the signal, enabling precise cellular responses. In some cascades, upstream MAP4Ks and downstream MAPK-activated protein kinases (MAPKAPKs) further tune signaling specificity and dynamics [110, 111]. The signal ultimately activates regulatory proteins in multiple subcellular compartments, including the nucleus, cytoplasm, Golgi apparatus, and mitochondria, thereby controlling gene expression, metabolism, and cell architecture. For example, MAPK cascades regulate gene expression by modifying transcription factors, chromatin remodelers, and other nuclear proteins [112]. In mammals, four primary MAPK pathways are distinguished by their terminal MAPKs: ERK1/2, which is frequently activated by growth factors and promotes proliferation; JNK and p38, which respond to stress and inflammatory cues; and ERK5, which contributes to cardiovascular development and neuronal survival [113]. Dysregulated MAPK signaling is associated with many diseases, notably cancer. Mutations or aberrant activation within these pathways can drive uncontrolled proliferation, apoptosis resistance, and enhanced survival, facilitating tumor progression [114]. The following sections outline key components, regulatory mechanisms, and cellular functions, with emphasis on tumorigenesis.

Components of MAPK Pathways

MAPK signaling is *via* a core cascade that relays extracellular signals to elicit intracellular responses, including proliferation, differentiation, cell death, and stress adaptation. The cascade has three principal tiers—MAP3K, MAP2K, and MAPK—and is conserved [115]. These components coordinate responses to diverse inputs, such as growth factors, cytokines, and environmental stressors including UV radiation, heat, and osmotic shock.

MAP3Ks initiate the cascade in response to these stimuli. Prominent examples include Raf, which activates the ERK1/2 pathway by phosphorylating mitogen-activated protein kinase kinase 1/2 (MEK1/2), and transforming growth factor---activated kinase 1 (TAK1), which initiates the JNK and p38 pathways in response to inflammatory signals. In addition, MEKK1–4 participate in the JNK and p38 pathways, enhancing stress responses and facilitating apoptosis [116, 117]. MAP3Ks contain a core kinase domain that mediates phosphorylation and regulatory regions that enable interactions with upstream activators such as Ras and downstream targets such as MAP2Ks [118]. Raf kinase, a major MAP3K, has three conserved regions: CR1, which binds Ras; CR2, enriched in serine/threonine residues for regulatory functions; and CR3, the kinase domain that phosphorylates MAP2Ks. Other MAP3Ks, such as TAK1 and MEKK1, possess analogous kinase domains but distinct regulatory regions that confer specific protein–protein interactions [119, 120].

MAP2Ks are activated by MAP3Ks and, in turn, phosphorylate MAPKs on threonine and tyrosine residues, thereby propagating the signal. Examples include MEK1/2, which activate ERK1/2 to control proliferation, and MKK3/6, which activate p38 MAPK in response to stress stimuli [121]. MKK4/7 similarly initiate the JNK pathway, essential for apoptosis and stress tolerance. MAP2Ks are defined by a catalytic domain that phosphorylates MAPKs and regulatory domains that interact with MAP3Ks and scaffold proteins such as KSR, aiding localization to upstream activators and downstream targets [122].

MAPKs are the terminal kinases that phosphorylate diverse downstream substrates, including transcription factors and enzymes. ERK5 is also crucial for cardiovascular development and cellular viability under oxidative stress [123]. MAPKs share a conserved serine/threonine kinase domain that binds ATP and catalyzes substrate phosphorylation. Phosphorylation of the activation loop by MAP2Ks induces the active conformation. In addition, MAPKs contain docking sites for interactions with MAP2Ks and substrates, as well as a substrate-binding groove that ensures accurate recognition of target proteins [124].

MAPKKK (*e.g.*, Raf): central kinase domain plus regulatory domains.

MAPKK (*e.g.*, MEK1/2): dual-specificity kinase domain plus regulatory domains.

MAPK (*e.g.*, ERK1/2): serine/threonine kinase domain plus docking and substrate-binding regions.

MAPK Pathways

MAPK pathways are central signaling cascades that control diverse cellular functions. Multiple MAPK modules operate in parallel, each defined by distinct components, functions, and stimuli.

Extracellular Signal-regulated Kinase (ERK1/2) Pathway

The ERK1/2 signaling pathway relays extracellular cues to the nucleus and regulates proliferation, differentiation, survival, death, and motility. Extracellular signals—including growth factors (*e.g.*, EGF, PDGF), hormones, and cytokines—activate RTKs at the cell surface. Ligand binding induces RTK autophosphorylation, creating docking sites for adaptor proteins such as Grb2 (Growth factor receptor-bound protein 2). Grb2 recruits Sos (Son of Sevenless), a guanine nucleotide exchange factor, to the plasma membrane to promote GDP–GTP exchange on Ras, activating this small GTPase. GTP-bound Ras recruits Raf, a serine/threonine kinase, to the plasma membrane. Raf-1 (c-Raf), together with the isoforms A-Raf and B-Raf, belongs to the MAP3K family. Upon recruitment, Raf is phosphorylated and activated through regulatory interactions with heat shock protein 90 (Hsp90), p50, and 14-3-3 proteins. 14-3-3 proteins stabilize Raf by binding specific phosphorylation sites, whereas Hsp90 and p50 preserve its stability to facilitate activation. These interactions tightly control Raf activation, permitting graded pathway outputs [125].

Upon activation, Raf phosphorylates and stimulates MEK1 and MEK2, members of the MAP2K family. MEK1/2, dual-specificity kinases, phosphorylate ERK1/2 on threonine (Thr) and tyrosine (Tyr) residues within the conserved TEY motif of the activation loop. This dual phosphorylation induces a conformational change that activates ERK1/2 and amplifies downstream signaling. Activated ERK1/2 translocates to the nucleus to phosphorylate substrates, including transcription factors such as Ets-like protein 1 (Elk-1), FBJ murine osteosarcoma viral oncogene homolog (c-Fos), Jun proto-oncogene (c-Jun), and myelocytomatosis viral oncogene homolog (c-Myc). These factors regulate genes associated with proliferation, differentiation, and survival. ERK1/2 also phosphorylates cytoplasmic proteins that control cytoskeletal remodeling, motility, and vesicle trafficking, underscoring its role in coordinating complex cellular responses [126, 127].

Multiple mechanisms regulate ERK1/2 signaling. Scaffold proteins such as Kinase Suppressor of Ras (KSR) and IQ Motif Containing GTPase Activating Protein (IQGAP) assemble pathway components into functional complexes to ensure spatial fidelity and efficiency. Moreover, dual-specificity phosphatases (DUSPs) dephosphorylate ERK1/2 to limit excessive or prolonged activation that could drive abnormal responses, such as uncontrolled proliferation [128]. The activation of Raf is also modulated by kinases including protein kinase C (PKC) and p21-activated kinase (PAK), which phosphorylate Raf on specific residues and thereby modulate its activity. These phosphorylation events allow graded Raf outputs that reflect input context. B-Raf and Raf-1 are regulated differently, particularly in their interactions with small GTPases: Raf-1 is mainly activated by H-Ras, K-Ras, and N-Ras, whereas B-Raf responds to Ras and Rap1. These differences have important biological implications, especially in neurons where B-Raf predominates [129].

JNK Signaling Pathway

The JNK, or stress-activated protein kinase (SAPK), pathway is a major MAPK module that shapes physiological and pathological processes, including immune responses, neurodegeneration, cancer, and metabolic disease [130].

Extracellular and intracellular stresses activate upstream small GTPases, including Ras-related C3 botulinum toxin substrate 1 (Rac1) and cell division control protein 42 (Cdc42) of the Rho family. These small GTPases act as molecular switches that engage MAP3Ks. Multiple MAP3Ks—including MEKK1–4, apoptosis signal-regulating kinase 1 (ASK1), TAK1, and mixed-lineage kinases (MLKs)—activate the JNK pathway by phosphorylating the downstream MAP2Ks MKK4 and MKK7.

Upon activation, MKK4 and MKK7 phosphorylate JNK on Thr183 and Tyr185 within the activation loop, producing full activation. MKK7 preferentially mediates cytokine-induced JNK activation, whereas MKK4 responds to a broader array of stress signals [131].

Activated JNK translocates to the nucleus and phosphorylates transcription factors, notably Activator Protein-1 (AP-1) family members c-Jun, Activating Transcription Factor 2 (ATF2), and c-Fos. JNK-mediated phosphorylation of c-Jun enhances its activity and promotes AP-1 assembly, which governs genes linked to proliferation, differentiation, and death. JNK also phosphorylates p53, c-Myc, and Elk-1, modulating diverse outcomes; for example, phosphorylation can enhance p53's pro-apoptotic activity and augment c-Myc's roles in proliferation and differentiation [132].

JNK's role in apoptosis is context dependent; it can promote death or support survival depending on stimulus and cell state. Under prolonged or intense stress, JNK often triggers the intrinsic mitochondrial apoptotic pathway by phosphorylating pro-apoptotic Bcl-2 family members, promoting cytochrome c release and caspase activation. Conversely, JNK can enhance survival by upregulating anti-apoptotic proteins such as B-cell lymphoma 2 (Bcl-2) and B-cell lymphoma-extra large (Bcl-xL) through cross-talk with survival pathways (*e.g.*, NF-κB). Feedback loops and interactions with PI3K–Akt and ERK further tune these outcomes [133, 134].

JNK also regulates inflammation: proinflammatory cytokines such as tumor necrosis factor-α (TNF-α) and interleukin-1 (IL-1) activate JNK, which drives expression of inflammatory genes. Cytokine-induced JNK activation promotes phosphorylation and activation of transcription factors, including NF-κB, to induce cytokines, chemokines, and adhesion molecules pertinent to the immune response. JNK is essential for T-helper cell differentiation and has been implicated in autoimmune and inflammatory diseases, where dysregulated signaling drives chronic inflammation and tissue damage. Multiple negative feedback mechanisms constrain JNK activity to prevent excessive or prolonged activation that can cause tissue damage or disease. DUSPs dephosphorylate and inactivate JNK, and scaffold proteins such as JNK-interacting proteins (JIPs) organize pathway components to ensure compartment-specific activation [135].

p38 MAPK Pathway

The p38 MAPK pathway is a core signaling cascade that maintains cellular homeostasis and orchestrates responses to extrinsic and intrinsic stress. It is activated by environmental stressors (*e.g.*, UV radiation, heat shock) and pro-inflammatory cytokines, including TNF-α and IL-1. Activation begins when these stimuli engage pattern-recognition or cytokine receptors, such as Toll-like receptors (TLRs). Receptor engagement triggers a kinase cascade. MAP3Ks—including ASK1, TAK1, and MLKs—activate the MAP2Ks MKK3 and MKK6, the direct upstream kinases for p38 MAPK. Phosphorylation of p38 at threonine 180 (Thr180) and tyrosine 182 (Tyr182) is required for full activation [136].

The p38 MAPK family comprises p38α, p38β, p38γ, and p38δ, which differ in function and tissue distribution. p38α is the most studied and regulates diverse cellular activities; p38β shares overlap but shows distinct expression; p38γ and p38δ are less prevalent and operate in specific physiological and pathological contexts. Activated p38 modulates numerous downstream targets. A principal role is the regulation of gene expression: p38 phosphorylates transcription factors such as ATF2 and Elk-1, driving expression programs linked to stress responses, inflammation, and differentiation [137, 138].

Beyond transcriptional control, p38 influences cytoskeletal dynamics by phosphorylating actin-regulatory proteins, thereby shaping cell morphology, motility, and adhesion—processes central to wound repair, immune function, and cancer dissemination. The pathway is tightly regulated to prevent excessive or prolonged activity that could cause dysfunction and disease (Fig. **5**). Dual-specificity phosphatases (DUSPs), including MAP kinase phosphatase 1 (MKP-1), dephosphorylate and inactivate p38. Other phosphatases, such as protein phosphatase 2A (PP2A), contribute as well. Scaffold proteins, for example JNK-interacting protein 1 (JIP1), organize pathway components to ensure precise signal transmission [139].

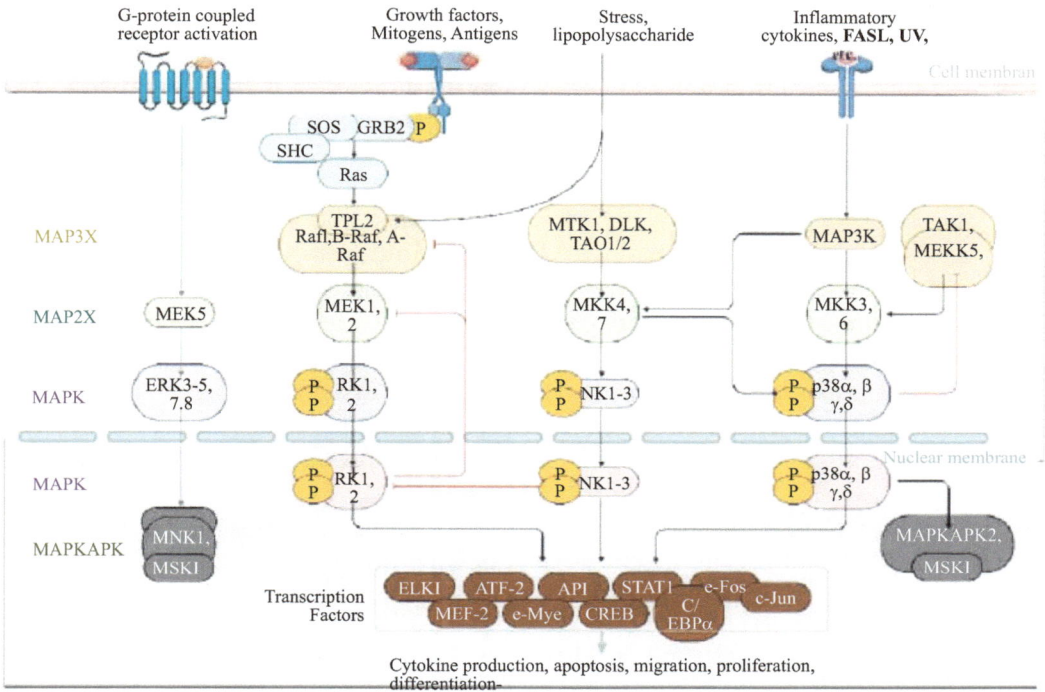

Fig. (5). Mitogen-activated protein kinase (MAPK) signaling pathways are activated by different stimuli.

Dysregulated MAPK Signaling in Disease

ERK, JNK, and p38 MAPK signaling is frequently dysregulated across cancers, promoting unchecked proliferation, survival, and metastasis. Mutations or hyperactivation within these modules drive oncogenesis and disease progression. For example, activating B-Raf mutations chronically stimulate ERK signaling, fostering abnormal growth; MEK and ERK inhibitors have shown therapeutic benefit in multiple malignancies [140]. Aberrant JNK signaling can amplify stress responses that support tumor growth, enhance proliferation and survival, and facilitate inflammation and metastasis; in some cancers, JNK also contributes to radio- and chemoresistance [141]. Dysregulated p38 signaling, which governs stress and inflammatory responses, can similarly increase survival and proliferation under stress and has been implicated in drug tolerance, making p38 a potential therapeutic target [142].

MAPK cascades also orchestrate inflammatory responses, and their disruption underlies chronic inflammatory diseases. ERK activation promotes production of pro-inflammatory cytokines and chemokines; sustained ERK activity is linked to rheumatoid arthritis and inflammatory bowel disease, where it exacerbates tissue damage [143]. JNK regulates cytokine expression and inflammatory proteins; persistent JNK activation contributes to conditions such as psoriasis and Crohn's disease, and its inhibition can reduce inflammation and improve outcomes [144]. p38 controls the synthesis of TNF-α, IL-1β, and IL-6; dysregulated p38 activity drives excessive inflammation associated with rheumatoid arthritis, asthma, and sepsis, motivating efforts to therapeutically target p38 [145].

In neurodegeneration, disturbed MAPK signaling promotes neuronal injury, death, and disease progression. In Alzheimer's and Parkinson's diseases, altered ERK signaling affects tau phosphorylation and amyloid-β production and compromises dopaminergic neuron survival [146]. JNK overactivation triggers neuronal stress responses and apoptosis, accelerates disease course in Alzheimer's and Huntington's diseases, and influences toxic protein aggregation and neuroinflammation [147]. p38 participates in neuronal stress and apoptotic pathways; in Alzheimer's disease it contributes to tau hyperphosphorylation and neuronal loss, and in amyotrophic lateral sclerosis and other disorders it correlates with neuroinflammation and motor neuron degeneration [148].

Targeted inhibitors of MAPK components, particularly MEK inhibitors, are central to treating cancers with MAPK dysregulation. Agents such as trametinib and cobimetinib selectively inhibit MEK1/2, blocking downstream ERK signaling that sustains tumor growth. These drugs are most effective in tumors with B-Raf V600E mutations and can markedly reduce tumor burden and prolong progression-free survival. B-Raf inhibitors, including vemurafenib and dabrafenib, further suppress ERK activation [149]. Emerging direct ERK inhibitors show promise for overcoming resistance to upstream agents. Combination regimens—either within the MAPK axis or together with immunotherapy—can enhance efficacy and limit resistance. Personalized strategies that match therapy to tumor features remain essential.

CROSS-TALK BETWEEN GPCRS AND THE MAPK PATHWAY

The MAPK pathway promotes cell proliferation through multiple effector molecules. While GPCR-linked nuclear responses arise from diverse pathways, activation of effectors within the RTK signaling cascade induces mitogenic programs and tumor growth [150]. Cross talk between the MAPK pathway and GPCRs, which also mediate growth stimulation, is well documented; evidence indicates that MAPK serves as a convergence node for signals from both RTKs and G proteins. RTKs activate MAPK by recruiting Shc, Grb2, and Sos proteins [151]. This recruitment initiates a cascade through Ras and the kinases Raf and MEK, culminating in MAPK activation. Activated MAPK translocates to the nucleus, where it phosphorylates transcription factors that drive cell growth.

GPCRs can induce p38 MAPK signaling in a ubiquitin-dependent manner, and deubiquitination is required to terminate inflammatory signaling [152]. Activation of PI3K/Akt and RAS/MAPK *via* receptors such as GPR75 has been implicated in aggressive prostate cancer phenotypes [153]. These pathways also activate NF-κB, a central regulator of cancer proliferation, apoptosis, and migration [154]. GPR75 binds 20-HETE, a proinflammatory ligand; ligand engagement triggers the IP3/DAG pathway, leading to PKC phosphorylation [155]. PKC subsequently activates the PI3K/Akt and MAPK pathways and, in turn, the NF-κB pathway [156]. Thus, GPR75 inhibition is an attractive target for chemotherapy. Blocking CCL5 binding—an activator of ERK1/2 and AKT—*via* GPR75 knockdown decreased EMT and cancer spread [157]. Another GPR75 ligand, RANTES, similarly stimulates MAPK *via* the PI3K pathway [158].

Beyond acting as passive scaffolds, β-arrestins activate c-Src and c-Raf, effector proteins in the MAPK–ERK pathway. GPCR-dependent β-arrestin activation enhances ERK2 phosphorylation and the downstream cascade [159].

Tang et al. reported that GPCR125, a receptor involved in tumor angiogenesis, promotes osteoclastogenesis and osteoclast-mediated bone resorption by upregulating the NF-κB and AKT pathways. Silencing this receptor significantly decreased MAPK–ERK and NF-κB–AKT signaling, indicating that GPCR125 engages these pathways to regulate osteoclasts [160].

REGULATION AND TERMINATION OF SIGNALING PATHWAYS

Regulation and timely termination of signaling are essential to prevent overstimulation and maintain organismal homeostasis. These processes rely on feedback mechanisms that modulate pathway strength and on the deactivation of signaling components to cease responses.

Negative feedback loops limit abnormal signaling by allowing downstream components to inhibit upstream steps, thereby controlling signal amplitude and duration and enabling context-appropriate responses.

Termination mechanisms stop signaling at appropriate times by degrading signaling molecules or reversing activating modifications. For example, ligand dissociation halts receptor engagement; intracellular enzymes reverse pathway modifications, such as phosphatases that dephosphorylate kinase targets. Small-molecule messengers, including cAMP, are hydrolyzed by specific enzymes, lowering their concentrations and stopping the associated responses. Together, these mechanisms ensure appropriate pathway function and prevent excessive or inappropriate cellular activity [61, 161].

Feedback Mechanisms

Negative feedback loops and receptor desensitization are central to maintaining homeostasis and preventing overstimulation. Acting together, they calibrate the strength and duration of cellular responses.

Negative feedback is embedded in most signaling pathways and constrains the output they produce. By routing a pathway's output back to its input, these loops dampen excessive stimulation and permit adaptation to environmental changes. In the epidermal growth factor receptor (EGFR) pathway, such feedback preserves signal integrity and prevents overstimulation that could otherwise promote disease [162].

Receptor desensitization reduces a cell's response after prolonged stimulation. Common mechanisms include receptor phosphorylation, internalization, and downregulation. In GPCRs, G protein–coupled receptor kinases (GRKs) phosphorylate receptors to inhibit signaling. Desensitization typically proceeds through three steps: uncoupling (loss of receptor–G protein coupling), internalization (endocytic uptake), and downregulation (post-internalization degradation). Collectively, these steps limit sensitivity to persistent stimuli and avert excessive responses.

Negative feedback loops and receptor desensitization regulate cellular signaling. Negative feedback preserves response fidelity, whereas desensitization prevents overstimulation, allowing cells to remain receptive to new inputs. Together, these mechanisms confer flexibility and resilience as cells adapt to changing environments [163, 164].

Role of Phosphatases and Ubiquitination

The interplay between phosphatases and ubiquitination is central to signal transduction, governing dephosphorylation and the turnover of signaling proteins. Serine/threonine phosphatases such as PP1 and PP2A modulate pathways by dephosphorylating key substrates, shaping processes including tumor angiogenesis. PP5 exemplifies dual regulation: phosphorylation and ubiquitination can enhance or suppress its activity, thereby influencing cancer-cell survival pathways. Ubiquitination dynamics, controlled by deubiquitinases such as USP15, maintain protein stability by reversing ubiquitin modifications and calibrating signaling output. Disruption of these regulatory layers can drive pathology and underscores their therapeutic potential. In angiogenesis, phosphatases tune endothelial signaling by dephosphorylating receptors and downstream effectors. In plants, PP2C phosphatases counterbalance kinases under stress, refining ion transport and signaling pathways to adapt to environmental change. More broadly, ubiquitination and deubiquitination determine protein stability and activity across signaling networks; for example, ubiquitination of PP5 targets it for degradation, altering survival pathways in cancer cells. Overall, precise regulation of phosphatases and ubiquitination is essential for signaling homeostasis, and its dysregulation highlights opportunities for targeted therapy [165–167].

Implications for Drug Development

Targeting signaling pathways offers substantial opportunities in cancer, metabolic disease, and immune disorders. Deeper delineation of these networks enables rational strategies that improve clinical outcomes. In oncology, modulating cytokine and chemokine pathways can shape immune responses and inhibit tumor progression, providing options to overcome resistance to conventional therapies. Inhibiting metabolic pathways such as PI3K and mTOR can suppress tumor growth and augment T cell–mediated immunity, particularly with dual-pathway inhibitors. In metabolic disease, immunometabolism is pivotal in conditions such as type 2 diabetes mellitus, where pathway-directed interventions can improve outcomes. In immune disorders, manipulating pathway crosstalk between metabolic and immune signaling offers avenues for interventions that strengthen anti-tumor immunity. Despite this promise, achieving specificity while minimizing adverse effects remains challenging, warranting further research to optimize therapeutic strategies [168, 169].

CONCLUSION

Emerging advances in signal transduction are accelerating personalized medicine and synthetic biology. Sala et al. are developing synthetic signal-transduction systems that mimic natural processes. Researchers are designing stimuli-responsive supramolecular constructs—noncovalent molecular assemblies—that embed within phospholipid bilayers. These constructs aim to replicate the behavior of ligand-gated ion channels, GPCRs, and RTKs. The lipid bilayer plays an essential role in these biomimetic systems because its anisotropic environment complicates both design and data analysis. This work seeks to elucidate the physical basis of natural signal transduction and could enable integration of artificial signaling pathways into cells [170]. Stefanov et al. are likewise developing advanced synthetic-biology platforms that leverage signal transduction to enable personalized medicines.

Key developments include closed-loop systems, designer cell implants, and molecular sensors. Closed-loop systems monitor disease markers and respond accordingly. For example, engineered cells can detect cytokines such as TNF and IL-22 and then release therapeutic proteins to treat conditions like psoriasis, in which skin cells build up to form scaly, itchy patches. Designer cell implants integrate signal-transduction pathways; for instance, cells can be engineered to sense luteinizing hormone, which is essential for ovulation, and release sperm at the appropriate time. These platforms incorporate molecular sensors for small molecules and cytokines, enabling real-time monitoring and response, including detection of histamine and other immune mediators to manage immune-related disease. Collectively, these engineered systems aim to deliver precise, signal-triggered therapies tailored to individual patients, thereby enhancing safety and efficacy [171].

CONSENT FOR PUBLICATION

All visual representations were created by the authors and were not sourced from other publications. No consent was required.

CONFLICT OF INTEREST

This work received no specific grant from any funding agency in the public, commercial, or not-for-profit sectors. The authors declare no conflicts of interest related to this publication.

ACKNOWLEDGEMENTS

We thank Kumaran K., Sanjana Dhayalan, and Sasitharan T. for their contributions to proofreading and editing.

REFERENCES

[1] Nair A, Chauhan P, Saha B, Kubatzky KF. Conceptual Evolution of Cell Signaling. Int J Mol Sci 2019; 20(13): 3292.
[http://dx.doi.org/10.3390/ijms20133292] [PMID: 31277491]

[2] Villalobo A, Gabius HJ. Signaling pathways for transduction of the initial message of the glycocode into cellular responses. Cells Tissues Organs 1998; 161(1-4): 110-29.
[http://dx.doi.org/10.1159/000046453] [PMID: 9780354]

[3] Cesaro G, Nagai JS, Gnoato N, *et al*. Advances and challenges in cell-cell communication inference: a comprehensive review of tools, resources, and future directions. Briefings in bioinformatics 2025; 26(3). 10.1093/bib/bbaf280

[4] Zaghloul Salem MS. Biological Networks: An Introductory Review. J Proteom Genom Res 2018; 2(1): 41-111.

[http://dx.doi.org/10.14302/issn.2326-0793.jpgr-18-2312]

[5] Rudel D, Sommer RJ. The evolution of developmental mechanisms. Dev Biol 2003; 264(1): 15-37.
[http://dx.doi.org/10.1016/S0012-1606(03)00353-1] [PMID: 14623229]

[6] Oligny LL. Human molecular embryogenesis: an overview. Pediatr Dev Pathol 2001; 4(4): 324-43.
[http://dx.doi.org/10.1007/s10024001-0033-2] [PMID: 11441334]

[7] Marmion M, Macori G, Ferone M, Whyte P, Scannell AGM. Survive and thrive: Control mechanisms that facilitate bacterial adaptation to survive manufacturing-related stress. Int J Food Microbiol 2022; 368109612
[http://dx.doi.org/10.1016/j.ijfoodmicro.2022.109612] [PMID: 35278797]

[8] Schneider JE. Energy balance and reproduction. Physiol Behav 2004; 81(2): 289-317.
[http://dx.doi.org/10.1016/j.physbeh.2004.02.007] [PMID: 15159173]

[9] Müller A, Oertli M, Arnold IC. H. pylori exploits and manipulates innate and adaptive immune cell signaling pathways to establish persistent infection. Cell Commun Signal 2011; 9(1): 25.
[http://dx.doi.org/10.1186/1478-811X-9-25] [PMID: 22044597]

[10] Tozzi A. Information processing in the CNS: a supramolecular chemistry? Cogn Neurodynamics 2015; 9(5): 463-77.
[http://dx.doi.org/10.1007/s11571-015-9337-1] [PMID: 26379797]

[11] Doğaner BA, Yan LKQ, Youk H. Autocrine Signaling and Quorum Sensing: Extreme Ends of a Common Spectrum. Trends in cell biology 2016; 26(4): 262-71.
[https://doi.org/10.1016/j.tcb.2015.11.002]

[12] Gnecchi M, Zhang Z, Ni A, Dzau VJ. Paracrine mechanisms in adult stem cell signaling and therapy. Circ Res 2008; 103(11): 1204-19.
[http://dx.doi.org/10.1161/CIRCRESAHA.108.176826] [PMID: 19028920]

[13] Gilbert R, Vickaryous M, Viloria-Petit A. Signalling by Transforming Growth Factor Beta Isoforms in Wound Healing and Tissue Regeneration. J Dev Biol 2016; 4(2): 21.
[http://dx.doi.org/10.3390/jdb4020021] [PMID: 29615587]

[14] Heyland A, Hodin J, Reitzel AM. Hormone signaling in evolution and development: a non-model system approachs. BioEssays 2005; 27(1): 64-75.
[http://dx.doi.org/10.1002/bies.20136] [PMID: 15612033]

[15] Gordon NK, Gordon R. The organelle of differentiation in embryos: the cell state splitter. Theor Biol Med Model 2016; 13(1): 11.
[http://dx.doi.org/10.1186/s12976-016-0037-2] [PMID: 26965444]

[16] Alberts B, Johnson A, Lewis J, et al. Molecular Biology of the Cell. 4th edition. Signaling through G-Protein-Linked Cell-Surface Receptors. New York: Garland Science; 2002.

[17] Cattaneo F, Guerra G, Parisi M, et al. Cell-surface receptors transactivation mediated by g protein-coupled receptors. Int J Mol Sci 2014; 15(11): 19700-28.
[http://dx.doi.org/10.3390/ijms151119700] [PMID: 25356505]

[18] Schlessinger J. Cell signaling by receptor tyrosine kinases. Cell 2000; 103(2): 211-25.
[http://dx.doi.org/10.1016/S0092-8674(00)00114-8] [PMID: 11057895]

[19] Heldin CH, Lu B, Evans R, Gutkind JS. Signals and Receptors. Cold Spring Harb Perspect Biol 2016; 8(4)a005900
[http://dx.doi.org/10.1101/cshperspect.a005900] [PMID: 27037414]

[20] Tokarz VL, MacDonald PE, Klip A. The cell biology of systemic insulin function. J Cell Biol 2018; 217(7): 2273-89.
[http://dx.doi.org/10.1083/jcb.201802095] [PMID: 29622564]

[21] Wang D, Jin S, Zou X. Crosstalk between pathways enhances the controllability of signalling networks. IET systems biology 2016; 10(1): 2-9.

[https://doi.org/10.1049/iet-syb.2014.0061]

[22] Soderling TR, Stull JT. Structure and regulation of calcium/calmodulin-dependent protein kinases. Chem Rev 2001; 101(8): 2341-52.
[http://dx.doi.org/10.1021/cr0002386] [PMID: 11749376]

[23] Wee P, Wang Z. Epidermal Growth Factor Receptor Cell Proliferation Signaling Pathways. Cancers (Basel) 2017; 9(5): 52.
[http://dx.doi.org/10.3390/cancers9050052] [PMID: 28513565]

[24] Cole K, Kohn E. Calcium-mediated signal transduction: biology, biochemistry, and therapy. Cancer Metastasis Rev 1994; 13(1): 31-44.
[http://dx.doi.org/10.1007/BF00690417] [PMID: 8143344]

[25] Elmore S. Apoptosis: a review of programmed cell death. Toxicol Pathol 2007; 35(4): 495-516.
[http://dx.doi.org/10.1080/01926230701320337] [PMID: 17562483]

[26] Kléber AG, Jin Q. Coupling between cardiac cells—An important determinant of electrical impulse propagation and arrhythmogenesis. Biophys Rev 2021; 2(3)031301.
[http://dx.doi.org/10.1063/5.0050192] [PMID: 34296210]

[27] Niessen CM, Leckband D, Yap AS. Tissue organization by cadherin adhesion molecules: dynamic molecular and cellular mechanisms of morphogenetic regulation. Physiol Rev 2011; 91(2): 691-731.
[http://dx.doi.org/10.1152/physrev.00004.2010] [PMID: 21527735]

[28] Zhao X, Guan JL. Focal adhesion kinase and its signaling pathways in cell migration and angiogenesis. Adv Drug Deliv Rev 2011; 63(8): 610-5.
[http://dx.doi.org/10.1016/j.addr.2010.11.001] [PMID: 21118706]

[29] Williams G. Aromatase up-regulation, insulin and raised intracellular oestrogens in men, induce adiposity, metabolic syndrome and prostate disease, *via* aberrant ER-α and GPER signalling. Mol Cell Endocrinol 2012; 351(2): 269-78.
[http://dx.doi.org/10.1016/j.mce.2011.12.017] [PMID: 22233684]

[30] Goldfine ID. The insulin receptor: molecular biology and transmembrane signaling. Endocr Rev 1987; 8(3): 235-55.
[http://dx.doi.org/10.1210/edrv-8-3-235] [PMID: 3308443]

[31] Burke K, Chidambaram L. How Much Bandwidth Is Enough? A Longitudinal Examination of Media Characteristics and Group Outcomes1. Manage Inf Syst Q 1999; 23(4): 557-79.
[http://dx.doi.org/10.2307/249489]

[32] Dinarello CA. Immunological and inflammatory functions of the interleukin-1 family. Annu Rev Immunol 2009; 27(1): 519-50.
[http://dx.doi.org/10.1146/annurev.immunol.021908.132612] [PMID: 19302047]

[33] Lan RY, Selmi C, Gershwin ME. The regulatory, inflammatory, and T cell programming roles of interleukin-2 (IL-2). J Autoimmun 2008; 31(1): 7-12.
[http://dx.doi.org/10.1016/j.jaut.2008.03.002] [PMID: 18442895]

[34] Mocellin S, Rossi C, Pilati P, Nitti D. Tumor necrosis factor, cancer and anticancer therapy. Cytokine Growth Factor Rev 2005; 16(1): 35-53.
[http://dx.doi.org/10.1016/j.cytogfr.2004.11.001] [PMID: 15733831]

[35] Martins-Green M, Petreaca M, Wang L. Chemokines and Their Receptors Are Key Players in the Orchestra That Regulates Wound Healing. Adv Wound Care (New Rochelle) 2013; 2(7): 327-47.
[http://dx.doi.org/10.1089/wound.2012.0380] [PMID: 24587971]

[36] Janeway CA Jr, Travers P, Walport M, *et al.* Immunobiology: The Immune System in Health and Disease. 5th edition. New York: Garland Science; 2001. Macrophage activation by armed CD4 TH1 cells. Available from: https://www.ncbi.nlm.nih.gov/books/NBK27153/

[37] Deutch AY, Roth RH. Pharmacology and Biochemistry of Synaptic Transmission: Classic Transmitters.From Molecules to Networks. Elsevier 2004; pp. 245-78.
[http://dx.doi.org/10.1016/B978-012148660-0/50010-X]

[38] Yousuf MS, Kerr BJ. The Role of Regulatory Transporters in Neuropathic Pain. In: Advances in Pharmacology. Elsevier 2016; vol. 75: pp. 245-71.
[http://dx.doi.org/10.1016/bs.apha.2015.12.003]

[39] Haddad JJ. N-methyl-d-aspartate (NMDA) and the regulation of mitogen-activated protein kinase (MAPK) signaling pathways: A revolving neurochemical axis for therapeutic intervention? Prog Neurobiol 2005; 77(4): 252-82.
[http://dx.doi.org/10.1016/j.pneurobio.2005.10.008] [PMID: 16343729]

[40] Fadel JR. Regulation of cortical acetylcholine release: Insights from *in vivo* microdialysis studies. Behav Brain Res 2011; 221(2): 527-36.
[http://dx.doi.org/10.1016/j.bbr.2010.02.022] [PMID: 20170686]

[41] Böhmová E, Machová D, Pechar M, *et al.* Cell-penetrating peptides: a useful tool for the delivery of various cargoes into cells. Physiol Res 2018; 67 (Suppl. 2): S267-79.
[http://dx.doi.org/10.33549/physiolres.933975] [PMID: 30379549]

[42] Lindsey S, Langhans SA. Epidermal Growth Factor Signalling in Transformed Cells. Int Rev Cell Mol Biol 2015; 314: 1-41.

[43] Eyster KM. The membrane and lipids as integral participants in signal transduction: lipid signal transduction for the non-lipid biochemist. Adv Physiol Educ 2007; 31(1): 5-16.
[http://dx.doi.org/10.1152/advan.00088.2006] [PMID: 17327576]

[44] García-Gómez E, Vázquez-Martínez ER, Reyes-Mayoral C, Cruz-Orozco OP, Camacho-Arroyo I, Cerbón M. Regulation of Inflammation Pathways and Inflammasome by Sex Steroid Hormones in Endometriosis. Front Endocrinol (Lausanne) 2020; 10: 935.
[http://dx.doi.org/10.3389/fendo.2019.00935] [PMID: 32063886]

[45] Bian K, Doursout MF, Murad F. Vascular system: role of nitric oxide in cardiovascular diseases. J Clin Hypertens (Greenwich) 2008; 10(4): 304-10.
[http://dx.doi.org/10.1111/j.1751-7176.2008.06632.x] [PMID: 18401228]

[46] Arango Duque G, Descoteaux A. Macrophage cytokines: involvement in immunity and infectious diseases. Front Immunol 2014; 5: 491.
[http://dx.doi.org/10.3389/fimmu.2014.00491] [PMID: 25339958]

[47] Ahmad HI, Jabbar A, Mushtaq N, *et al.* Immune Tolerance vs. Immune Resistance: The Interaction Between Host and Pathogens in Infectious Diseases. Front Vet Sci 2022; 9827407
[http://dx.doi.org/10.3389/fvets.2022.827407] [PMID: 35425833]

[48] Jackson D, Purcell A, Fitzmaurice C, Zeng W, Hart D. The central role played by peptides in the immune response and the design of peptide-based vaccines against infectious diseases and cancer. Curr Drug Targets 2002; 3(2): 175-96.
[http://dx.doi.org/10.2174/1389450024605436] [PMID: 11958299]

[49] Dallas SL, Alliston T, Bonewald LF. Transforming Growth Factor-β.Principles of Bone Biology. Elsevier 2008; pp. 1145-66.
[http://dx.doi.org/10.1016/B978-0-12-373884-4.00067-7]

[50] Lee SY, Ju MK, Jeon HM, *et al.* Regulation of Tumor Progression by Programmed Necrosis. Oxid Med Cell Longev. 2018 Jan 31;2018(1).
[http://dx.doi.org/10.1155/2018/3537471]

[51] Barrientos S, Stojadinovic O, Golinko MS, Brem H, Tomic-Canic M. PERSPECTIVE ARTICLE: Growth factors and cytokines in wound healing. Wound Repair Regen 2008; 16(5): 585-601.
[http://dx.doi.org/10.1111/j.1524-475X.2008.00410.x] [PMID: 19128254]

[52] Bode AM, Dong Z. The paradox of arsenic: molecular mechanisms of cell transformation and chemotherapeutic effects. Crit Rev Oncol Hematol 2002; 42(1): 5-24.
[http://dx.doi.org/10.1016/S1040-8428(01)00215-3] [PMID: 11923065]

[53] Alberts B. Molecular Biology of the Cell (6th ed.). W.W. Norton & Company, 2015.
[https://doi.org/10.1201/9781315735368]

[54] Kiel C, Serrano L. Challenges ahead in signal transduction: MAPK as an example. Curr Opin Biotechnol 2012; 23(3): 305-14.
[http://dx.doi.org/10.1016/j.copbio.2011.10.004] [PMID: 22036710]

[55] Su J, Song Y, Zhu Z, et al. Cell–cell communication: new insights and clinical implications. Signal Transduct Target Ther 2024; 9(1): 196.
[http://dx.doi.org/10.1038/s41392-024-01888-z] [PMID: 39107318]

[56] Lee MJ, Yaffe MB. Protein Regulation in Signal Transduction. Cold Spring Harb Perspect Biol 2016; 8(6)a005918.
[http://dx.doi.org/10.1101/cshperspect.a005918] [PMID: 27252361]

[57] Yue J, López JM. Understanding MAPK Signaling Pathways in Apoptosis. Int J Mol Sci 2020; 21(7): 2346.
[http://dx.doi.org/10.3390/ijms21072346] [PMID: 32231094]

[58] Soares-Silva M, Diniz FF, Gomes GN, Bahia D. The Mitogen-Activated Protein Kinase (MAPK) Pathway: Role in Immune Evasion by Trypanosomatids. Front Microbiol 2016; 7: 183.
[http://dx.doi.org/10.3389/fmicb.2016.00183] [PMID: 26941717]

[59] Koyama-Honda I, Fujiwara TK, Kasai RS, et al. High-speed single-molecule imaging reveals signal transduction by induced transbilayer raft phases. J Cell Biol 2020; 219(12)e202006125.
[http://dx.doi.org/10.1083/jcb.202006125] [PMID: 33053147]

[60] Pan S, Zhang W. Molecules in Signal Pathways.Clinical Molecular Diagnostics. Singapore: Springer Singapore 2021; pp. 139-54.
[http://dx.doi.org/10.1007/978-981-16-1037-0_11]

[61] Valls PO, Esposito A. Signalling dynamics, cell decisions, and homeostatic control in health and disease. Curr Opin Cell Biol 2022; 75102066.
[http://dx.doi.org/10.1016/j.ceb.2022.01.011] [PMID: 35245783]

[62] Trowbridge IS, Omary MB. Human cell surface glycoprotein related to cell proliferation is the receptor for transferrin. Proc Natl Acad Sci USA 1981; 78(5): 3039-43.
[http://dx.doi.org/10.1073/pnas.78.5.3039] [PMID: 6265934]

[63] Limbird LE. Cell Surface Receptors: A Short Course on Theory and Methods. Boston, MA: Springer US 1996.
[http://dx.doi.org/10.1007/978-1-4613-1255-0]

[64] Liggett SB, McGraw DW. G-protein-coupled receptors. Encyclopedia of Respiratory Medicine. Elsevier 2006; pp. 248-51.
[http://dx.doi.org/10.1016/B0-12-370879-6/00169-1]

[65] Teng X, Chen S, Wang Q, et al. Structural insights into G protein activation by D1 dopamine receptor. Sci Adv 2022; 8(23)eabo4158.
[http://dx.doi.org/10.1126/sciadv.abo4158] [PMID: 35687690]

[66] Saengsawang W, Chukaew P, Rasenick MM. G-Protein Coupled Receptors.Encyclopedia of Cell Biology. Elsevier 2023; pp. 62-9.
[http://dx.doi.org/10.1016/B978-0-12-821618-7.00123-1]

[67] Omble A, Kulkarni K. GPCRs that *Rh* oar the Guanine nucleotide exchange factors. Small GTPases 2022; 13(1): 84-99.
[http://dx.doi.org/10.1080/21541248.2021.1896963] [PMID: 33849392]

[68] Cheng L, Xia F, Li Z, *et al.* Structure, function and drug discovery of GPCR signaling. Molecular Biomedicine 2023; 4(1): 46.
[http://dx.doi.org/10.1186/s43556-023-00156-w] [PMID: 38047990]

[69] Zhang M, Chen T, Lu X, Lan X, Chen Z, Lu S. G protein-coupled receptors (GPCRs): advances in structures, mechanisms and drug discovery. Signal Transduct Target Ther 2024; 9(1): 88.
[http://dx.doi.org/10.1038/s41392-024-01803-6] [PMID: 38594257]

[70] Batool Z, Azfal A, Liaquat L, *et al.* Receptor tyrosine kinases (RTKs).Receptor Tyrosine Kinases in Neurodegenerative and Psychiatric Disorders. Elsevier 2023; pp. 117-85.
[http://dx.doi.org/10.1016/B978-0-443-18677-6.00012-9]

[71] Karl K, Light TP, Hristova K. Receptor Tyrosine Kinases.Comprehensive Pharmacology. Elsevier 2022; pp. 10-36.
[http://dx.doi.org/10.1016/B978-0-12-820472-6.00135-3]

[72] Jawad Ahmad F. Receptor Tyrosine Kinases Mediated Cell Signalling Pathways. Pakistan BioMedical Journal. 2022 Jul 31;01–01.
[http://dx.doi.org/10.54393/pbmj.v5i7.572]

[73] Starbird C, Shao B, Hincapie-Otero M. Investigating the roles of oligomerization and lipid interaction in TAM receptor activation. Biophys J 2024; 123(3): 36a.
[http://dx.doi.org/10.1016/j.bpj.2023.11.307]

[74] Tovar KR, Westbrook GL. Ligand-Gated Ion Channels. In: Sperelakis N (Ed) Cell Physiology Source Book. Elsevier 2012; pp. 549-62.
[http://dx.doi.org/10.1016/B978-0-12-387738-3.00031-7]

[75] Schmauder R, Eick T, Schulz E, *et al.* Fast functional mapping of ligand-gated ion channels. Commun Biol 2023; 6(1): 1003.
[http://dx.doi.org/10.1038/s42003-023-05340-w] [PMID: 37783870]

[76] McFarlane A, Fyfe PK, Moraga I. Cytokine Receptors.Comprehensive Pharmacology. Elsevier 2022; pp. 37-64.
[http://dx.doi.org/10.1016/B978-0-12-820472-6.00112-2]

[77] Pogozheva ID, Cherepanov S, Park SJ, Raghavan M, Im W, Lomize AL. Structural Modeling of Cytokine-Receptor-JAK2 Signaling Complexes Using AlphaFold Multimer. J Chem Inf Model 2023; 63(18): 5874-95.
[http://dx.doi.org/10.1021/acs.jcim.3c00926] [PMID: 37694948]

[78] McFarlane A, Fyfe PK, Moraga I. Cytokine Receptors.Comprehensive Pharmacology. Elsevier 2022; pp. 37-64.
[http://dx.doi.org/10.1016/B978-0-12-820472-6.00112-2]

[79] Mierke DF, Pellegrini M. Receptor–Ligand Interactions. Encyclopedia of Endocrine Diseases. Elsevier 2004; pp. 163-6.
[http://dx.doi.org/10.1016/B0-12-475570-4/01127-6]

[80] García-Peñarrubia P, Gálvez JJ, Gálvez J. Mathematical modelling and computational study of two-dimensional and three-dimensional dynamics of receptor–ligand interactions in signalling response mechanisms. J Math Biol 2014; 69(3): 553-82.
[http://dx.doi.org/10.1007/s00285-013-0712-4] [PMID: 23893005]

[81] Tuteja N. Signaling through G protein coupled receptors. Plant Signal Behav 2009; 4(10): 942-7.
[http://dx.doi.org/10.4161/psb.4.10.9530] [PMID: 19826234]

[82] Sriram K, Insel PA. G Protein-Coupled Receptors as Targets for Approved Drugs: How Many Targets and How Many Drugs? Mol Pharmacol 2018; 93(4): 251-8.
[http://dx.doi.org/10.1124/mol.117.111062] [PMID: 29298813]

[83] Luo Y, Sun l, Peng Y, The structural basis of the G protein–coupled receptor and ion channel axis, Current Research in Structural Biology 2025; 9: 100165.
[https://doi.org/10.1016/j.crstbi.2025.100165]

[84] Chen Z, Ren X, Zhou Y, Huang N. Exploring structure-based drug discovery of GPCRs beyond the orthosteric binding site, hLife, 2024; 2(5).
[https://doi.org/10.1016/j.hlife.2024.01.002]

[85] Syrovatkina V, Alegre KO, Dey R, Huang XY. Regulation, Signaling, and Physiological Functions of G-Proteins. J Mol Biol 2016; 428(19): 3850-68.
[http://dx.doi.org/10.1016/j.jmb.2016.08.002] [PMID: 27515397]

[86] Kobilka BK. G protein coupled receptor structure and activation. Biochim Biophys Acta Biomembr 2007; 1768(4): 794-807.
[http://dx.doi.org/10.1016/j.bbamem.2006.10.021]

[87] Maruta N, Trusov Y, Jones AM, Botella JR, Heterotrimeric G. Heterotrimeric G Proteins in Plants: Canonical and Atypical Gα Subunits. Int J Mol Sci 2021; 22(21): 11841.
[http://dx.doi.org/10.3390/ijms222111841] [PMID: 34769272]

[88] Carman CV, Benovic JL. G-protein-coupled receptors: turn-ons and turn-offs. Curr Opin Neurobiol 1998; 8(3): 335-44.
[http://dx.doi.org/10.1016/S0959-4388(98)80058-5] [PMID: 9687355]

[89] Yang D, Zhou Q, Labroska V, et al. G protein-coupled receptors: structure- and function-based drug discovery. Signal Transduct Target Ther 2021; 6(1): 7.
[http://dx.doi.org/10.1038/s41392-020-00435-w] [PMID: 33414387]

[90] Liu Y, Wang X, Dong D, et al. Research Advances in Heterotrimeric G-Protein α Subunits and Uncanonical G-Protein Coupled Receptors in Plants. Int J Mol Sci 2021; 22(16): 8678.
[http://dx.doi.org/10.3390/ijms22168678] [PMID: 34445383]

[91] Sigismund S, Confalonieri S, Ciliberto A, Polo S, Scita G, Di Fiore PP. Endocytosis and signaling: cell logistics shape the eukaryotic cell plan. Physiol Rev 2012; 92(1): 273-366.
[http://dx.doi.org/10.1152/physrev.00005.2011] [PMID: 22298658]

[92] Jean-Charles PY, Kaur S, Shenoy SK. G Protein–Coupled Receptor Signaling Through β-Arrestin–Dependent Mechanisms. J Cardiovasc Pharmacol 2017; 70(3): 142-58.
[http://dx.doi.org/10.1097/FJC.0000000000000482] [PMID: 28328745]

[93] Rysiewicz B, Błasiak E, Mystek P, Dziedzicka-Wasylewska M, Polit A. Beyond the G protein α subunit: investigating the functional impact of other components of the Gαi$_3$ heterotrimers. Cell Commun Signal 2023; 21(1): 279.
[http://dx.doi.org/10.1186/s12964-023-01307-w] [PMID: 37817242]

[94] Bock A, Irannejad R, Scott JD. cAMP signaling: a remarkably regional affair. Trends in biochemical sciences 2024; 49(4): 305-17.
[https://doi.org/10.1016/j.tibs.2024.01.004]

[95] Newton AC, Bootman MD, Scott JD. Second Messengers. Cold Spring Harb Perspect Biol 2016; 8(8)a005926
[http://dx.doi.org/10.1101/cshperspect.a005926] [PMID: 27481708]

[96] Sassone-Corsi P. The cyclic AMP pathway. Cold Spring Harb Perspect Biol 2012; 4(12): a011148-8.
[http://dx.doi.org/10.1101/cshperspect.a011148] [PMID: 23209152]

[97] Hutchings CJ. Mini-review: antibody therapeutics targeting G protein-coupled receptors and ion channels. Antib Ther 2020; 3(4): 257-64.
[http://dx.doi.org/10.1093/abt/tbaa023] [PMID: 33912796]

[98] McCudden CR, Hains MD, Kimple RJ, Siderovski DP, Willard FS. G-protein signaling: back to the future. Cell Mol Life Sci 2005; 62(5): 551-77.

[http://dx.doi.org/10.1007/s00018-004-4462-3] [PMID: 15747061]

[99] Zhang H, Liu Y, Liu J, *et al.* cAMP-PKA/EPAC signaling and cancer: the interplay in tumor microenvironment. J Hematol Oncol 2024; 17(1): 5.
[http://dx.doi.org/10.1186/s13045-024-01524-x] [PMID: 38233872]

[100] Williamson JR, Cooper RH, Joseph SK, Thomas AP. Inositol trisphosphate and diacylglycerol as intracellular second messengers in liver. Am J Physiol Cell Physiol 1985; 248(3): C203-16.
[http://dx.doi.org/10.1152/ajpcell.1985.248.3.C203] [PMID: 2579567]

[101] Falzone ME, MacKinnon R. *Gβγ* activates *PIP2* hydrolysis by recruiting and orienting *PLCβ* on the membrane surface. Proc Natl Acad Sci USA 2023; 120(20)e2301121120.
[http://dx.doi.org/10.1073/pnas.2301121120] [PMID: 37172014]

[102] Catterall WA. Voltage-gated calcium channels. Cold Spring Harb Perspect Biol 2011; 3(8): a003947-7.
[http://dx.doi.org/10.1101/cshperspect.a003947] [PMID: 21746798]

[103] Eisner DA, Caldwell JL, Kistamás K, Trafford AW. Calcium and Excitation-Contraction Coupling in the Heart. Circ Res 2017; 121(2): 181-95.
[http://dx.doi.org/10.1161/CIRCRESAHA.117.310230] [PMID: 28684623]

[104] Park J, Selvam B, Sanematsu K, Shigemura N, Shukla D, Procko E. Structural architecture of a dimeric class C GPCR based on co-trafficking of sweet taste receptor subunits. J Biol Chem 2019; 294(13): 4759-74.
[http://dx.doi.org/10.1074/jbc.RA118.006173] [PMID: 30723160]

[105] Kumar A, Plückthun A. *In vivo* assembly and large-scale purification of a GPCR - Gα fusion with Gβγ, and characterization of the active complex. PLoS One 2019; 14(1)e0210131.
[http://dx.doi.org/10.1371/journal.pone.0210131] [PMID: 30620756]

[106] Verbakel SK, van Huet RAC, Boon CJF, *et al.* Non-syndromic retinitis pigmentosa. Prog Retin Eye Res 2018; 66: 157-86.
[http://dx.doi.org/10.1016/j.preteyeres.2018.03.005] [PMID: 29597005]

[107] Zhen F, Zou T, Wang T, Zhou Y, Dong S, Zhang H. Rhodopsin-associated retinal dystrophy: Disease mechanisms and therapeutic strategies. Front Neurosci 2023; 171132179.
[http://dx.doi.org/10.3389/fnins.2023.1132179] [PMID: 37077319]

[108] Sack DA, Sack RB, Nair GB, Siddique AK. Cholera. Lancet 2004; 363(9404): 223-33.
[http://dx.doi.org/10.1016/S0140-6736(03)15328-7] [PMID: 14738797]

[109] Seger R, Krebs EG. The MAPK signaling cascade. FASEB J 1995; 9(9): 726-35.
[http://dx.doi.org/10.1096/fasebj.9.9.7601337] [PMID: 7601337]

[110] Pandey P, Verma M, Siddiqui S, *et al.* Modulation of Mitogen-Activated Protein Kinase (MAPK) Signaling Pathway in Gastrointestinal Cancers by Phytochemicals. Pharm Res 2025 Nov 20.
[doi: 10.1007/s11095-025-03977-2]

[111] Zhang W, Liu HT. MAPK signal pathways in the regulation of cell proliferation in mammalian cells. Cell Res 2002; 12(1): 9-18.
[http://dx.doi.org/10.1038/sj.cr.7290105] [PMID: 11942415]

[112] Tamemoto H, Kadowaki T, Tobe K, *et al.* Biphasic activation of two mitogen-activated protein kinases during the cell cycle in mammalian cells. J Biol Chem 1992; 267(28): 20293-7.
[http://dx.doi.org/10.1016/S0021-9258(19)88700-8] [PMID: 1400347]

[113] Widmann C, Gibson S, Jarpe MB, Johnson GL. Mitogen-activated protein kinase: conservation of a three-kinase module from yeast to human. Physiol Rev 1999; 79(1): 143-80.
[http://dx.doi.org/10.1152/physrev.1999.79.1.143] [PMID: 9922370]

[114] Santarpia L, Lippman SM, El-Naggar AK. Targeting the MAPK–RAS–RAF signaling pathway in cancer therapy. Expert Opin Ther Targets 2012; 16(1): 103-19.

[http://dx.doi.org/10.1517/14728222.2011.645805] [PMID: 22239440]

[115] Cargnello M, Roux PP. Activation and function of the MAPKs and their substrates, the MAPK-activated protein kinases. Microbiology and molecular biology reviews: MMBR 2011; 75(1): 50-83.
[https://doi.org/10.1128/MMBR.00031-10]

[116] Cristina M, Petersen M, Mundy J. Mitogen-activated protein kinase signaling in plants. Annu Rev Plant Biol 2010; 61(1): 621-49.
[http://dx.doi.org/10.1146/annurev-arplant-042809-112252] [PMID: 20441529]

[117] Peti W, Page, R. Molecular basis of MAP kinase regulation. Protein science: a publication of the Protein Society 2013; 22(12): 1698-710.
[https://doi.org/10.1002/pro.2374]

[118] Stronach B, Lennox AL, Garlena RA. Domain specificity of MAP3K family members, MLK and Tak1, for JNK signaling in Drosophila. Genetics 2014; 197(2): 497-513.
[https://doi.org/10.1534/genetics.113.160937]

[119] Crews CM, Alessandrini A, Erikson RL. The Primary Structure of MEK, a Protein Kinase that Phosphorylates the ERK Gene Product. Science (1979). 1992 Oct 16;258(5081):478-80.

[120] Craig EA, Stevens MV, Vaillancourt RR, Camenisch TD. MAP3Ks as central regulators of cell fate during development. Dev Dyn 2008; 237(11): 3102-14.
[http://dx.doi.org/10.1002/dvdy.21750] [PMID: 18855897]

[121] Moodie SA, Willumsen BM, Weber MJ, Wolfman A. Complexes of Ras□GTP with Raf-1 and Mitogen-Activated Protein Kinase Kinase. Science (1979). 1993 Jun 11;260(5114):1658-61.

[122] Min X, Akella R, He H, *et al.* The structure of the MAP2K MEK6 reveals an autoinhibitory dimer. Structure 2009; 17(1): 96-104.
[http://dx.doi.org/10.1016/j.str.2008.11.007] [PMID: 19141286]

[123] Meister M, Tomasovic A, Banning A, Tikkanen R. Mitogen-Activated Protein (MAP) Kinase Scaffolding Proteins: A Recount. Int J Mol Sci 2013; 14(3): 4854-84.
[http://dx.doi.org/10.3390/ijms14034854] [PMID: 23455463]

[124] Pumiglia KM, Decker SJ. Cell cycle arrest mediated by the MEK/mitogen-activated protein kinase pathway. Proceedings of the National Academy of Sciences of the United States of America 1997; 94(2): 448-52.
[https://doi.org/10.1073/pnas.94.2.448]

[125] Roskoski R Jr. ERK1/2 MAP kinases: Structure, function, and regulation. Pharmacol Res 2012; 66(2): 105-43.
[http://dx.doi.org/10.1016/j.phrs.2012.04.005] [PMID: 22569528]

[126] Roskoski R Jr. MEK1/2 dual-specificity protein kinases: Structure and regulation. Biochem Biophys Res Commun 2012; 417(1): 5-10.
[http://dx.doi.org/10.1016/j.bbrc.2011.11.145] [PMID: 22177953]

[127] Courcelles M, Frémin C, Voisin L, Lemieux S, Meloche S, Thibault P. Phosphoproteome dynamics reveal novel ERK1/2 MAP kinase substrates with broad spectrum of functions. Molecular systems biology 2013; 9: 669.
https://doi.org/10.1038/msb.2013.25

[128] Fujioka A, Terai K, Itoh RE, *et al.* Dynamics of the Ras/ERK MAPK cascade as monitored by fluorescent probes. J Biol Chem 2006; 281(13): 8917-26.
[http://dx.doi.org/10.1074/jbc.M509344200] [PMID: 16418172]

[129] Boulton TG, Nye SH, Robbins DJ, *et al.* ERKs: A family of protein-serine/threonine kinases that are activated and tyrosine phosphorylated in response to insulin and NGF. Cell 1991; 65(4): 663-75.
[http://dx.doi.org/10.1016/0092-8674(91)90098-J] [PMID: 2032290]

[130] Chen J, Ye C, Wan C, *et al.* The Roles of c-Jun N-Terminal Kinase (JNK) in Infectious Diseases. Int J Mol Sci 2021; 22(17): 9640.
[http://dx.doi.org/10.3390/ijms22179640] [PMID: 34502556]

[131] Davis RJ. 1. Signal transduction by the c-Jun N-terminal kinase.Cellular Responses to Stress. Princeton University Press 1999; pp. 1-12.
[http://dx.doi.org/10.1515/9781400865048.1]

[132] Bogoyevitch MA, Kobe B. Uses for JNK: the many and varied substrates of the c-Jun N-terminal kinases. Microbiol Mol Biol Rev 2006; 70(4): 1061-95.
[http://dx.doi.org/10.1128/MMBR.00025-06] [PMID: 17158707]

[133] Davis RJ. 1. Signal transduction by the c-Jun N-terminal kinase.Cellular Responses to Stress. Princeton University Press 1999; pp. 1-12.
[http://dx.doi.org/10.1515/9781400865048.1]

[134] King LE, Hohorst L, García-Sáez AJ. Expanding roles of BCL-2 proteins in apoptosis execution and beyond. Journal of cell science 2023; 136(22): jcs260790.
[https://doi.org/10.1242/jcs.260790]

[135] Bogoyevitch MA, Ngoei KRW, Zhao TT, Yeap YYC, Ng DCH. c-Jun N-terminal kinase (JNK) signalling: Recent advances and challenges. Biochimica et Biophysica Acta (BBA) -. Proteins and Proteomics 2010; 1804(3): 463-75.
[http://dx.doi.org/10.1016/j.bbapap.2009.11.002]

[136] Zarubin T, Han J. Activation and signaling of the p38 MAP kinase pathway. Cell Res 2005; 15(1): 11-8.
[http://dx.doi.org/10.1038/sj.cr.7290257] [PMID: 15686620]

[137] New L, Han J. The p38 MAP kinase pathway and its biological function. Trends Cardiovasc Med 1998; 8(5): 220-8.
[http://dx.doi.org/10.1016/S1050-1738(98)00012-7] [PMID: 14987568]

[138] Wang J, Liu Y, Guo Y, *et al.* Function and inhibition of P38 MAP kinase signaling: Targeting multiple inflammation diseases. Biochemical pharmacology 2024; 220: 115973.
[https://doi.org/10.1016/j.bcp.2023.115973]

[139] Cuadrado A, Nebreda AR. Mechanisms and functions of p38 MAPK signalling. Biochem J 2010; 429(3): 403-17.
[http://dx.doi.org/10.1042/BJ20100323] [PMID: 20626350]

[140] Kohno M, Pouyssegur J. Targeting the ERK signaling pathway in cancer therapy. Ann Med 2006; 38(3): 200-11.
[http://dx.doi.org/10.1080/07853890600551037] [PMID: 16720434]

[141] Bubici C, Papa S. JNK signalling in cancer: in need of new, smarter therapeutic targets. Br J Pharmacol 2014; 171(1): 24-37.
[http://dx.doi.org/10.1111/bph.12432] [PMID: 24117156]

[142] Bradham C, McClay DR. p38 MAPK in development and cancer. Cell Cycle 2006; 5(8): 824-8.
[http://dx.doi.org/10.4161/cc.5.8.2685] [PMID: 16627995]

[143] Huang G, Shi LZ, Chi H. Regulation of JNK and p38 MAPK in the immune system: Signal integration, propagation and termination. Cytokine 2009; 48(3): 161-9.
[http://dx.doi.org/10.1016/j.cyto.2009.08.002] [PMID: 19740675]

[144] Hammouda M, Ford A, Liu Y, Zhang J. The JNK Signaling Pathway in Inflammatory Skin Disorders and Cancer. Cells 2020; 9(4): 857.
[http://dx.doi.org/10.3390/cells9040857] [PMID: 32252279]

[145] Yong HY, Koh MS, Moon A. The p38 MAPK inhibitors for the treatment of inflammatory diseases and cancer. Expert Opin Investig Drugs 2009; 18(12): 1893-905.
[http://dx.doi.org/10.1517/13543780903321490] [PMID: 19852565]

[146] Keshri PK, Singh SP. Unraveling the AKT/ERK cascade and its role in Parkinson disease. Arch Toxicol 2024; 98(10): 3169-90.
[http://dx.doi.org/10.1007/s00204-024-03829-9] [PMID: 39136731]

[147] Tiziana Borsello , Gianluigi Forloni . JNK signalling: a possible target to prevent neurodegeneration. Curr Pharm Des 2007; 13(18): 1875-86.
[http://dx.doi.org/10.2174/138161207780858384] [PMID: 17584114]

[148] Munoz L, Ammit AJ. Targeting p38 MAPK pathway for the treatment of Alzheimer's disease. Neuropharmacology 2010; 58(3): 561-8.
[http://dx.doi.org/10.1016/j.neuropharm.2009.11.010] [PMID: 19951717]

[149] Zhou L, Xu N, Sun Y, Liu XM. Targeted biopharmaceuticals for cancer treatment. Cancer Lett 2014; 352(2): 145-51.
[http://dx.doi.org/10.1016/j.canlet.2014.06.020] [PMID: 25016064]

[150] Jain R, Watson U, Vasudevan L, Saini DK. ERK Activation Pathways Downstream of GPCRs. > Int Rev Cell Mol Biol 2018: 338: 79-109.

[151] Liebmann C, Böhmer F. Signal transduction pathways of G protein-coupled receptors and their cross-talk with receptor tyrosine kinases: lessons from bradykinin signaling. Curr Med Chem 2000; 7(9): 911-43.
[http://dx.doi.org/10.2174/0929867003374589] [PMID: 10911023]

[152] Cheng N, Trejo J. An siRNA library screen identifies CYLD and USP34 as deubiquitinases that regulate GPCR-p38 MAPK signaling and distinct inflammatory responses. J Biol Chem 2023; 299(12)105370.
[http://dx.doi.org/10.1016/j.jbc.2023.105370] [PMID: 37865315]

[153] Dashti MR, Gorbanzadeh F, Jafari-Gharabaghlou D, Farhoudi Sefidan Jadid M, Zarghami N. G Protein-Coupled Receptor 75 (GPR75) As a Novel Molecule for Targeted Therapy of Cancer and Metabolic Syndrome. Asian Pac J Cancer Prev 2023; 24(5): 1817-25.
[http://dx.doi.org/10.31557/APJCP.2023.24.5.1817] [PMID: 37247305]

[154] Bonizzi G, Karin M. The two NF-κB activation pathways and their role in innate and adaptive immunity. Trends Immunol 2004; 25(6): 280-8.
[http://dx.doi.org/10.1016/j.it.2004.03.008] [PMID: 15145317]

[155] Pascale JV, Park EJ, Adebesin AM, Falck JR, Schwartzman ML, Garcia V. Uncovering the signalling, structure and function of the 20-HETE-GPR75 pairing: Identifying the chemokine CCL5 as a negative regulator of GPR75. Br J Pharmacol 2021; 178(18): 3813-28.
[http://dx.doi.org/10.1111/bph.15525] [PMID: 33974269]

[156] Fan F, Roman RJ. GPR75 Identified as the First 20-HETE Receptor. Circ Res 2017; 120(11): 1696-8.
[http://dx.doi.org/10.1161/CIRCRESAHA.117.311022] [PMID: 28546348]

[157] Dedoni S, Campbell LA, Harvey BK, Avdoshina V, Mocchetti I. The orphan G□protein□coupled receptor 75 signaling is activated by the chemokine CCL 5. J Neurochem 2018; 146(5): 526-39.
[http://dx.doi.org/10.1111/jnc.14463] [PMID: 29772059]

[158] Ignatov A, Robert J, Gregory-Evans C, Schaller HC. RANTES stimulates Ca^{2+} mobilization and inositol trisphosphate (IP_3) formation in cells transfected with G protein□coupled receptor 75. Br J Pharmacol 2006; 149(5): 490-7.
[http://dx.doi.org/10.1038/sj.bjp.0706909] [PMID: 17001303]

[159] Kahsai AW, Shah KS, Shim PJ, *et al.* Signal transduction at GPCRs: Allosteric activation of the ERK MAPK by β-arrestin. Proc Natl Acad Sci USA 2023; 120(43)e2303794120.
[http://dx.doi.org/10.1073/pnas.2303794120] [PMID: 37844230]

[160] Tang CY, Wang H, Zhang Y, *et al.* GPR125 positively regulates osteoclastogenesis potentially through AKT-NF-κB and MAPK signaling pathways. Int J Biol Sci 2022; 18(6): 2392-405.
[http://dx.doi.org/10.7150/ijbs.70620] [PMID: 35414778]

[161] Milanesi R, Coccetti P, Tripodi F. The Regulatory Role of Key Metabolites in the Control of Cell Signaling. Biomolecules 2020; 10(6): 862.
[http://dx.doi.org/10.3390/biom10060862] [PMID: 32516886]

[162] Lemmon MA, Freed DM, Schlessinger J, Kiyatkin A. The Dark Side of Cell Signaling: Positive Roles for Negative Regulators. Cell 2016; 164(6): 1172-84.
[http://dx.doi.org/10.1016/j.cell.2016.02.047] [PMID: 26967284]

[163] Kim WK, Lee Y, Jang SJ, Hyeon C. Kinetic Model for the Desensitization of G Protein-Coupled Receptor. J Phys Chem Lett 2024; 15(23): 6137-45.
[http://dx.doi.org/10.1021/acs.jpclett.4c00967] [PMID: 38832827]

[164] Rajagopal S, Shenoy SK. GPCR desensitization: Acute and prolonged phases. Cell Signal 2018; 41: 9-16.
[http://dx.doi.org/10.1016/j.cellsig.2017.01.024] [PMID: 28137506]

[165] Fonódi M, Nagy L, Boratkó A. Role of Protein Phosphatases in Tumor Angiogenesis: Assessing PP1, PP2A, PP2B and PTPs Activity. Int J Mol Sci 2024; 25(13): 6868.
[http://dx.doi.org/10.3390/ijms25136868] [PMID: 38999976]

[166] Sanyal SK, Rajasheker G, Kishor PBK, *et al.* Role of Protein Phosphatases in Signalling, Potassium Transport, and Abiotic Stress Responses. Protein Phosphatases and Stress Management in Plants. Cham: Springer International Publishing 2020; pp. 203-32.
[http://dx.doi.org/10.1007/978-3-030-48733-1_11]

[167] Das T, Song EJ, Kim EE. The Multifaceted Roles of USP15 in Signal Transduction. Int J Mol Sci 2021; 22(9): 4728.
[http://dx.doi.org/10.3390/ijms22094728] [PMID: 33946990]

[168] Kaur KK, Allahbadia GN. The mechanistic modes of Targeting immunometabolism in cancer: An innovative strategy: A narrative review. GSC Advanced Research and Reviews. 2024 Jul 30;20(1):005–24.

[169] Ali Ahmad H, Seemab K, Wahab F, Imran Khan M. Signaling Pathways in Drug Development [Internet]. Pharmaceutical Science. IntechOpen; 2024. Available from: http://dx.doi.org/10.5772/intechopen.114041

[170] della Sala F, Tilly DP, Webb SJ. Approaches Towards Synthetic Signal Transduction in Phospholipid Bilayers.New Trends in Macromolecular and Supramolecular Chemistry for Biological Applications. Cham: Springer International Publishing 2021; pp. 1-24.
[http://dx.doi.org/10.1007/978-3-030-57456-7_1]

[171] Stefanov BA, Fussenegger M. Biomarker-driven feedback control of synthetic biology systems for next-generation personalized medicine. Front Bioeng Biotechnol 2022; 10986210.
[http://dx.doi.org/10.3389/fbioe.2022.986210] [PMID: 36225597]

CHAPTER 13

Cell Death: Mechanisms and Mysteries beyond Apoptosis

Shambhavi Jha[1], Rohan Vyas[1], S. Manvi[1], Vasanth Kanth T.L.[1], Keerthivasu Ramasamy[1], K.N. Aruljothi[1] and **Ramya Lakshmi Rajendran[2,3,4,*]**

[1] *Department of Genetic Engineering, SRM Institute of Science and Technology, Kattankulathur, Chengalpattu, Tamil Nadu, India*

[2] *Department of Nuclear Medicine, School of Medicine, Kyungpook National University, Daegu, Republic of Korea*

[3] *Cardiovascular Research Institute, Kyungpook National University, Daegu, Republic of Korea*

[4] *BK21 FOUR KNU Convergence Educational Program of Biomedical Sciences for Creative Future Talents, School of Medicine, Kyungpook National University, Daegu, Republic of Korea*

Abstract: This book chapter on cell death explores the mechanisms, significance, and implications of various forms of cellular demise in health and disease. It delves into classic pathways like apoptosis, which ensures cellular homeostasis, immunity, and necrosis, traditionally viewed as accidental cell death but now recognized for its role in inflammation and tissue damage. Emerging forms of cell death, such as autophagy, pyroptosis, ferroptosis, and NETosis, highlight the complexity of cellular life and death decisions. The chapter underscores how each form of death is tightly regulated by specific signaling pathways and proteins, contributing to tissue development, immune responses, and the progression of diseases like cancer and neurodegenerative disorders. A particular focus is placed on the molecular crosstalk between these pathways and their potential as therapeutic targets. The historical context of cell death research—from identifying apoptosis to modern-day advancements in targeted therapies and gene-editing technologies—provides a comprehensive view of how our understanding of cellular death has evolved.

Keywords: Apoptosis, Autophagy, Cancer, Disorders, Ferroptosis, Immune responses, Necrosis, NETosis, Neurodegenerative, Pyroptosis, Signalling pathways, Therapeutic targets.

---

[*] **Corresponding author Ramya Lakshmi Rajendran:** Department of Nuclear Medicine, School of Medicine, Kyungpook National University, Daegu, Republic of Korea; E-mail: ramyag@knu.ac.kr

K.N. Aruljothi, Prakash Gangadaran, Krishnan Anand, Satish Ramalingam, K. Kumaran & Kruthika Prakash (Ed.)
All rights reserved-© 2026 Bentham Science Publishers

INTRODUCTION

In the intricate tapestry of life, cell death stands as a cornerstone, shaping the very essence of biological existence. It is a process as ancient as life itself, a fundamental mechanism intricately woven into the fabric of development, homeostasis, and disease. Cell death is not a mere cessation of cellular activity; instead, it is a meticulously regulated, multifaceted series of events that culminate in the demise of a living entity. Beyond its apparent simplicity lies a world of complexity, where cellular components intricately communicate and orchestrate a symphony of self-destruction. The study of cell death has transcended the realms of mere biological inquiry; it has become a frontier where life sciences, genetics, and medicine converge. At its core, cell death represents nature's ultimate balancing act. It is the balance of nature, which calls for everything that is ever born to meet its end, its death. Every human, every microbe has a certain life span, after which death is inevitable. This concept, at the most cellular level, is more frequent and much more complex. There are countless cells in our body dying every day, and countless still that are newly generated. It is a process of profound importance, eliminating cells that have outlived their usefulness, have suffered irreparable damage, or have succumbed to genetic aberrations. This process is indispensable in sculpting and developing organisms and ensuring the proper formation of tissues, organs, and intricate structures [1]. Additionally, cell death acts as a guardian, protecting the organism from potential threats. It is the body's natural response to infections, injuries, and internal dysregulation that underlie various diseases, preventing the spread of damage to neighbouring cells and tissues. Yet, the significance of cell death extends far beyond its role as a biological safeguard. It is a process deeply entwined with the very essence of evolution. The ability to control cell death has played a pivotal role in the emergence of complex life forms, allowing for the development of intricate biological systems. In essence, the study of cell death is a study of life itself—a journey into the heart of existence, exploring the delicate balance between creation and destruction, growth and decay [2, 3].

In this chapter, we embark on a comprehensive exploration of cell death. We delve into the mechanisms that govern this intricate process, uncovering the molecular intricacies that define its various forms. From the classic pathways of apoptosis and necrosis to the emerging realms of autophagy, pyroptosis, and ferroptosis, we unravel the mysteries that underlie each type of cellular demise. Moreover, we examine the physiological significance of cell death, its pivotal role in development, tissue maintenance, and immunity, as well as its implications in diseases ranging from cancer to neurodegenerative disorders. As we venture deeper into the labyrinth of cell death, we also cast our gaze towards the future. Recent advancements in the field have opened new avenues of research, promising innovative therapies and treatments. From targeted therapies tailored to specific cell death pathways to the revolutionary potentials of gene-editing technologies, the landscape of cell death research is constantly evolving [4].

The Historical Tapestry of Cell Death Research

The annals of cell death research are rich with the endeavours of pioneering scientists whose meticulous observations and ingenious experiments have illuminated the enigmatic pathways of cellular demise. Our journey through time takes us back to the 19th century, when rudimentary microscopes provided the first glimpses into the microscopic world. Eminent scientists like Rudolf Virchow and Carl Vogt observed cellular structures and noted instances where cells seemed to disintegrate, laying the foundation for the intriguing field of cell death studies. In the early 20th century, the term "apoptosis" made its debut in scientific discourse, thanks to the work of Walther Flemming. This term, embodying the Greek notion of a "falling off" or "dropping," encapsulated the elegant yet complex process of programmed cell death. Flemming's observations sparked curiosity, fuelling a wave of research that sought to unravel the mysteries of apoptosis [1]. Amidst these explorations, the mid-20th century heralded significant breakthroughs. Pioneering electron microscopy techniques provided unprecedented resolution, enabling scientists to delve deeper into cellular structures and observe the orchestrated disassembly of dying cells. It was during this period that the morphological hallmarks of apoptosis—cell shrinkage, membrane blebbing, chromatin condensation, and formation of apoptotic bodies—were meticulously documented, offering a profound understanding of this fundamental process (Table **1**) [5].

In the latter half of the 20th century, the scientific community witnessed the emergence of groundbreaking discoveries. A key milestone came in the form of the identification of caspases, the enzymatic executioners orchestrating apoptosis. The elucidation of these proteases' role in cleaving vital cellular substrates offered a molecular perspective, transforming apoptosis from a mere morphological phenomenon into a finely regulated biochemical cascade. As the 21st century dawned, the exploration of cell death expanded beyond apoptosis. Necrosis, once deemed a chaotic and uncontrolled form of cell death, was redefined. It became evident that necrosis, too, had regulatory elements, especially in the context of ischemic injury and inflammation. Concurrently, autophagy, a cellular recycling mechanism, garnered increasing attention. Its role in maintaining cellular homeostasis, particularly during periods of stress and starvation, unfolded as researchers decoded the intricate machinery of autophagosome formation and cargo recognition. The modern era of cell death research is marked by a profound transformation catalysed by advances in molecular biology, genetics, and high-throughput technologies. Scientists now delve into the genetic signatures associated with various forms of cell death, unravelling intricate signalling pathways and unveiling novel regulators. This deeper understanding has paved the way for innovative therapeutic interventions, ranging from targeted cancer therapies to approaches aimed at modulating cell death in neurodegenerative diseases [1, 4].

Table 1. Timeline of discoveries.

Year	Discovery	Discoverer(s)
1842	Identification of cell death during development	Karl Vogt
1858	Coining the term "necrosis"	Rudolf Virchow
1965	Identification of apoptosis as a distinct process	John Kerr, Andrew Wyllie, and Alastair Currie
1972	Discovery of programmed cell death in tadpoles	Sydney Brenner
1974	Electron microscopy studies on apoptosis	John F. R. Kerr and Andrew H. Wyllie
1986	Cloning of the first apoptosis-related gene	Robert Horvitz
1991	Discovery of the Bcl-2 protein	Stanley Korsmeyer
1993	Identification of caspases as key mediators	Xiaodong Wang and Junying Yuan
2000	Discovery of PARP-1's role in parthanatos	Valina L. Dawson and Ted M. Dawson
2003	Discovery of the ferroptosis cell death pathway	Brent R. Stockwell
2017	Identification of the NETosis cell death pathway	Arturo Zychlinsky and team

Exploring Diverse Cell Death Pathways

As scientific techniques advanced, so did our understanding of cell death. It became evident that apoptosis was not the sole player in the cellular demise orchestra. Necrosis, once considered a chaotic and unregulated form of cell death, was revealed to have its nuances, often triggered by external factors such as infections or toxins [6]. Autophagy, the cell's self-cannibalization process, emerged as a crucial mechanism for recycling cellular components and maintaining cellular health [7]. In recent years, newer players have joined the stage, including pyroptosis, an inflammatory form of programmed cell death, and ferroptosis, characterized by iron-dependent lipid peroxidation. Let us first understand the difference between programmed and non-programmed cell death mechanisms (Fig. 1) [8].

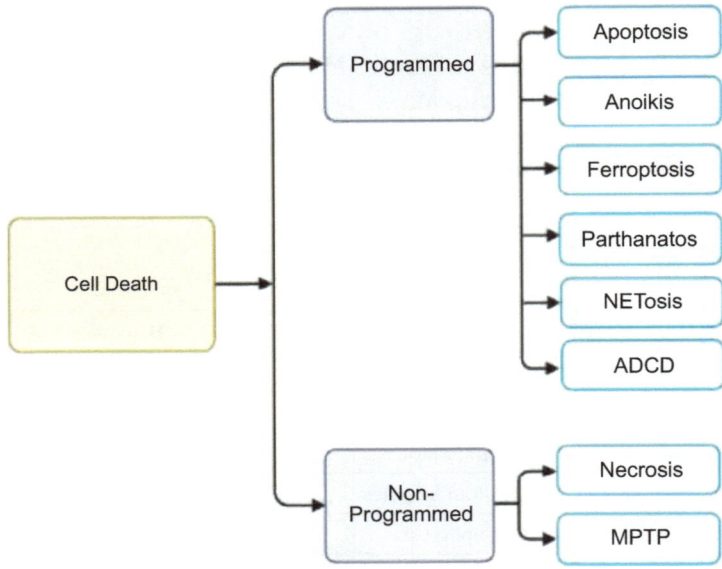

Fig. (1). Different cell death mechanisms and their various types.

PROGRAMMED CELL DEATH

Programmed cell death, a fundamental biological process, encompasses several distinct types, each finely orchestrated and crucial for various physiological functions in multicellular organisms. Apoptosis, the most well-known form, involves a series of tightly regulated steps, including cell shrinkage, membrane blebbing, and DNA fragmentation. Apoptosis plays a vital role in normal development, immune responses, and tissue homeostasis, ensuring the elimination of damaged or unnecessary cells without triggering inflammation. Autophagy, another programmed cell death mechanism, serves as nature's recycling process [6]. During autophagy, cells engulf and degrade their components within autophagosomes, promoting cellular renewal and survival during nutrient scarcity or stress [9, 10].

Pyroptosis, on the other hand, represents a form of programmed cell death that is closely linked to the immune response. It is characterized by the activation of inflammasomes, leading to the cleavage of Gasdermin D (GSDMD) and the subsequent release of pro-inflammatory cytokines [11]. Pyroptosis acts as a defence mechanism against microbial infections, enabling the elimination of invading pathogens while enhancing the immune response [12]. Necroptosis, a regulated necrotic cell death pathway, is initiated under conditions where apoptosis is blocked, often in response to viral infections. It involves the activation of Receptor-Interacting Protein Kinases (RIPKs) and Mixed-Lineage Kinase domain-like protein (MLKL), resulting in cellular swelling, membrane rupture, and inflammation [13]. Many more forms of programmed cell death are continuously being discovered with advancing research. These distinct forms of programmed cell death highlight the remarkable complexity of cellular processes, underscoring the importance of their tight regulation in maintaining tissue integrity and ensuring the proper functioning of the immune system. We will be further discussing the types of programmed cell death and how they occur, along with recent progress in the understanding of their operation [9, 14].

Delving into Apoptosis

Apoptosis, a Greek term meaning "falling off" or "dropping," is a highly regulated form of programmed cell death essential for the development, homeostasis, and survival of multicellular organisms [15]. This intricate process is a natural mechanism to eliminate unwanted or damaged cells without causing inflammation or harm to neighbouring tissues. For example, sometimes, cells need to retire, especially when they are old, damaged, or no longer needed. Apoptosis is like a planned retirement for these cells. Instead of causing chaos and harm, as in some other forms of cell death, apoptosis is like a graceful farewell. When a cell undergoes apoptosis, it follows a set of instructions, just like retiring employees might follow a company protocol. These instructions ensure that the cell carefully and tidily dismantles itself. Think of it as packing up belongings, shutting down the operations, and leaving a clean workspace behind. The cell's components are recycled or safely disposed of, ensuring that there's no mess or disruption to the surrounding "neighbourhood" of cells. It is a natural and essential process, ensuring the body stays healthy and functional. It's like a well-orchestrated departure, where cells gracefully exit the stage, allowing new and healthy cells to take their place, keeping the body running smoothly, much like a city evolving and renewing itself [1, 16].

Apoptosis is meticulously orchestrated by a complex interplay of signalling pathways and molecular events, ensuring the removal of cells during various physiological processes, including embryonic development, immune responses, and tissue remodelling. Apoptosis can be triggered by internal or external signals [17, 18]. External signals may include growth factor withdrawal, DNA damage, or the binding of specific ligands to death receptors on the cell membrane [7]. Internal signals can arise from cellular stress, metabolic imbalances, or the presence of damaged DNA. In response to these signals, a cascade of events unfolds, leading to the activation of caspases, the central executioners of apoptosis. Caspases are a family of protease enzymes that cleave specific target proteins, leading to the characteristic changes observed in apoptotic cells [19].

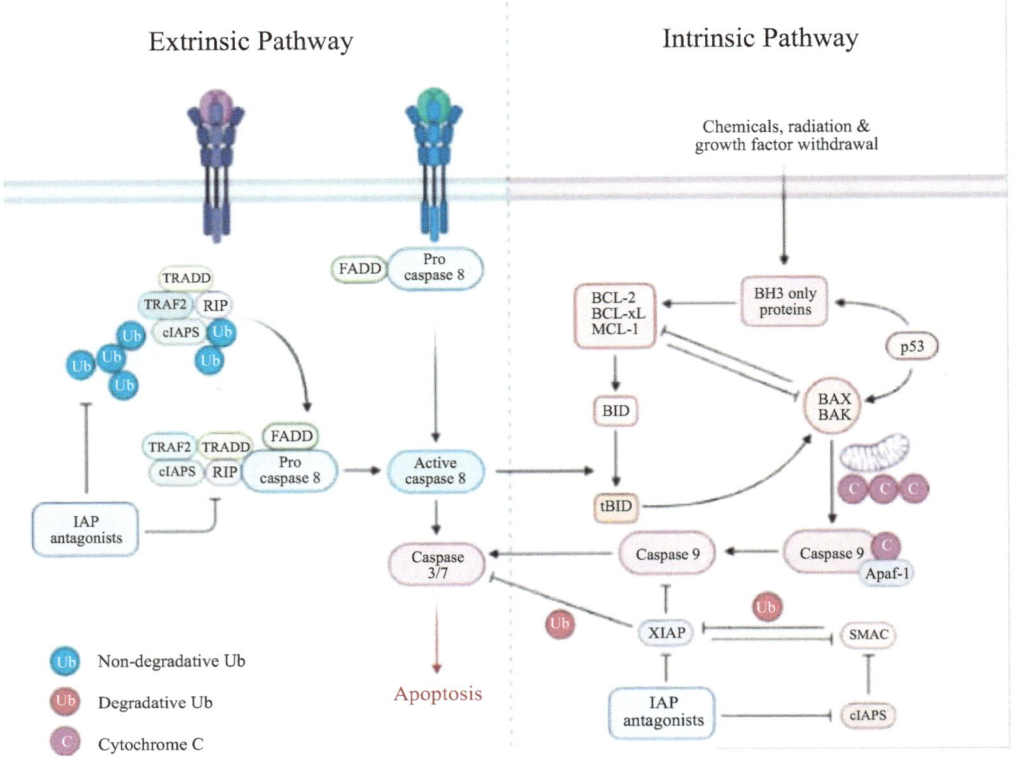

Fig. (2). Extrinsic and intrinsic pathways of apoptosis.

Extrinsic and Intrinsic Pathways in Apoptosis: Dual Routes to Cellular Demise

Apoptosis, or programmed cell death, can be initiated through two main pathways: the extrinsic pathway and the intrinsic pathway. Both pathways converge at a point where caspases, a family of protease enzymes, are activated, leading to the systematic dismantling of the cell (Fig. **2**).

Extrinsic Pathway

In the complex world of cell communication, the extrinsic pathway acts as a sentinel, guarding the body against potentially harmful cells. Imagine this pathway as a sophisticated security system. When specific external signals, often carried by proteins known as ligands, encounter their corresponding receptors like Fas or TNFR1 on the cell surface, it's akin to a security breach alarm being triggered. These receptors, part of the TNF receptor superfamily, are strategically placed on the surface of cells, especially immune cells. When ligands such as Fas ligand (FasL) or tumour necrosis factor-alpha (TNF-alpha) bind to these receptors, it's like receiving a distress call, indicating that something is wrong with the cell. This binding event initiates the formation of the Death-Inducing Signalling Complex (DISC). Think of the DISC as a specialized control centre that assembles in response to these distress signals [20].

Within the DISC, a critical enzyme called procaspase-8 is activated. Procaspase-8 is like a key officer in the security force. Once activated, it takes charge and starts the caspase cascade directly. Caspases are enzymes that act like molecular scissors, capable of slicing and activating various cellular proteins. In this context, caspase-8 cleaves and activates downstream caspases, including caspase-3, which serves as a major executioner in the apoptosis process. This amplification of the apoptotic signal is vital. It ensures that a single distress signal triggers a robust and efficient response, ultimately leading to the controlled dismantling of the cell. This pathway is not only essential for eliminating virus-infected or mutated cells but also plays a crucial role in maintaining tissue homeostasis, especially in the immune system. In essence, the extrinsic pathway acts as a first responder, swiftly identifying and neutralizing cells that could pose a threat to the body, thus contributing significantly to our overall health and well-being [1].

Intrinsic Pathway

The cell is a highly organized component in our bodies. Here, the mitochondria act like power plants, generating energy to fuel various cellular activities. However, there are internal challenges that come in the form of DNA damage, oxidative stress, or nutrient deprivation – stresses that can severely disrupt the cell's functions and well-being. When these internal signals of distress arise, it's as if the cell has encountered a blackout or a severe shortage of resources. In response to these signals, the mitochondria, the power plants of the cell, release pro-apoptotic proteins like cytochrome c into the cytoplasm. Cytochrome c, like a messenger, travels from the mitochondria to the cytoplasm, carrying news of cellular distress. Upon arrival in the cytoplasm, cytochrome c joins forces with a key cellular coordinator called apoptotic protease-activating factor 1 (Apaf-1) and procaspase-9, forming a formidable trio known as the apoptosome. Think of the apoptosome as a crisis management team, gathering to assess the situation and make critical decisions for the cell's fate [15].

Under the guidance of the apoptosome, procaspase-9, akin to an executive decision-maker, is activated. This activation sets off a cascading reaction, initiating the activation of downstream effector caspases, especially caspase-3. Caspase-3, the chief executioner, carries out precise cellular disassembly, leading to apoptosis. In essence, this internal pathway ensures that, when the cell faces insurmountable challenges from within, it can make the difficult decision to gracefully exit the stage, preserving the overall harmony of the biological system. Crucially, the intrinsic pathway is under the vigilant watch of the Bcl-2 protein family, the guardians of cellular fate. This family includes both pro-apoptotic (*e.g.*, Bax, Bak) and anti-apoptotic (*e.g.*, Bcl-2, Bcl-xL) proteins. The balance between these proteins acts as a molecular scale, tilting the cell's fate toward life or death. If pro-apoptotic signals dominate and overwhelm the anti-apoptotic defences, a critical event known as mitochondrial outer membrane permeabilization (MOMP) occurs. This breach leads to the release of cytochrome c, further amplifying the apoptotic signal and ensuring that the cell bows out gracefully when its internal challenges become insurmountable [21].

Convergence and Caspase Activation

Imagine the activation of caspase-3 as the climax of a meticulously choreographed pathway where caspase-3 plays the lead role, the star of the show, and the master regulator of cellular demise. When both the extrinsic and intrinsic pathways converge at caspase-3, it's akin to the entire pathway seamlessly uniting in perfect harmony. Activated caspase-3 is not just any enzyme; it's the executioner, the molecular scissors that precisely cut specific cellular substrates, initiating a cascade of events that define apoptosis. These substrates include structural and regulatory proteins within the cell. As caspase-3 snips through these proteins, the cell undergoes remarkable transformations.

Cell Shrinkage

Caspase-3 cleaves proteins involved in maintaining the cell's shape and structure. As these proteins are dismantled, the cell undergoes shrinkage. This reduction in size is a characteristic feature of apoptotic cells and ensures the dying cell is neatly compacted, ready for removal.

Membrane Blebbing

The cell's outer membrane, which separates it from its surroundings, starts to form characteristic protrusions known as blebs. Caspase-3 orchestrates these changes by cleaving membrane-associated proteins. Blebbing allows the cell to break into smaller, membrane-bound fragments, making it easier for neighbouring cells to engulf and digest these apoptotic remnants without causing inflammation.

DNA Fragmentation

One of the most crucial actions of activated caspase-3 is its ability to fragment the cell's DNA. This cleavage occurs at specific sites within the genome, leading to the characteristic ladder-like pattern seen during DNA gel electrophoresis. DNA fragmentation ensures that the cell's genetic material is neatly packaged into smaller fragments, preventing any potential harm to surrounding cells and tissues. These orchestrated changes facilitate the efficient removal of the dying cell without triggering inflammation or damage to neighbouring cells [22].

Moreover, the beauty of apoptosis lies in its adaptability. The intricate interplay between the extrinsic and intrinsic pathways allows cells to respond flexibly to a myriad of signals, ensuring the precise regulation of cell death in diverse physiological and pathological contexts [23]. Understanding these pathways not only reveals the intricacies of fundamental biological processes but also unveils potential therapeutic targets for diseases where apoptosis regulation is compromised, such as cancer or neurodegenerative disorders.

Apart from the majorly functioning Apoptosis pathways, there are some other pathways that also take place side by side in this function of cell death. These include: ER Stress Pathway [24, 25], Granzyme-Perforin Pathway (in Cytotoxic T lymphocytes and Natural Killer Cells), Mitochondrial Outer Membrane Permeabilization (MOMP) [24].

These pathways are not mutually exclusive and can crosstalk, allowing the cell to integrate various signals and make decisions regarding its fate. The balance between pro-apoptotic and anti-apoptotic molecules, along with the integration of signals from these pathways, determines whether a cell undergoes apoptosis, highlighting the complexity and precision of this fundamental biological process [26].

Delving into the intricate realm of apoptosis, a fascinating symphony of cellular events comes to light, orchestrated with precision and purpose. As cells navigate their life cycle, they encounter an array of signals, both internal and external, that can trigger programmed cell death. This process, crucial for tissue homeostasis and immune regulation, involves a series of meticulously choreographed morphological and biochemical changes. Apoptotic cells undergo orchestrated transformations, including cell shrinkage, membrane blebbing, chromatin condensation, and ultimately, fragmentation of the nucleus. The fragmentation results in the formation of apoptotic bodies, containing neatly packaged cellular components to prevent the release of potentially harmful contents into the extracellular space. Remarkably, the plasma membrane's integrity is maintained until the late stages of apoptosis, preventing leakage of cellular contents and subsequent inflammation [1, 27].

Central to this intricate dance are the delicate balances struck between pro-apoptotic and anti-apoptotic proteins. Pro-apoptotic proteins like Bax and Bak promote cell death by permeabilizing the mitochondrial outer membrane, releasing cytochrome c into the cytoplasm. This released cytochrome c acts as a trigger, assembling the apoptosome, a complex that activates caspase-9. On the contrary, anti-apoptotic proteins such as Bcl-2 and Bcl-xL counteract these actions, preserving cell survival by inhibiting the release of cytochrome c and caspase activation. This balance is pivotal for cell survival and is often disrupted in diseases like cancer, autoimmune disorders, and neurodegenerative diseases. In the context of disease, dysregulation of apoptosis plays a central role. In cancer, cells adept at evading apoptosis gain uncontrolled growth advantages, contributing to tumorigenesis and resistance to therapy. Understanding these molecular mechanisms has spurred the development of targeted therapies. For instance, BH3 mimetics, designed to restore the apoptotic balance, have shown promise in promoting cancer cell destruction. Moreover, the removal of apoptotic cells is an artful process orchestrated by phagocytes such as macrophages and specialized immune cells like dendritic cells. These cells recognize specific "eat-me" signals on apoptotic cells, leading to their swift engulfment and subsequent degradation. Importantly, this phagocytosis occurs silently, without inciting an immune response, thereby maintaining tissue homeostasis and immune tolerance [22].

In essence, apoptosis represents a finely tuned biological phenomenon, essential for maintaining the delicate equilibrium of cellular life. Its dysregulation underpins the pathogenesis of numerous diseases. Through ongoing research and targeted interventions, scientists strive to unravel its complexities, paving the way for innovative therapeutic strategies and a deeper understanding of the intricate dance between life and death at the cellular level.

Recent Advancements

In recent years, research in the field of apoptosis has witnessed remarkable advancements, driven by innovative technologies and a deeper understanding of cellular signalling pathways. One significant area of progress involves unravelling the intricate network of regulators that modulate apoptosis. Researchers have identified various proteins, both pro-apoptotic and anti-apoptotic, that fine-tune the apoptotic process. This knowledge has paved the way for the development of targeted therapies, especially in the treatment of cancers [28]. Furthermore, recent studies have shed light on the role of apoptosis in tissue regeneration and development. Apoptosis is not merely a process of cellular destruction; it also serves as a mechanism for sculpting tissues and organs during embryonic development. Understanding the delicate balance between cell survival and apoptosis has provided insights into how complex multicellular organisms form and grow. Additionally, advancements in single-cell analysis techniques have allowed scientists to study apoptosis at a level of granularity previously unattainable. Single-cell RNA sequencing and imaging technologies have provided unprecedented insights into the heterogeneity of apoptotic responses within tissues, offering a more nuanced understanding of how individual cells within a population respond to apoptotic signals. Moreover, the emerging field of immunotherapy, particularly in cancer treatment, has capitalized on apoptosis. Immunotherapeutic strategies, such as immune checkpoint inhibitors, harness the body's immune system to recognize and eliminate cancer cells, often by inducing apoptosis. This approach has shown promising results in various cancer types, revolutionizing the landscape of cancer therapy [29].

Anoikis: Cell Death due to Detachment

Anoikis stands as a sentinel, guarding the delicate balance between cell adhesion and detachment in multicellular organisms. Rooted in Greek, where "an" means without and "oikos" translates to home, anoikis is a form of programmed cell death triggered by the loss of cell adhesion to the extracellular matrix (ECM) or neighbouring cells. This process is not merely a cellular response; it's a fundamental biological mechanism ensuring tissue integrity, preventing the dissemination of rogue cells, and maintaining the proper architecture of organs and tissues [30]. In simpler terms, imagine that your cells are like the people residing in various neighbourhoods. Each cell has its own specific place to stay, ensuring that the city functions smoothly. Now, think of the extracellular matrix, a network of proteins and molecules, as the city's solid ground. Just like buildings need strong foundations to stand tall, cells rely on this matrix to stay in their designated spots. Anoikis is like the city planner ensuring that each resident stays in their assigned home. It's the mechanism that ensures cells are anchored to their specific place in the body. When cells lose their grip due to various reasons, like injury or disease, anoikis steps in; it's akin to a safety protocol, preventing cells from wandering off and causing trouble elsewhere. At a deeper level, cells have special receptors called integrins that act like locks fitting into the city's architectural structure. When these locks don't connect properly, it sends a signal that something's wrong. Anoikis then triggers a process inside the cell, leading to a controlled self-destruction. It's like cells have an innate sense of belonging, and when that's disrupted, anoikis ensures they die and make room for new, healthy cells [5].

At its core, anoikis is orchestrated by a complex interplay of molecular events. Integrins, transmembrane receptors connecting the cell cytoskeleton to the ECM, play a pivotal role. These receptors, which mediate adhesion, also act as signalling hubs [31]. When cells lose their anchorage, integrin-mediated signalling is disrupted, leading to the activation of pro-apoptotic proteins. Bim, a member of the Bcl-2 family, is a key mediator of anoikis. It gets activated when integrin signalling wanes, triggering the intrinsic apoptotic pathway. Mitochondria, often referred to as the powerhouse of the cell, release cytochrome c into the cytoplasm, initiating the caspase cascade that ultimately dismantles the cell [32].

Several key mechanisms underlie anoikis: Cell Adhesion Molecules (CAMs), Activation of Caspases [33], Bcl-2 Family Proteins [32], p53 Activation, and Mitochondrial Dysfunction.

Extrinsic and Intrinsic Pathways in Anoikis

Anoikis, the programmed cell death induced by the loss of cell adhesion, integrates both extrinsic and intrinsic apoptotic pathways, ensuring that detached cells are efficiently eliminated from tissues. These pathways serve as molecular mechanisms to safeguard tissue integrity and prevent the spread of potentially harmful cells to distant sites (Fig. **3**) [34].

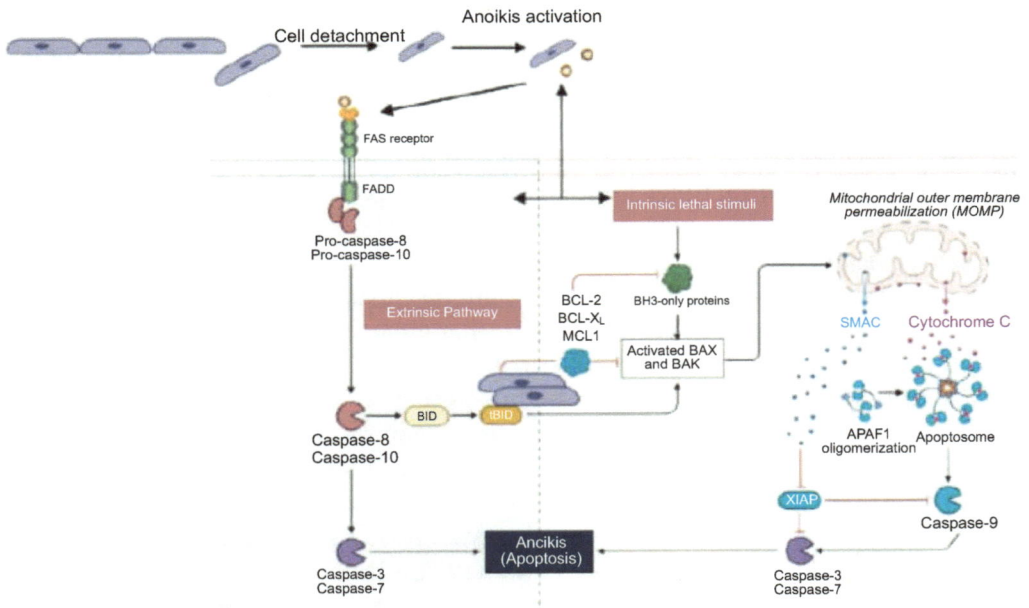

Fig. (3). Extrinsic and intrinsic pathways of anoikis.

The extrinsic pathway, often dubbed the death receptor pathway, serves as a vigilant sentinel in the intricate world of cell biology, particularly in processes like anoikis. Picture cells as vigilant inhabitants of a bustling community, each with a designated place to stay, firmly anchored to their surroundings. In this scenario, death receptors like Fas and TNFR1 act as vigilant gatekeepers. These receptors are like security sensors on the cell's surface, constantly scanning for signals. When ligands, specific molecules in the cellular environment, bind to these receptors, it's akin to a distress signal [35]. Think of it as a call for help when something goes wrong. Upon ligand binding, a complex event unfolds. This binding initiates the formation of what scientists call the Death-Inducing Signalling Complex (DISC). Imagine this complex as a central command centre in a crisis. At the DISC, procaspase-8, a dormant enzyme, is activated. This activation is akin to empowering a key officer in the security forces, giving them the authority to take charge. Activated caspase-8 then becomes the commander, cleaving and activating downstream effector caspases, particularly caspase-3. Caspase-3 is a chief executor, akin to molecular scissors, precisely cutting through cellular proteins [36].

Now, consider a scenario where cells lose their anchorage due to factors like injury or disease, disrupting their connection to the extracellular matrix. This disruption prevents the engagement of survival signals through death receptors. It's similar to cutting the power supply to the city's security system. In the absence of these survival signals, the balance tips in favour of apoptosis. Cells, realizing they are detached and vulnerable, activate the extrinsic pathway as an early responder. This swift response ensures that cells promptly undergo apoptosis upon loss of adhesion, preventing them from wandering aimlessly and causing harm to the body. In the grand scheme of cellular life, this extrinsic pathway acts as a first responder, swiftly identifying cells that have lost their foothold and orchestrating their graceful exit, maintaining the overall harmony of the biological system [36].

The intrinsic pathway, often known as the mitochondrial pathway, offers a profound insight into the intricate world of cell survival and demise, especially in phenomena like anoikis. Picture cells as vigilant inhabitants of a bustling city, each firmly anchored to its surroundings, ensuring the smooth functioning of the community. In this scenario, the mitochondria within the cell act as power plants, providing the necessary energy for cellular activities. Now, consider a situation where cells face internal challenges, such as stress or loss of attachment. In these moments, a crucial event occurs: the weakening of integrin-mediated survival signals. Integrins are like cellular anchors, keeping cells firmly attached to the extracellular matrix, the cellular neighbourhood. When these signals diminish, it's akin to a sudden power outage in the city, disrupting the energy flow [37].

Here's where a key player, Bim, steps into the spotlight. Bim is a pro-apoptotic member of the Bcl-2 protein family, and in the realm of anoikis, it plays a pivotal role. Upon cell detachment, Bim becomes activated and upregulated, responding to the distress signal of lost adhesion. Activated Bim interacts with anti-apoptotic proteins within the cell, disrupting their harmony. This interaction destabilizes the mitochondrial outer membrane, akin to breaching a city's security perimeter. The breach, in this case, leads to the permeabilization of the mitochondrial outer membrane, allowing the escape of cytochrome c and other pro-apoptotic factors into the cell's cytoplasm. Imagine these factors as messengers carrying news of the city's unrest. Their presence triggers the formation of the apoptosome complex, a molecular meeting ground. The apoptosome, akin to a crisis management team, activates procaspase-9, an essential decision-maker. Once activated, procaspase-9 sets off a cascading reaction, initiating the caspase cascade, the cellular equivalent of emergency protocols. This cascade, led by caspase-9, leads to the activation of downstream effector caspases, ultimately resulting in the execution of apoptosis. It's like a carefully orchestrated dismantling of the cityscape, ensuring that cells, faced with insurmountable challenges, gracefully exit the stage without causing chaos or harm to the surrounding cellular community. In the realm of anoikis, the loss of integrin-mediated signals acts as a trigger, unleashing the pro-apoptotic forces within the cell, setting in motion the intrinsic pathway, and culminating in the cell's peaceful demise [38].

Both the extrinsic and intrinsic pathways in anoikis converge at the activation of effector caspases, particularly caspase-3. The integration of these pathways ensures a robust and coordinated response to cell detachment, safeguarding tissue integrity and preventing the dissemination of detached cells [39].

Several other key signalling pathways are involved in regulating anoikis, ensuring that cells respond appropriately to the loss of anchorage and preventing detached cells from surviving in inappropriate locations. These include:

Integrin-Mediated Signalling Pathway

Integrins, transmembrane receptors that bind cells to the extracellular matrix (ECM), are central players in anoikis. In adherent cells, integrins transmit survival signals, promoting cell attachment and inhibiting apoptosis. However, when cells lose their attachment, integrin-mediated signalling is disrupted. This disruption leads to the activation of pro-apoptotic proteins, including Bim, a member of the Bcl-2 family. Bim promotes Mitochondrial Outer Membrane Permeabilization (MOMP), releasing cytochrome c into the cytoplasm, a hallmark of the intrinsic apoptotic pathway.

PI3K/Akt Pathway

The phosphoinositide 3-kinase (PI3K)/Akt pathway is a central regulator of cell survival and apoptosis. In adherent cells, activated integrins stimulate this pathway, promoting cell survival. Upon detachment, PI3K/Akt signalling is inhibited, diminishing the anti-apoptotic signals. This pathway regulates the expression and activity of Bcl-2 family proteins and caspases, influencing the commitment of cells to anoikis [40].

MAPK/ERK Pathway

The Mitogen-Activated Protein Kinase/Extracellular Signal-Regulated Kinase (MAPK/ERK) pathway, involved in cell proliferation and survival, is also modulated during anoikis. Disruption of integrin-mediated signalling leads to decreased ERK activation, contributing to anoikis induction. ERK can influence the expression of pro-survival and pro-apoptotic proteins, affecting the cellular response to detachment [41].

Focal Adhesion Kinase (FAK) Signalling

Focal Adhesion Kinase (FAK) is a non-receptor tyrosine kinase that plays a crucial role in integrin signalling. In adherent cells, FAK is activated, promoting cell survival and preventing apoptosis. Upon loss of adhesion, FAK activity decreases, allowing pro-apoptotic signals to prevail. FAK inhibition sensitizes cells to anoikis, highlighting its significance in regulating cell survival in anchorage-dependent conditions [41, 42].

SRC Kinase Pathway

SRC kinases are involved in regulating cell adhesion and migration. They influence integrin-mediated signalling and are activated in response to cell detachment. Src activation can lead to the phosphorylation of pro-apoptotic proteins, promoting anoikis. Src kinase inhibitors have been studied for their potential to sensitize anoikis-resistant cancer cells to detachment-induced cell death [34, 41].

Understanding the intricate interplay between these signalling pathways is crucial for deciphering the mechanisms underlying anoikis and its dysregulation in diseases such as cancer. Targeting these pathways presents potential therapeutic opportunities to sensitize cancer cells to anoikis, thereby preventing metastatic spread and improving treatment outcomes.

Anoikis and Cancer

In the context of cancer, anoikis holds immense significance. Cancer cells are adept at resisting anoikis to gain a survival advantage. When these cells detach from the primary tumour site, they evade anoikis, becoming Circulating Tumour Cells (CTCs). These CTCs can spread to distant tissues and organs, leading to the formation of metastases [43]. Understanding the mechanisms behind anoikis resistance has become crucial in cancer research [44]. Oncogenes, like Ras and Src, and tumour suppressor genes, including p53, can influence anoikis sensitivity, highlighting the complex interplay between genetic alterations and cellular survival. Several signalling pathways contribute to anoikis resistance in cancer cells. The phosphoinositide 3-kinase (PI3K)/Akt pathway, which promotes cell survival and inhibits apoptosis, is often upregulated in anoikis-resistant cells. Additionally, Epithelial-Mesenchymal Transition (EMT), a process where cells lose their epithelial characteristics and acquire a migratory, mesenchymal phenotype, is associated with anoikis resistance. EMT allows cells to detach from their primary site and evade anoikis, facilitating metastatic spread [45]. Recent advancements in cancer therapeutics have focused on targeting anoikis resistance. Small molecules and biological agents are being developed to sensitize cancer cells to anoikis, thus preventing metastatic dissemination. These agents often target components of the PI3K/Akt pathway, integrin signalling, or EMT-associated molecules. In preclinical studies, these strategies have shown promise, presenting potential avenues for therapeutic interventions [46].

Recent Advancements

Recent studies have delved into the molecular intricacies of anoikis and its significance in cancer biology. Advances in high-throughput screening technologies have facilitated the identification of novel regulators of anoikis, revealing potential targets for therapeutic intervention [47]. Researchers have also explored the role of microRNAs and long non-coding RNAs in modulating anoikis, shedding light on the intricate gene regulatory networks that govern this process [48]. Furthermore, advancements in 3D cell culture systems have allowed scientists to create more physiologically relevant models for studying anoikis. These 3D culture systems better mimic the *in vivo* microenvironment, providing valuable insights into the behaviour of cancer cells during detachment and metastasis [49]. Additionally, techniques such as single-cell RNA sequencing have enabled the characterization of heterogeneity within cancer cell populations, offering a deeper understanding of the molecular mechanisms driving anoikis resistance in specific subsets of cells [50].

Therapeutically, recent research efforts have explored the development of small molecules and biological agents targeting key components of the anoikis pathway. These compounds aim to sensitize cancer cells to anoikis, preventing their survival during detachment and inhibiting metastatic spread. Clinical trials and preclinical studies evaluating these novel agents are ongoing, holding promise for future cancer therapies [51].

Ferroptosis: Iron-Dependent Cell Death Mechanism

Unlike apoptosis or necrosis, ferroptosis represents a unique cell death pathway with significant implications in various physiological and pathological processes. Let's understand this in an easier way. Cells are like busy workers managing various tasks. Just as any city needs to maintain a delicate balance to function smoothly, your cells also need to maintain a balance of different substances, including lipids, which are essential for their structure and function. Now, think of ferroptosis as a kind of disruption, a sudden imbalance in this delicate system. In this analogy, ferroptosis is like a city under attack by a rogue element - in this case, a buildup of harmful molecules called reactive oxygen species (ROS). ROS are like aggressive vandals wreaking havoc within the city, causing damage to vital structures. Ordinarily, your cells have defence mechanisms, like antioxidants, to keep these vandals in check. However, in ferroptosis, the balance tips, and the antioxidants are overwhelmed [52, 53].

Here's where iron, an essential mineral in your body, becomes a key player. Iron is essential for many cellular functions, but in ferroptosis, it becomes a double-edged sword. It reacts with the excess ROS, creating a toxic environment within the cell. This toxic environment damages lipids, the building blocks of your cell's membranes, leading to their disintegration [54]. Imagine these lipids as the architectural framework of the city, holding everything together. When they break down, the city's structures collapse. In ferroptosis, the cell faces a sort of 'architectural crisis.' The once-sturdy buildings, representing your cells, start falling apart due to lipid damage. As this destruction continues, the cell undergoes a process of controlled self-destruction, akin to a city deciding to demolish a damaged building to prevent it from collapsing and causing harm to nearby structures. Scientists are studying ferroptosis to understand how to prevent this process, especially in diseases where it plays a harmful role, such as in certain types of cancer and neurodegenerative disorders. By understanding and managing ferroptosis, researchers are essentially working to restore peace and order in the cellular city, ensuring that its structures remain intact and functional (Fig. **4**) [55, 56].

In ferroptosis, iron accumulates in cells, promoting the production of lipid hydroperoxides from polyunsaturated fatty acids. These lipid hydroperoxides can propagate and amplify, damaging cell membranes and organelles. The glutathione-dependent lipid peroxidase, glutathione peroxidase 4 (GPX4), acts as a crucial defence mechanism against ferroptosis by reducing lipid hydroperoxides to non-toxic lipid alcohols. Inhibition or depletion of GPX4 disrupts this balance, leading to ferroptotic cell death [57]. Ferroptosis is implicated in various pathological conditions, including cancer, neurodegenerative diseases, and ischemia-reperfusion injury. In cancer, ferroptosis resistance contributes to tumour progression and therapy resistance. In neurodegenerative disorders, such as Parkinson's and Alzheimer's diseases, ferroptosis-induced neuronal death exacerbates the progression of these conditions [58]. Additionally, ferroptosis plays a role in ischemic injuries, where the accumulation of iron and lipid peroxides contributes to tissue damage [52, 55].

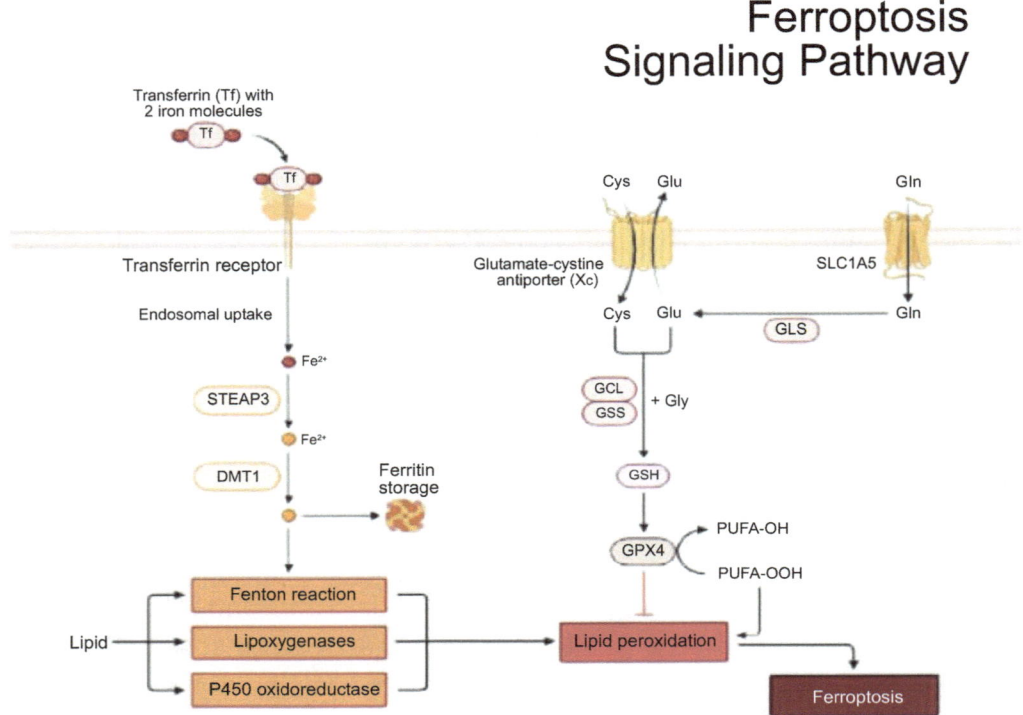

Fig. (4). Ferroptosis signalling pathway.

Some important characteristics of Ferroptosis include

- **Lipid Peroxidation:** The hallmark feature of ferroptosis is the accumulation of lipid peroxides within cell membranes, particularly in phospholipids containing polyunsaturated fatty acids (PUFAs). This peroxidation of lipids disrupts the integrity of cell membranes and impairs their function. At the core of ferroptosis is the accumulation of lipid peroxides, particularly phospholipid hydroperoxides. This process is driven by the presence of iron and reactive oxygen species (ROS), which initiate lipid peroxidation cascades in cellular membranes [55].

- **Mitochondrial Dysfunction:** Ferroptosis is associated with alterations in mitochondrial function, including reduced mitochondrial membrane potential and impaired mitochondrial respiration. Dysfunctional mitochondria contribute to the production of reactive oxygen species (ROS), further promoting lipid peroxidation [59, 60].
- **Iron Dependence:** Ferroptosis gets its name from its dependence on iron. The intracellular iron pool plays a crucial role in generating ROS through Fenton reactions, which drive lipid peroxidation. Therefore, excess iron can promote ferroptosis.
- **Inhibition by Lipid Peroxidation Inhibitors:** Ferroptosis can be inhibited by various compounds, including lipophilic antioxidants like ferrostatins and liproxstatins. These compounds counteract lipid peroxidation and protect cells from ferroptotic cell death [61].
- **Inhibition of Glutathione Peroxidase 4 (GPX4):** GPX4 is an enzyme that detoxifies lipid peroxides. In ferroptosis, GPX4 activity is inhibited, further promoting the accumulation of lipid peroxides and cell death [62].

The regulation of this condition may be either genetic, wherein several genes and signalling pathways are involved, especially the KEAP1-NRF2 pathway and p53, which are among the regulators that influence cellular responses to oxidative stress and lipid peroxidation [63]. Moreover, the enzymes involved in lipid metabolism, such as ACSL4 (acyl-CoA synthetase long-chain family member 4, play a crucial role in ferroptosis by promoting the incorporation of polyunsaturated fatty acids into cellular membranes. Finally, the cellular regulation of iron uptake, transport, and storage is tightly controlled to prevent ferroptosis. Proteins like ferritin help sequester iron and reduce its availability for Fenton reactions [64].

Unlike apoptosis, necroptosis, or other well-characterized forms of cell death, ferroptosis does not have a clearly defined extrinsic or intrinsic pathway akin to those seen in apoptosis. However, the process of ferroptosis involves intricate molecular interactions that can be broadly categorized into the following pathways:

Iron Metabolism and Iron Regulatory Proteins

Ferroptosis is closely linked to iron metabolism. Iron, a vital cellular component, can catalyse the production of Reactive Oxygen Species (ROS) through the Fenton reaction. Iron Regulatory Proteins (IRPs) And Iron-Responsive Element-Binding Proteins (IREBPs) control the expression of proteins involved in cellular iron homeostasis. Dysregulation of these proteins can lead to iron accumulation, promoting ferroptosis [65].

Lipid Metabolism and Lipid Peroxidation

Polyunsaturated fatty acids (PUFAs) are essential components of cellular membranes. During ferroptosis, PUFAs are susceptible to peroxidation, leading to the generation of lipid hydroperoxides. This process is catalysed by enzymes like lipoxygenases (LOX) and involves iron as a cofactor. Accumulation of lipid peroxides disrupts membrane integrity, leading to cell death [66, 67].

Glutathione and Glutathione Peroxidase 4 (GPX4)

Glutathione, a potent antioxidant, plays a crucial role in preventing ferroptosis. Glutathione peroxidase 4 (GPX4) is an enzyme that reduces lipid hydroperoxides to their corresponding alcohols, preventing lipid peroxidation and ferroptosis. Depletion of cellular glutathione or inhibition of GPX4 removes this protective barrier, leading to ferroptotic cell death [67, 68].

Nuclear Factor Erythroid 2-Related Factor 2 (NRF2) Pathway

The NRF2 pathway is a cellular defence mechanism against oxidative stress. NRF2 regulates the expression of Antioxidant Response Element (ARE)-containing genes, including enzymes involved in glutathione synthesis and detoxification. Activation of the NRF2 pathway can protect cells from ferroptosis by enhancing the cellular antioxidant capacity [68].

Ferroptosis Regulators

Several proteins have been identified as direct regulators of ferroptosis. For example, ferroptosis suppressor protein 1 (FSP1) acts as a cofactor for ubiquinone (CoQ10), which can reduce lipid peroxides and protect cells from ferroptosis. Additionally, heat shock protein beta-1 (HSPB1) has been implicated in ferroptosis regulation, although its precise mechanism is still under investigation [62, 69].

The major causes and triggers of Ferroptosis include:

1. **Depletion of Glutathione:** Glutathione is an important antioxidant that helps protect cells from oxidative stress. Depletion of intracellular glutathione levels can trigger ferroptosis, often due to the inhibition of the enzyme system that regenerates glutathione.
2. **Inhibition of Glutathione Peroxidase 4 (GPX4):** GPX4 is an enzyme that reduces lipid peroxides and prevents lipid peroxidation. Inhibition of GPX4 activity, either genetically or pharmacologically, can lead to ferroptosis.

3. **Excessive Iron Accumulation:** An increase in intracellular iron levels can drive ferroptosis by promoting lipid peroxidation. This can occur due to various factors, including altered iron metabolism.
4. **Oxidative Stress:** Oxidative stress from various sources, including the accumulation of ROS, can contribute to ferroptosis by initiating lipid peroxidation processes [62].

The oxidative damage and neuronal death associated with these conditions may involve ferroptosis. Inducing ferroptosis in cancer cells has emerged as a potential therapeutic strategy, where some chemotherapeutic agents and targeted therapies are designed to trigger ferroptosis in cancer cells, leading to tumour regression. In conditions such as stroke and myocardial infarction, where tissues undergo oxygen deprivation followed by reperfusion, ferroptosis has been proposed to contribute to cell death and tissue damage. It may also play a role in modulating immune responses during infection and inflammation and can influence the activation and death of immune cells in response to pathogens. Its iron-dependent nature and involvement in various diseases make it a subject of growing interest in cell biology and biomedical research, and a deeper understanding of ferroptosis may open new avenues for therapeutic interventions and provide insights into the complex interplay between oxidative stress, lipid metabolism, and cell death.

Recent Advancements

Recent research efforts have focused on unravelling the regulatory mechanisms and signalling pathways associated with ferroptosis. One of the key advancements involves the identification of small molecules and compounds that modulate ferroptosis sensitivity. Ferrostatins, liproxstatins, and other ferroptosis inhibitors have been discovered, showing promise as potential therapeutic agents in diseases where ferroptosis is implicated. Researchers have also explored the role of ferroptosis in the tumour microenvironment, uncovering its impact on immune responses and cancer therapy outcomes [70]. Furthermore, studies have elucidated the crosstalk between ferroptosis and other forms of cell death, such as apoptosis and autophagy. Understanding these interactions provides insights into the complex network of cellular processes governing cell fate decisions [71]. Additionally, advancements in imaging techniques, metabolomics, and genetic screenings have enabled scientists to dissect the molecular events leading to ferroptosis in a more precise and comprehensive manner [72].

The recognition of ferroptosis as a distinct cell death pathway has opened new avenues for therapeutic interventions. Researchers are exploring innovative strategies, including the development of ferroptosis-inducing agents [73] and combination therapies targeting ferroptosis resistance mechanisms [74]. Clinical trials and preclinical studies are underway to evaluate the efficacy and safety of these novel approaches.

Parthanatos: Death by PARP-1 Overactivation

Parthanatos, a unique form of regulated cell death, stands at the intersection of DNA damage, energy depletion, and the activation of poly (ADP-ribose) polymerase (PARP) enzymes. Parthanatos is not dependent on caspase-activated cell death, although it is dependent on the nuclear translocation of the mitochondrial-associated Apoptosis-Inducing Factor (AIF). AIF is, primarily, a mitochondrial protein, and when it enters the nucleus, it triggers parthanatos [75]. In cases of excessive DNA damage, overactivation of PARP-1 leads to over-PARylation in the nucleus, which induces mitochondrial AIF translocation to the nucleus and causes non-caspase-dependent cell death. Additionally, overactivation of PARP-1 by its activity leads to the consumption of both NAD+ and ATP, causing an energy collapse in the cell and leading to necrosis [76, 77].

Parthanatos is characterized by the overactivation of PARP enzymes, primarily PARP-1, in response to severe DNA damage. PARP-1 is involved in DNA repair processes, particularly in the base excision repair pathway. When DNA damage overwhelms the cell's repair capacity, PARP-1 is hyperactivated, leading to the excessive synthesis of poly(ADP-ribose) (PAR) polymers. These PAR polymers, in turn, lead to the translocation of apoptosis-inducing factor (AIF) from the mitochondria to the nucleus. Once in the nucleus, AIF induces chromatin condensation and large-scale DNA fragmentation, hallmark features of parthanatos. Some important mechanisms of Parthanatos include:

- **PARP Activation:** Parthanatos is initiated by the hyperactivation of PARP enzymes in response to DNA damage. PARP enzymes play a role in DNA repair, and their activation is a cellular response to repair single-strand DNA breaks. However, in conditions of excessive DNA damage, PARP-1 becomes overactivated.
- **Poly (ADP-ribose) (PAR) Polymer Synthesis:** Overactivated PARP-1 consumes cellular NAD+ and utilizes it to synthesize PAR polymers. These polymers serve as a signal for the recruitment of apoptosis-inducing factor (AIF) from the mitochondria to the nucleus [78].
- **AIF Translocation:** AIF, a mitochondrial protein, translocates from the mitochondria to the nucleus in response to PAR polymers. Once in the nucleus, AIF induces large-scale DNA fragmentation, chromatin condensation, and ultimately, cell death [79].

Parthanatos has been implicated in several pathological conditions. In neurodegenerative diseases, such as Parkinson's and Alzheimer's, parthanatos-mediated neuronal death contributes to the progression of these disorders [80]. Ischemia-reperfusion injury, where tissues experience damage upon restoration of blood supply, involves parthanatos as a significant component of cell death [81]. Additionally, cancer cells, particularly those with DNA repair deficiencies, are susceptible to parthanatos induction, making this pathway a potential target for cancer therapy [75, 82].

There is limited evidence to suggest that parthanatos, a form of regulated cell death, follows distinct intrinsic and extrinsic pathways akin to apoptosis. Some other important signalling pathways associated with parthanatos are mentioned below:

PARP Activation

Parthanatos begins with the activation of PARP enzymes, especially PARP-1, in response to severe DNA damage. PARP-1 is a nuclear enzyme that detects DNA strand breaks and facilitates DNA repair. However, under conditions of excessive DNA damage, PARP-1 is hyperactivated.

Poly(ADP-ribose) (PAR) Polymer Synthesis

Hyperactivated PARP-1 consumes cellular NAD+ and utilizes it to synthesize PAR polymers. These polymers serve as a signal for the translocation of apoptosis-inducing factor (AIF) from the mitochondria to the nucleus.

AIF Translocation and Nuclear Events

Once in the nucleus, AIF induces large-scale DNA fragmentation and chromatin condensation, leading to cell death. The exact mechanism through which AIF causes these nuclear events is still an area of active research.

Involvement of Other Proteins

Several other proteins, such as apoptosis-inducing factor mitochondria-associated 2 (AIFM2) and histone H1, have been implicated in the nuclear events of parthanatos. AIFM2 has been suggested to interact with AIF during parthanatos.

Mitochondrial Dysfunction

Parthanatos is associated with mitochondrial dysfunction, including the release of AIF from the mitochondria. Mitochondrial impairment and the generation of reactive oxygen species (ROS) are considered important factors in parthanatos induction.

Interactions with Apoptotic Pathways

Parthanatos has been suggested to interact with apoptosis in certain contexts. The crosstalk between these pathways is a topic of ongoing research, and the distinction between these cell death pathways is not always clear-cut [76].

Recent Advancements

Recent research in the field of parthanatos has unveiled novel insights into its regulatory mechanisms and potential therapeutic interventions. Scientists have identified small molecules and inhibitors that target PARP enzymes, preventing their hyperactivation and subsequent parthanatos induction. These compounds, known as PARP inhibitors, have shown promise in cancer therapy, especially in cancers with defects in DNA repair pathways, like BRCA-mutated tumours [83]. Moreover, studies have focused on the crosstalk between parthanatos and other cell death pathways, such as apoptosis and necroptosis. Understanding these interactions provides a comprehensive view of the cell's decision-making processes in response to various stresses [84]. Researchers have also explored the role of mitochondrial dynamics, oxidative stress, and mitochondrial DNA damage in parthanatos, unravelling the complexity of this cell death mechanism [85, 86].

The recognition of parthanatos as a distinct form of regulated cell death has opened avenues for therapeutic interventions. PARP inhibitors, in addition to their application in cancer therapy, are being explored for their potential in neurodegenerative diseases and ischemia-reperfusion injury. Clinical trials are underway to evaluate the efficacy and safety of these inhibitors, offering hope for patients suffering from conditions where parthanatos plays a detrimental role [87].

NETosis: The Unique Cell Death Pathway

NETosis is a type of cell death in neutrophils characterized by the release of extracellular traps (NETs) composed of DNA, histones, and antimicrobial proteins, which capture and kill invading pathogens. Discovered in 2004, this kind of cell death is triggered in response to various signals and stimuli, primarily as a defence mechanism by neutrophils to combat infections and pathogens. Neutrophils are the first line of defence against pathogens and have huge amounts of antimicrobial granules since they are professional phagocytes [88].

Mechanism of NETosis

The induction of NETosis is closely linked to the production of reactive oxygen species (ROS) by NADPH oxidase and mitochondria. NADPH oxidase activation is primarily driven by phosphorylation through protein kinase C (PKC) and other kinases like c-Raf, MEK, Akt, and ERK, which assemble the enzyme on the membrane to produce ROS. In conditions like chronic granulomatous disease (CGD), where NADPH oxidase is deficient, exogenous ROS can still promote NET formation. Different pathogens, such as Staphylococcus aureus, Helicobacter pylori, and Entamoeba histolytica, trigger NETosis *via* distinct signalling pathways, with the c-Raf-MEK-ERK cascade playing a significant role. Notably, NETosis is associated with the suppression of apoptosis, enhancing the neutrophils' antimicrobial function. Activation of NETosis can be triggered by signals from the G-protein-coupled fMLP receptor, which leads to the release of calcium (Ca^{2+}) from intracellular stores and activates PI3K. Both Ca^{2+} and PI3K are important for activating NADPH oxidase in neutrophils, which produces ROS. Additionally, the generation of mitochondrial ROS (mtROS) plays a key role in NETosis, particularly when triggered by Ca^{2+} ionophores like A23187 or ionomycin. Studies have shown that inhibiting mtROS or NADPH oxidase can block NETosis induced by these stimuli [89].

ADCC: The Immune Response Cell Death

In recent years, a fascinating aspect of cell death, known as "Antigen-Dependent Cellular cytotoxicity" (ADCC), has gained significant attention. ADCC is a specialized form of apoptosis in which dying cells actively engage the immune system by presenting antigens to immune cells, thereby contributing to immune responses and immunological memory. It is a mechanism of immune response cell death that plays a vital role in the immune system's defence against infected or abnormal cells, particularly in the context of antibody-based immune responses. ADCC involves the destruction of target cells by immune cells, primarily natural killer (NK) cells and certain types of white blood cells, in a process mediated by antibodies. In this section, we will explore ADCC in detail [90, 91].

ADCC shares many features with classical apoptosis, including DNA fragmentation, caspase activation, and membrane blebbing (Fig. **5**). However, ADCC is distinguished by specific mechanisms that allow dying cells to interact with the immune system:

- **Antibody Binding:** ADCC begins when antibodies, such as immunoglobulin G (IgG), bind to specific antigens present on the surface of target cells. These

antigens can be derived from pathogens (*e.g.*, viruses, bacteria) or abnormal host cells (*e.g.*, cancer cells).
- **Recognition by Effector Cells:** Once antibodies have bound to the target cells, immune effector cells, primarily NK cells and certain white blood cells (*e.g.*, macrophages, neutrophils), recognize the antibody-coated target cells through receptors on their surfaces. These receptors interact with the Fc (constant fragment) portion of the antibodies, leading to immune cell activation [90].

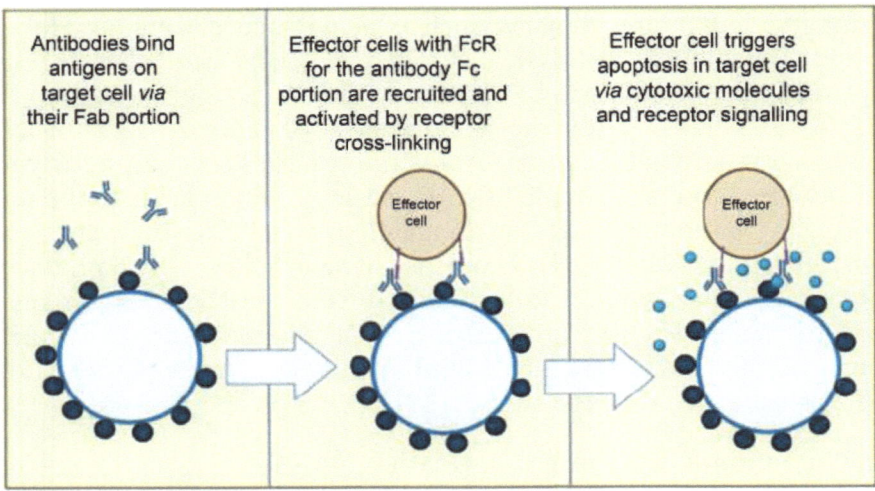

Fig. (5). Antigen-mediated cellular cytotoxicity.

Some important uses of ADCC include:

1. **Vaccination Strategies:** Understanding the mechanisms of ADCC has implications for vaccine development. Vaccines can be designed to induce immunogenic cell death, enhancing the immune response and the production of long-lasting immunity [92].
2. **Autoimmune Diseases:** Dysregulation of ADCC processes can contribute to autoimmune diseases, where the immune system mistakenly targets healthy cells. Understanding ADCC may help develop targeted therapies for autoimmune conditions.
3. **Cancer Immunotherapy:** Manipulating ADCC in the context of cancer can enhance the immune response against tumour cells. Certain cancer treatments aim to induce immunogenic cell death, promoting the clearance of cancerous cells.
4. **Transplantation and Organ Preservation:** Modulating ADCC may have implications for transplantation, as it can help reduce immune responses against transplanted organs and improve organ preservation techniques [93].

ADCC plays a crucial role in maintaining immune tolerance by facilitating the clearance of cells undergoing normal apoptosis. This helps prevent autoimmune responses against self-antigens. Dying cells in ADCC contribute to the presentation of antigens to T cells, leading to the activation of specific immune responses against pathogens or cancerous cells. It can contribute to the formation of immunological memory, as the immune system "learns" to recognize specific antigens presented by apoptotic cells. This memory aids in the rapid and robust immune response upon subsequent encounters with the same antigens. The understanding of ADCC has led to the development of therapeutic antibodies, such as monoclonal antibodies, for the treatment of various diseases. These antibodies can be designed to specifically target cancer cells or pathogenic cells and enhance ADCC as part of the therapeutic response. The dysregulation of ADCC mechanisms can have implications in autoimmune diseases, where the immune system may mistakenly target healthy cells. Therefore, understanding ADCC can provide insights into the underlying mechanisms of autoimmune disorders.

NON-PROGRAMMED CELL DEATH

Necrosis and its Implications for Cell Death

Necrosis is a complex process that can be triggered by a variety of factors, including physical injury, toxins, hypoxia (lack of oxygen), and infection (Fig. **6**). Unlike apoptosis, which is an orderly and programmed process, necrosis typically results from cellular stress or damage that overwhelms the cell's ability to maintain its structural and functional integrity [94]. The various types of Necrosis include:

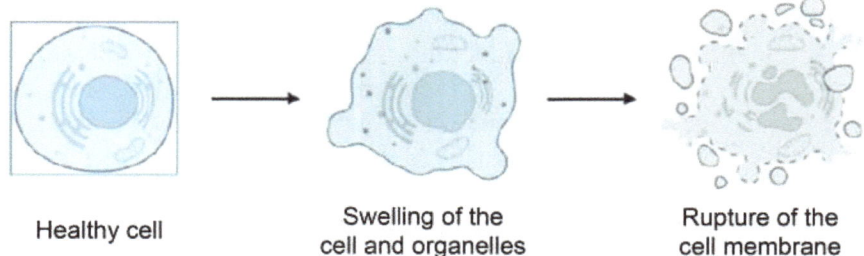

Fig. (6). A simple depiction of necrosis.

Coagulative Necrosis

This is the most common form of necrosis, characterized by the denaturation of cellular proteins and the preservation of tissue architecture. Coagulative necrosis is a specific type of necrosis characterized by a well-defined outline of necrotic tissue that retains the basic architecture and shape of the affected organ or tissue. This type of necrosis is commonly associated with ischemic injuries, where there is a lack of blood flow to a particular area, leading to oxygen and nutrient deprivation. Coagulative necrosis is often seen in various organs, including the heart, kidneys, and liver, and it typically manifests as a consequence of severe hypoxia or ischemia.

Some major characteristics of Coagulative Necrosis include:

Preservation of Tissue Structure

One of the hallmark features of coagulative necrosis is the preservation of the overall architecture of the affected tissue. The cellular outlines and basic tissue structure remain intact, at least initially.

Denaturation of Proteins

Coagulative necrosis is marked by the denaturation of cellular proteins due to the lack of oxygen and metabolic substrates. This results in the loss of enzymatic activity and renders the affected cells non-functional.

Nuclear Changes

The nuclei of cells undergoing coagulative necrosis typically undergo pyknosis (shrinkage), followed by karyorrhexis (fragmentation). This nuclear deterioration is a characteristic feature of necrotic cells.

It may be caused by Ischemic Injuries, Hypoxia, or Chemical and Thermal Injuries. It can impair the function of the affected organ or tissue poorly, for example, in the case of myocardial infarction (heart attack), coagulative necrosis of heart muscle cells can lead to reduced cardiac function. It may also impair the function of the organ and lead to reduced cardiac function. If not managed promptly, coagulative necrosis can lead to complications, including abscess formation or tissue loss. In some cases, coagulative necrosis can progress to liquefactive necrosis, especially if there is a secondary infection. Proper diagnosis and timely intervention are essential to minimize the impact of coagulative necrosis on affected organs and tissues [95].

Liquefactive Necrosis

Liquefactive necrosis is a distinct form of cell death characterized by the dissolution of dead tissue, resulting in the formation of a liquid, viscous mass. This type of necrosis occurs predominantly in soft tissues and is commonly associated with bacterial or fungal infections, as well as certain types of brain injuries. Some major characteristics of Liquefactive Necrosis include:

Liquid Formation

Liquefactive necrosis is named for its most prominent feature, which is the transformation of the necrotic tissue into a liquid substance. This liquid is often composed of dead cells, cellular debris, and inflammatory cells.

Rapid Tissue Destruction

Unlike coagulative necrosis, where the tissue structure is preserved, liquefactive necrosis leads to rapid tissue destruction and loss of architecture. The affected area appears as a pus-filled cavity or abscess.

Inflammatory Response

Liquefactive necrosis is associated with a robust inflammatory response, as the release of cellular contents and the presence of invading pathogens trigger the recruitment of immune cells to the site of injury.

It may be caused by bacterial, fungal, or parasitic infections wherein microorganisms release enzymes and toxins that not only kill host cells but also liquefy the surrounding tissue to facilitate their spread and survival. Another cause may include brain tissue injury, which is also called "liquefactive encephalomalacia." Liquefactive necrosis can result in the loss of functional tissue and can have severe consequences, depending on the affected organ. For example, brain abscesses can lead to neurological deficits, while abscesses in other organs can impair their function. It may also lead to a robust inflammatory response, with the influx of immune cells to combat infection. This inflammatory response can cause local swelling, pain, and tissue damage. Pus is the characteristic product of liquefactive necrosis in response to infection. It consists of dead white blood cells, cellular debris, and pathogens. Pus-filled abscesses may need to be drained surgically to remove the accumulated material. Timely diagnosis and appropriate treatment, such as antibiotics or surgical drainage of abscesses, are essential to manage the consequences of liquefactive necrosis and prevent further tissue damage [96].

Caseous Necrosis

Caseous necrosis is a distinctive form of cell death characterized by a cheese-like, friable, and granular appearance of the necrotic tissue. This type of necrosis is typically associated with certain types of chronic infections, particularly mycobacterial infections like tuberculosis, as well as fungal infections. In this section, we will explore the characteristics, causes, and implications of caseous necrosis. Some major characteristics of Caseous Necrosis include:

Cheese-Like Appearance

The term "caseous" is derived from the Latin word "caseous," meaning cheese. Caseous necrosis results in the formation of a necrotic tissue that has a crumbly, cheese-like texture and appearance.

Loss of Tissue Structure

Similar to liquefactive necrosis, caseous necrosis is characterized by the loss of tissue structure. The affected area does not retain the typical architecture of the organ or tissue.

Distinct Granulomas

In many cases, caseous necrosis is associated with the formation of granulomas—organized clusters of immune cells, particularly macrophages, that encase the necrotic material. These granulomas are a hallmark of certain chronic infections.

It may be caused by Mycobacteria, such as *Mycobacterium tuberculosis*. The immune system attempts to wall off the bacteria, leading to the formation of granulomas and the characteristic cheese-like necrotic tissue. Another causative agent includes fungi like *Histoplasma* and *Aspergillus* species. Caseous necrosis often signifies a chronic and persistent infection. The formation of granulomas and the cheesy necrotic material are part of the body's attempt to contain and isolate the infectious agent. This can also lead to systemic infections and complications as the infectious agents can escape from the granulomas and spread to other parts of the body. Identifying caseous necrosis in tissue samples can help healthcare professionals narrow down the potential causes of an infection. Chronic infections associated with caseous necrosis can be challenging to treat, as the necrotic material may not be easily accessible to antibiotics or immune responses. Prolonged and complex treatment regimens are often required [95].

Gangrenous Necrosis: Gangrenous necrosis is a type of cell death that occurs primarily in body tissues, particularly in extremities like the limbs, and is characterized by tissue decay and death. It is often caused by severe bacterial infections, lack of blood supply (ischemia), or a combination of both. In this section, we will explore the characteristics, causes, and implications of gangrenous necrosis. Some important characteristics of Gangrenous Necrosis include:

Tissue Discoloration

Gangrenous tissue typically exhibits a characteristic discoloration. There are three common types of gangrene, each with its own colour:

• **Dry Gangrene (black or brown)** - It is often seen in conditions where blood flow is compromised, but bacterial invasion is limited.

• **Wet Gangrene:** (swollen, discoloured, greenish-black, or grey) - It is associated with bacterial infection and characterized by tissue liquefaction due to the presence of bacteria and excess fluid accumulation.

• **Gas Gangrene:** Gas gangrene results from infection by anaerobic bacteria, which produce gas within the affected tissues. This leads to a crackling sound when the tissue is touched and a foul odour.

1. **Tissue Decomposition:** Gangrenous tissue undergoes extensive decomposition and liquefaction due to the action of enzymes produced by bacteria (in the case of wet and gas gangrene) or autolytic processes (in the case of dry gangrene).

2. **Loss of Function:** As gangrenous tissue deteriorates, it loses its function, which can lead to severe impairment of the affected body part, potentially necessitating amputation to prevent the spread of infection.

It may be caused by Ischemia when blood flow to a tissue is significantly compromised, resulting in arterial blockage, embolism, or severe frostbite, leading to dry gangrene. Another causative factor includes gangrenous necrosis accompanied by bacterial infections, particularly by anaerobic bacteria like Clostridium species. These infections can lead to wet or gas gangrene and contribute to rapid tissue decomposition. This condition can result in the complete loss of the affected tissue, which can be a limb in the case of severe gangrene. Amputation may be required to prevent the spread of infection and save the patient's life. Severe infections associated with gangrenous necrosis can lead to systemic effects, such as sepsis, which is a life-threatening condition characterized by a systemic inflammatory response to infection. Gangrenous necrosis requires immediate medical attention. Treatment typically involves antibiotics, surgical debridement (removal of dead tissue), and, in some cases, amputation. Timely intervention is crucial to prevent the spread of infection [95].

Another cell death pathway called Necroptosis behaves very similarly to necrosis, but it is very important to understand the difference between the two for a comprehensive understanding of the subject. Imagine your body as a bustling city, where cells are the citizens working together to maintain harmony. In this cityscape, necrosis can be likened to a sudden, chaotic event—like an unexpected earthquake that damages buildings and disrupts the city's peace. Necrosis occurs randomly, without any organized plan, and often leaves a mess behind, causing inflammation and distress among the city's inhabitants. On the other hand, necroptosis is akin to a well-coordinated emergency response. Picture it as a team of skilled professionals who step in when needed, working together with precision and purpose. While there might be similarities, such as buildings collapsing (representing cell swelling and membrane rupture), the difference lies in the approach. Necroptosis follows a carefully orchestrated plan, with specific teams and tools involved, ensuring that the impact on the city is controlled and limited. Unlike the chaos of necrosis, necroptosis is like a structured, organized effort to address a crisis, offering hope for targeted solutions in the face of cellular challenges, much like experts managing an emergency in a bustling city [97].

MECHANISMS OF NECROSIS

Several mechanisms underlie necrosis, often involving loss of cell membrane integrity and disruption of cellular homeostasis:

1. **Ion Imbalance and Membrane Permeabilization:** Necrosis often involves a sudden influx of calcium ions into the cell, disrupting cellular processes. This influx can lead to the activation of enzymes like proteases and lipases, causing the degradation of cellular structures. Increased calcium levels can also affect mitochondria, leading to mitochondrial dysfunction and energy depletion. The excessive ion influx can eventually compromise the integrity of the cell membrane, causing it to rupture [98].
2. **Mitochondrial Dysfunction:** Mitochondria, the powerhouse of the cell, play a crucial role in necrosis. Cellular stressors can impair mitochondrial function, leading to the production of Reactive Oxygen Species (ROS) and the opening of the Mitochondrial Permeability Transition Pores (MPTP). The opening of MPTP disrupts the mitochondrial membrane potential and further contributes to cellular damage.
3. **Inflammation and Cellular Breakdown:** Once the cell membrane ruptures, cellular contents, including enzymes and pro-inflammatory molecules, are released into the extracellular space. These molecules act as Damage-Associated Molecular Patterns (DAMPs), triggering an immune response and inflammation. Inflammation in necrosis is often a consequence of the spillage of cellular contents, attracting immune cells to clear the debris and initiate tissue repair.
4. **Enzymatic Digestion:** After membrane rupture, lysosomal enzymes are released into the cytoplasm, leading to autolysis – a process where the cell

digests itself. Lysosomal enzymes, such as proteases and nucleases, break down cellular components, contributing to the disintegration of the cell structure.

5. **Osmotic Imbalance:** Disruption of ion balance and membrane integrity can lead to an influx of water into the cell, causing cellular swelling (oncosis). The increased intracellular pressure, combined with weakened membrane integrity, eventually causes cell rupture and the release of cellular contents [95, 99].

It's important to note that while necrosis is traditionally viewed as unregulated and passive, emerging research suggests that certain forms of necrosis can involve specific molecular pathways, blurring the lines between regulated and unregulated cell death processes. These nuances in the mechanisms of necrosis continue to be a subject of active investigation in the field of cell biology (Table **2**).

Table 2. Differences between apoptosis, necrosis, and necroptosis.

Aspect	Apoptosis	Necrosis	Necroptosis
Definition	Controlled, programmed cell death	Uncontrolled, accidental cell death	Regulated, programmed cell death
Regulation	Regulated by intra- and extracellular signals	Not regulated; passive process	Regulated by specific signalling pathways
Energy Requirement	Requires large amounts of ATP	Does not require ATP	Partially requires ATP, as it involves signalling pathways and molecular events.
Cell Membrane	Cell shrinks and fragments, releasing apoptotic bodies	Cell swells and bursts; inflammation follows	Cell membrane remains intact, minimal inflammation
Involvement of Enzymes	Involves the activation of caspases	No involvement of specific enzymes	Involves enzymes like RIPK1, RIPK3, and MLKL
Pathway Activation	Activated by DNA damage, hypoxia, cytokines, ligands, oncogenes, *etc*	Not a specific pathway	Activated by TNF-α, Fas ligand, or other death receptors
Role in Disease	No role in disease, although defective apoptotic processes can lead to abnormalities, autoimmune diseases, cancer, infections, *etc*	Associated with tissue damage and inflammation	Implicated in various diseases like inflammatory disorders and neurodegenerative diseases
Treatment Targets	Apoptosis itself helps prevent cancer	Focus on managing symptoms and inflammation	Potential targets include RIPK1 and other necroptosis-specific molecules for therapeutic intervention.

Recent Advancements

In recent years, significant strides have been made in understanding the intricate mechanisms of necrosis, a once-thought chaotic and unregulated form of cell death. New studies have illuminated the molecular pathways underlying necrotic cell demise, challenging the traditional view of necrosis as a purely accidental event. Researchers have identified specific signalling cascades and molecular players that govern regulated necrosis, blurring the lines between programmed and uncontrolled cell death processes [100]. For instance, recent research by Linkermann and colleagues (2013) highlighted the role of RIPK1 and RIPK3 kinases in regulated necrosis, leading to the discovery of necroptosis, a regulated necrotic pathway. Additionally, studies by Christofferson and Yuan (2010) have shed light on the importance of mitochondria in necrotic cell death, unveiling the intricate interplay between mitochondrial dysfunction and necrosis. These advancements not only expand our knowledge of necrosis but also offer potential therapeutic avenues. Targeting specific molecular components in necrotic pathways has emerged as a promising strategy, holding implications for conditions where regulated necrosis plays a detrimental role, such as ischemia-reperfusion injury and neurodegenerative disorders [101].

Exploring mPTP-Mediated Necrosis

mPTP-dependent cell death is a necrotic form of cell death. Mitochondria are the microscopic powerhouses of our cells, regulating most of our functioning, and when they start malfunctioning, it inevitably leads to cell death. Mitochondrial Outer Membrane Permeability (MOMP) and Mitochondrial permeability transition pore (mPTP) have been recognized to induce cell death, under certain conditions, by affecting the permeability of the outer and inner mitochondrial membranes, respectively [102].

The mitochondrial Permeability Transition Pore (mPTP) is a dynamic feature in the inner mitochondrial membrane. In situations where there is an excess of calcium within the mitochondria, particularly when combined with oxidative stress, increased phosphate levels, and a shortage of adenine nucleotides, a generalized pore called the mitochondrial Permeability Transition Pore (mPTP) opens in the inner mitochondrial membrane. The opening of the mPTP allows small molecules, including proteins, weighing less than 1.5 kDa, to freely enter the mitochondria. This disrupts the coupling of oxidative phosphorylation, resulting in a depletion of ATP and ultimately leading to necrotic cell death [103, 104].

Key events in mPTP-induced cell death are:

1. **mPTP Opening:** Certain pathological conditions or stressors can cause the mPTP to open, such as

   º Ca^{2+} overload and flux of other ions (Na^+, *etc.*)

   º Alkalinization of matrix pH

   º Oxidative stress

   º Metabolic dysfunction

2. **Calcium Dysregulation:** One of the key factors involved in mPTP-mediated cell death is the accumulation and dysregulation of calcium ions (Ca^{2+}). When mPTPs are open for a prolonged time, there is an uncontrolled influx of calcium into the mitochondrial matrix. This causes a change in mitochondrial transmembrane potential, further leading to a cascade of events.

3. **Mitochondrial Dysfunction:** The excessive calcium within the mitochondria can lead to mitochondrial dysfunction, characterized by several critical changes.

4. **Release of Pro-Apoptotic Factors:** mPTP-mediated mitochondrial dysfunction triggers the release of pro-apoptotic factors from the mitochondria into the cytoplasm. One crucial factor released is cytochrome c.

5. **Apoptosis Activation:** Cytochrome c initiates the intrinsic apoptotic pathway by interacting with other proteins in the cytoplasm. This leads to the activation of caspase enzymes, which are central regulators of apoptosis. Caspases, in turn, activate a cascade of proteolytic events that culminate in the controlled dismantling of the cell.

6. **Cell Death:** The activation of apoptosis, driven by mPTP-mediated mitochondrial dysfunction, ultimately leads to programmed cell death. Apoptosis is characterized by a series of cellular changes, including cell shrinkage, DNA fragmentation, and the formation of apoptotic bodies, which are then cleared by phagocytic cells.

CELL DEATH: KEY FINDINGS AND FUTURE PROSPECTS

In the realm of cell death, a journey through history reveals a tapestry of discoveries that has deepened our understanding of the intricate mechanisms governing life and death at the cellular level. From the early recognition of necrosis by Rudolf Virchow to the modern elucidation of various cell death pathways, including apoptosis, which illuminated the carefully orchestrated dismantling of a cell into discrete, membrane-bound apoptotic bodies, to the recognition of necroptosis as a regulated, programmed form of necrosis, scientists have journeyed through a myriad of pathways, each unravelling unique facets of this essential biological phenomenon. The discovery of parthanatos, a pathway deeply interwoven with DNA damage response and poly(ADP-ribose) metabolism, opened new dimensions in understanding the connections between genomic stability and cellular demise. The emergence of ferroptosis, emphasizing the role of iron and lipid peroxidation, has challenged traditional notions of cell death and offered innovative avenues for therapeutic intervention in diseases such as cancer. Moreover, the unveiling of NETosis, a mechanism through which neutrophils deploy neutrophil extracellular traps (NETs) to ensnare and eliminate pathogens, showcases the intricate interplay between immune defence and programmed cellular death. These findings have not only shed light on essential physiological processes, such as development and immune responses, but also offered insights into the pathogenesis of diseases like cancer and neurodegenerative disorders [105].

Looking forward, the future of cell death research is both promising and challenging. With cutting-edge technologies such as single-cell RNA sequencing and advanced imaging techniques, scientists are poised to explore the heterogeneity of cell death responses within tissues, unravelling the subtleties that govern cell fate decisions in complex microenvironments. The intersection of cell death pathways and autophagy, the cell's recycling mechanism, promises to uncover novel regulatory networks that orchestrate cellular homeostasis. Additionally, understanding the impact of these pathways in the context of ageing and age-related diseases is becoming a focal point, offering potential avenues for extending healthy lifespans [106].

As researchers delve deeper into the intricate molecular and biochemical mechanisms underpinning cell death, the implications for human health are vast. Targeted therapies for cancer, neurodegenerative disorders, autoimmune diseases, and even infectious diseases are on the horizon. Moreover, the fundamental knowledge gained from cell death research continues to inspire innovation in regenerative medicine and tissue engineering, offering hope for patients with organ failure or degenerative conditions.

In this ever-expanding frontier of knowledge, the study of cell death not only deepens our understanding of the fundamental principles of life but also holds the key to unlocking novel therapeutic strategies, pushing the boundaries of medical science, and transforming the landscape of healthcare for generations to come. Hence, as we delve deeper into the mysteries of cellular demise, we embark on a journey of discovery that not only deepens our understanding of life's fragility but also holds the promise of innovative solutions for human health and disease.

CONCLUSION

Cell death plays a crucial role in maintaining biological balance, from development and tissue maintenance to immunity and disease prevention. The chapter explored various types of cell death, including apoptosis, necrosis, autophagy, pyroptosis, and ferroptosis, emphasizing their importance in both health and disease, such as cancer and neurodegenerative disorders. Understanding these pathways not only deepens our knowledge of cellular processes but also opens doors to new therapies, including targeted treatments and gene-editing technologies. As research advances, the study of cell death remains key to unlocking innovative medical approaches and enhancing human health.

CONSENT FOR PUBLICATION

All the visual representations were created by the authors involved in this publication, and it was not sourced from any other publications. No consent required.

CONFLICT OF INTEREST

This book chapter received no specific grant from any funding agency in the public, commercial, or not-for-profit sector. The authors have no conflict of interest related to this publication.

ACKNOWLEDGEMENTS

We would like to acknowledge Mr. Kumaran. K, Ms. Kruthika P, Ms. Raksa Arun, Ms. Srisri SatishKartik, Ms. Sanjana Dhayalan, and Mr. Sasitharan T for their contribution in proofreading and editing works.

REFERENCES

[1] O'Brien MA, Kirby R. Apoptosis: A review of pro-apoptotic and anti-apoptotic pathways and dysregulation in disease. J Vet Emerg Crit Care (San Antonio) 2008; 18(6): 572-85.
[http://dx.doi.org/10.1111/j.1476-4431.2008.00363.x]

[2] Park W, Wei S, Kim BS, et al. Diversity and complexity of cell death: a historical review. Exp Mol Med 2023; 55(8): 1573-94.
[http://dx.doi.org/10.1038/s12276-023-01078-x] [PMID: 37612413]

[3] Shen S, Shao Y, Li C. Different types of cell death and their shift in shaping disease. Cell Death Discov 2023; 9(1): 284.
[http://dx.doi.org/10.1038/s41420-023-01581-0] [PMID: 37542066]

[4] Wang H, Li Q, Alam P, et al. Aggregation-Induced Emission (AIE), Life and Health. ACS Nano 2023; 17(15): 14347-405.
[http://dx.doi.org/10.1021/acsnano.3c03925] [PMID: 37486125]

[5] Zhang R, Banik NL, Ray SK. Combination of all-trans retinoic acid and interferon-gamma upregulated p27kip1 and down regulated CDK2 to cause cell cycle arrest leading to differentiation and apoptosis in

human glioblastoma LN18 (PTEN-proficient) and U87MG (PTEN-deficient) cells. Cancer Chemother Pharmacol 2008; 62(3): 407-16.
[http://dx.doi.org/10.1007/s00280-007-0619-0] [PMID: 17960384]

[6] Tonnus W, Meyer C, Paliege A, *et al.* The pathological features of regulated necrosis Journal of Pathology. John Wiley and Sons Ltd 2019; pp. 697-707.

[7] Elaasser B, Arakil N, Mohammad KS. Bridging the Gap in Understanding Bone Metastasis: A Multifaceted Perspective. Int J Mol Sci 2024; 25(5): 2846.
[http://dx.doi.org/10.3390/ijms25052846] [PMID: 38474093]

[8] Wu Z, Deng J, Zhou H, Tan W, Lin L, Yang J. Programmed Cell Death in Sepsis Associated Acute Kidney Injury. Front Med (Lausanne) 2022; 9883028.
[http://dx.doi.org/10.3389/fmed.2022.883028] [PMID: 35655858]

[9] Nichols LA, Grunz-Borgmann EA, Wang X, Parrish AR. A role for the age-dependent loss of α(E)-catenin in regulation of N-cadherin expression and cell migration. Physiol Rep 2014; 2(6)e12039.
[http://dx.doi.org/10.14814/phy2.12039] [PMID: 24920123]

[10] Qian S, Long Y, Tan G, *et al.* Programmed cell death: molecular mechanisms, biological functions, diseases, and therapeutic targets. Beijing: MedComm 2024; p. 5.

[11] Fang Y, Tian S, Pan Y, *et al.* Pyroptosis: A new frontier in cancer. Biomedicine and Pharmacotherapy. Elsevier Masson SAS 2020.

[12] Yu P, Zhang X, Liu N, Tang L, Peng C, Chen X. Pyroptosis: mechanisms and diseases. Signal Transduct Target Ther 2021; 6(1): 128.
[http://dx.doi.org/10.1038/s41392-021-00507-5] [PMID: 33776057]

[13] Yan J, Wan P, Choksi S, Liu ZG. Necroptosis and tumor progression. Trends Cancer 2022; 8(1): 21-7.
[http://dx.doi.org/10.1016/j.trecan.2021.09.003] [PMID: 34627742]

[14] Anosike NL, Adejuwon JF, Emmanuel GE, *et al.* Necroptosis in the developing brain: role in neurodevelopmental disorders. Metab Brain Dis 2023; 38(3): 831-7.
[http://dx.doi.org/10.1007/s11011-023-01203-9] [PMID: 36964816]

[15] Elena-Real CA, Díaz-Quintana A, González-Arzola K, *et al.* Cytochrome c speeds up caspase cascade activation by blocking 14-3-3ε-dependent Apaf-1 inhibition. Cell Death Dis 2018; 9(3): 365.
[http://dx.doi.org/10.1038/s41419-018-0408-1] [PMID: 29511177]

[16] Morana O, Wood W, Gregory CD. The Apoptosis Paradox in Cancer. Int J Mol Sci 2022; 23(3): 1328.
[http://dx.doi.org/10.3390/ijms23031328] [PMID: 35163253]

[17] Frisch SM, Francis H. Disruption of epithelial cell-matrix interactions induces apoptosis. J Cell Biol 1994; 124(4): 619-26.
[http://dx.doi.org/10.1083/jcb.124.4.619] [PMID: 8106557]

[18] Goyal MR. Scientific and Technical Terms in Bioengineering and Biological Engineering. Apple Academic Press: New York 2018: p. 662.
[http://dx.doi.org/10.1201/b22469]

[19] Kim NH, Kang PM. Apoptosis in cardiovascular diseases: mechanism and clinical implications. Korean Circ J 2010; 40(7): 299-305.
[http://dx.doi.org/10.4070/kcj.2010.40.7.299] [PMID: 20664736]

[20] Faherty CS, Maurelli AT. Staying alive: bacterial inhibition of apoptosis during infection. Trends Microbiol 2008; 16(4): 173-80.
[http://dx.doi.org/10.1016/j.tim.2008.02.001] [PMID: 18353648]

[21] Redi C. Essentials of apoptosis - A guide for basic and clinical research. Eur J Histochem 2010; 54(1): 11.
[http://dx.doi.org/10.4081/ejh.2010.e11]

[22] Yan G, Elbadawi M, Efferth T. Multiple cell death modalities and their key features (Review). World

[23] Zakeri Z, Lockshin RA Cell Death: History and Future. In: Programmed Cell Death in Cancer Progression and Therapy. Advances in Experimental Medicine and Biology 2008; vol 615. Springer, Dordrecht.
[http://dx.doi.org/10.1007/978-1-4020-6554-5_1] [PMID: 37612413]

[24] Bonora M, Giorgi C, Pinton P. Molecular mechanisms and consequences of mitochondrial permeability transition. Nat Rev Mol Cell Biol 2022; 23(4): 266-85.
[http://dx.doi.org/10.1038/s41580-021-00433-y] [PMID: 34880425]

[25] 2012. Favaloro B, Allocati N, Graziano V, Di Ilio C, De Laurenzi V. Role of apoptosis in disease. Aging (Albany NY) 2012; 4(5): 330-49.

[26] Voss AK, Strasser A. The essentials of developmental apoptosis. F1000Research 2020; 9: F1000.
[http://dx.doi.org/10.12688/f1000research.21571.1]

[27] Deng H, Kuang P, Cui H, et al. Sodium fluoride induces apoptosis in cultured splenic lymphocytes from mice. Oncotarget 2016; 7(42): 67880-900.
[http://dx.doi.org/10.18632/oncotarget.12081] [PMID: 27655720]

[28] Singh P, Lim B. Targeting Apoptosis in Cancer. Curr Oncol Rep 2022; 24(3): 273-84.
[http://dx.doi.org/10.1007/s11912-022-01199-y] [PMID: 35113355]

[29] Pfeffer C, Singh A. Apoptosis: A Target for Anticancer Therapy. Int J Mol Sci 2018; 19(2): 448.
[http://dx.doi.org/10.3390/ijms19020448] [PMID: 29393886]

[30] Raeisi M, Zehtabi M, Velaei K, Fayyazpour P, Aghaei N, Mehdizadeh A. Anoikis in cancer: The role of lipid signaling Cell Biol Int. John Wiley and Sons Inc 2022; pp. 1717-28.

[31] Mei J, Jiang XY, Tian HX, et al. Anoikis in cell fate, physiopathology, and therapeutic interventions MedComm. Beijing: John Wiley and Sons Inc 2024.

[32] Adeshakin FO, Adeshakin AO, Afolabi LO, Yan D, Zhang G, Wan X. Mechanisms for Modulating Anoikis Resistance in Cancer and the Relevance of Metabolic Reprogramming. Front Oncol 2021; 11626577.
[http://dx.doi.org/10.3389/fonc.2021.626577] [PMID: 33854965]

[33] Encyclopedic Reference of Genomics and Proteomics in Molecular Medicine. Berlin, Heidelberg: Springer Berlin Heidelberg 2006; pp. 29-9.

[34] Taddei ML, Giannoni E, Fiaschi T, Chiarugi P. Anoikis: an emerging hallmark in health and diseases. J Pathol 2012; 226(2): 380-93.
[http://dx.doi.org/10.1002/path.3000] [PMID: 21953325]

[35] Han YH, Wang Y, Lee SJ, Jin MH, Sun HN, Kwon T. Regulation of anoikis by extrinsic death receptor pathways. Cell Communication and Signaling. BioMed Central Ltd 2023.
[http://dx.doi.org/10.1186/s12964-023-01247-5]

[36] Elmore S. Apoptosis: a review of programmed cell death. Toxicol Pathol 2007; 35(4): 495-516.
[http://dx.doi.org/10.1080/01926230701320337] [PMID: 17562483]

[37] Vachon PH. Integrin signaling, cell survival, and anoikis: distinctions, differences, and differentiation. J Signal Transduct 2011; 2011: 1-18.
[http://dx.doi.org/10.1155/2011/738137] [PMID: 21785723]

[38] Nepali PR, Kyprianou N. Anoikis in phenotypic reprogramming of the prostate tumor microenvironment. Front Endocrinol (Lausanne) 2023; 141160267.
[http://dx.doi.org/10.3389/fendo.2023.1160267] [PMID: 37091854]

[39] Kiechle FL, Zhang X. Apoptosis: biochemical aspects and clinical implications. Clin Chim Acta 2002; 326(1-2): 27-45.
[http://dx.doi.org/10.1016/S0009-8981(02)00297-8] [PMID: 12417095]

[40] Liu B, Yan S, Qu L, Zhu J. Celecoxib enhances anticancer effect of cisplatin and induces anoikis in osteosarcoma *via* PI3K/Akt pathway. Cancer Cell Int 2017; 17(1): 1.
[http://dx.doi.org/10.1186/s12935-016-0378-2] [PMID: 28053596]

[41] Paoli P, Giannoni E, Chiarugi P. Anoikis molecular pathways and its role in cancer progression. Biochim Biophys Acta Mol Cell Res 2013; 1833(12): 3481-98.
[http://dx.doi.org/10.1016/j.bbamcr.2013.06.026] [PMID: 23830918]

[42] Liu G, Meng X, Jin Y, *et al.* Inhibitory role of focal adhesion kinase on anoikis in the lung cancer cell A549. Cell Biol Int 2008; 32(6): 663-70.
[http://dx.doi.org/10.1016/j.cellbi.2008.01.292] [PMID: 18343694]

[43] Kim YN, Koo KH, Sung JY, Yun UJ, Kim H. Anoikis resistance: an essential prerequisite for tumor metastasis. Int J Cell Biol 2012: 2012: 306879.
[http://dx.doi.org/10.1155/2012/306879]

[44] Wang Y, Cheng S, Fleishman JS, *et al.* Targeting anoikis resistance as a strategy for cancer therapy. Drug Resist Updat 2024; 75: 101099.
[http://dx.doi.org/10.1016/j.drup.2024.101099] [PMID: 38850692]

[45] Khan SU, Fatima K, Malik F. Understanding the cell survival mechanism of anoikis-resistant cancer cells during different steps of metastasis. Clin Exp Metastasis. Springer Science and Business Media B.V. 2022; pp. 715-26.

[46] Shaw P, Bhowmik AD, Pillai MSG, *et al.* Anoikis resistance in Cancer: Mechanisms, therapeutic strategies, potential targets, and models for enhanced understanding. Cancer Letters 2025; 624: 217750.
[http://dx.doi.org/10.1016/j.canlet.2025.217750] [PMID: 33854965]

[47] Dai Y, Zhang X, Ou Y, *et al.* Anoikis resistance—protagonists of breast cancer cells survive and metastasize after ECM detachment. Cell Commun Signal 2023; 21(1): 190.
[http://dx.doi.org/10.1186/s12964-023-01183-4] [PMID: 37537585]

[48] Lee HY, Son SW, Moeng S, Choi SY, Park JK. The role of noncoding RNAs in the regulation of anoikis and anchorage-independent growth in cancer. Int J Mol Sci. MDPI AG 2021; pp. 1-34.

[49] Patankar M, Eskelinen S, Tuomisto A, Mäkinen MJ, Karttunen TJ. KRAS and BRAF mutations induce anoikis resistance and characteristic 3D phenotypes in Caco-2 cells. Mol Med Rep 2019; 20(5): 4634-44.
[http://dx.doi.org/10.3892/mmr.2019.10693] [PMID: 31545494]

[50] Sattari Fard F, Jalilzadeh N, Mehdizadeh A, Sajjadian F, Velaei K. Understanding and targeting anoikis in metastasis for cancer therapies. Cell Biol Int 2023; 47(4): 683-98.
[http://dx.doi.org/10.1002/cbin.11970] [PMID: 36453448]

[51] Neuendorf HM, Simmons JL, Boyle GM. Therapeutic targeting of anoikis resistance in cutaneous melanoma metastasis. Front Cell Dev Biol 2023; 11: 1183328.
[http://dx.doi.org/10.3389/fcell.2023.1183328]

[52] Li J, Cao F, Yin H, *et al.* Ferroptosis: past, present and future. Cell Death Dis 2020; 11(2): 88.
[http://dx.doi.org/10.1038/s41419-020-2298-2] [PMID: 32015325]

[53] Jiang X, Stockwell BR, Conrad M. Ferroptosis: mechanisms, biology and role in disease. Nat Rev Mol Cell Biol 2021; 22(4): 266-82.
[http://dx.doi.org/10.1038/s41580-020-00324-8] [PMID: 33495651]

[54] Xie Y, Hou W, Song X, *et al.* Ferroptosis: Process and function. Cell Death Differ. Nature Publishing Group 2016; pp. 369-79.

[55] Fang Y, Chen X, Tan Q, Zhou H, Xu J, Gu Q. Inhibiting Ferroptosis through Disrupting the NCOA4–FTH1 Interaction: A New Mechanism of Action. ACS Cent Sci 2021; 7(6): 980-9.
[http://dx.doi.org/10.1021/acscentsci.0c01592] [PMID: 34235259]

[56] Tang D, Chen X, Kang R, Kroemer G. Ferroptosis: molecular mechanisms and health implications. Cell Res 2021; 31(2): 107-25.
[http://dx.doi.org/10.1038/s41422-020-00441-1] [PMID: 33268902]

[57] Newton K, Strasser A, Kayagaki N, Dixit VM. Cell death. Cell 2024; 187(2): 235-56.
[http://dx.doi.org/10.1016/j.cell.2023.11.044] [PMID: 38242081]

[58] Cao JY, Dixon SJ. Mechanisms of ferroptosis Cellular and Molecular Life Sciences. Birkhauser Verlag AG 2016; pp. 2195-209.

[59] Tian HY, Huang BY, Nie HF, *et al.* The Interplay between Mitochondrial Dysfunction and Ferroptosis during Ischemia-Associated Central Nervous System Diseases Brain Sci Multidisciplinary Digital Publishing Institute. MDPI 2023.

[60] Gan B. Mitochondrial regulation of ferroptosis. J Cell Biol 2021; 220(9)e202105043.
[http://dx.doi.org/10.1083/jcb.202105043] [PMID: 34328510]

[61] Angeli JPF, Shah R, Pratt DA, Conrad M. Ferroptosis Inhibition: Mechanisms and Opportunities Trends Pharmacol Sci. Elsevier Ltd 2017; pp. 489-98.

[62] Hunsaker EW, Franz KJ. Emerging opportunities to manipulate metal trafficking for therapeutic benefit. Inorg Chem 2019; 58(20): 13528-45.
[http://dx.doi.org/10.1021/acs.inorgchem.9b01029] [PMID: 31247859]

[63] Pope LE, Dixon SJ. Regulation of ferroptosis by lipid metabolism Trends Cell Biol. Elsevier Ltd 2023; pp. 1077-87.

[64] Zhang XD, Liu ZY, Sen Wang M, *et al.* Mechanisms and regulations of ferroptosis. Frontiers Media, SA: Front Immunol 2023.
[http://dx.doi.org/10.3389/fimmu.2023.1269451]

[65] Chen X, Yu C, Kang R, Tang D. Iron Metabolism in Ferroptosis. Front Cell Dev Biol. Frontiers Media S.A. 2020.
[http://dx.doi.org/10.3389/fcell.2020.590226]

[66] Lee JY, Kim WK, Bae KH, Lee SC, Lee EW. Lipid metabolism and ferroptosis Biology. Basel: MDPI AG 2021; pp. 1-16.

[67] Seibt TM, Proneth B, Conrad M. Role of GPX4 in ferroptosis and its pharmacological implication Free Radic Biol Med. Elsevier Inc. 2019; pp. 144-52.

[68] Huang X, Yan X, Chen G, *et al.* Insufficient autophagy enables the nuclear factor erythroid 2-related factor 2 (NRF2) to promote ferroptosis in morphine-treated SH-SY5Y cells. Psychopharmacology (Berl) 2024; 241(2): 291-304.
[http://dx.doi.org/10.1007/s00213-023-06485-6] [PMID: 38049617]

[69] Ma TL, Chen JX, Zhu P, Zhang C. Bin, Zhou Y, Duan JX Focus on ferroptosis regulation: Exploring novel mechanisms and applications of ferroptosis regulator Life Sci. Elsevier Inc. 2022.

[70] Mbah NE, Lyssiotis CA. Metabolic regulation of ferroptosis in the tumor microenvironment Journal of Biological Chemistry. American Society for Biochemistry and Molecular Biology Inc. 2022.

[71] Lee YS, Lee DH, Choudry HA, Bartlett DL, Lee YJ. Ferroptosis-induced endoplasmic reticulum stress: Cross-talk between ferroptosis and apoptosis Molecular Cancer Research. American Association for Cancer Research Inc. 2018; pp. 1073-6.

[72] Zeng F, Nijiati S, Tang L, Ye J, Zhou Z, Chen X. Ferroptosis Detection: From Approaches to Applications. Angew Chem Int Ed 2023; 62(35)e202300379.
[http://dx.doi.org/10.1002/anie.202300379] [PMID: 36828775]

[73] Liang C, Zhang X, Yang M, Dong X. Recent Progress in Ferroptosis Inducers for Cancer Therapy. Advanced Materials. Wiley-VCH Verlag 2019.
[http://dx.doi.org/10.1002/adma.201904197]

[74] Sun S, Shen J, Jiang J, Wang F, Min J. Targeting ferroptosis opens new avenues for the development of novel therapeutics. Signal Transduct Target Ther. Springer Nature 2023.
[http://dx.doi.org/10.1038/s41392-023-01606-1]

[75] Huang P, Chen G, Jin W, Mao K, Wan H, He Y. Molecular Mechanisms of Parthanatos and Its Role in Diverse Diseases. Int J Mol Sci 2022; 23(13): 7292.
[http://dx.doi.org/10.3390/ijms23137292] [PMID: 35806303]

[76] Rodríguez-Vargas JM, Rodríguez MI, Majuelos-Melguizo J, et al. Autophagy requires poly(adp-ribosyl)ation-dependent AMPK nuclear export. Cell Death Differ 2016; 23(12): 2007-18.
[http://dx.doi.org/10.1038/cdd.2016.80] [PMID: 27689873]

[77] Moura RD, Mattos PD, Valente PF, Hoch NC. Molecular mechanisms of cell death by parthanatos: More questions than answers. Genet Mol Biol 2024; 47(47) (Suppl. 1).e20230357.
[http://dx.doi.org/10.1590/1678-4685-gmb-2023-0357] [PMID: 39356140]

[78] Liu S, Luo W, Wang Y. Emerging role of PARP-1 and PARthanatos in ischemic stroke J Neurochem. John Wiley and Sons Inc 2022; pp. 74-87.

[79] Liu L, Li J, Ke Y, et al. The key players of parthanatos: opportunities for targeting multiple levels in the therapy of parthanatos-based pathogenesis. Cell Mol Life Sci 2022; 79(1): 60.
[http://dx.doi.org/10.1007/s00018-021-04109-w] [PMID: 35000037]

[80] Wang X, Ge P. Parthanatos in the pathogenesis of nervous system diseases Neuroscience. Elsevier Ltd 2020; pp. 241-50.

[81] Koehler RC, Dawson VL, Dawson TM. Targeting Parthanatos in Ischemic Stroke. Front Neurol. Frontiers Media S.A. 2021.
[http://dx.doi.org/10.3389/fneur.2021.662034]

[82] Gupta G, Afzal M, Moglad E, et al. Parthanatos and apoptosis: unraveling their roles in cancer cell death and therapy resistance. EXCLI J 2025; 24: 351-80.
[http://dx.doi.org/10.17179/excli2025-8251]

[83] Bondar D, Karpichev Y. Poly(ADP-Ribose) Polymerase (PARP) Inhibitors for Cancer Therapy: Advances, Challenges, and Future Directions Biomolecules. Multidisciplinary Digital Publishing Institute 2024.

[84] Robinson N, Ganesan R, Hegedűs C, Kovács K, Kufer TA, Virág L. Programmed necrotic cell death of macrophages: Focus on pyroptosis, necroptosis, and parthanatos. Redox Biol. Elsevier B.V. 2019.

[85] Chen A. PARP inhibitors: its role in treatment of cancer. Chin J Cancer 2011; 30(7): 463-71.
[http://dx.doi.org/10.5732/cjc.011.10111] [PMID: 21718592]

[86] Yang L, Guttman L, Dawson VL, Dawson TM. Parthanatos: Mechanisms, modulation, and therapeutic prospects in neurodegenerative disease and stroke. Biochem Pharmacol 2024; 228: 116174
[http://dx.doi.org/10.1016/j.bcp.2024.116174] [PMID: 38552851]

[87] Zhou Y, Liu L, Tao S, et al. Parthanatos and its associated components: Promising therapeutic targets for cancer. Pharmacol Res 2021; 163: 105299.
[http://dx.doi.org/10.1016/j.phrs.2020.105299] [PMID: 33171306]

[88] Chen T, Li Y, Sun R, et al. Receptor-Mediated NETosis on Neutrophils. Front Immunol 2021; 12: 775267.
[http://dx.doi.org/10.3389/fimmu.2021.775267] [PMID: 34804066]

[89] Huang J, Hong W, Wan M, Zheng L. Molecular mechanisms and therapeutic target of NETosis in diseases. Beijing: MedComm 2022; p. 3.

[90] Lo Nigro C, Macagno M, Sangiolo D, Bertolaccini L, Aglietta M, Merlano MC. NK-mediated antibody-dependent cell-mediated cytotoxicity in solid tumors: biological evidence and clinical perspectives. Ann Transl Med 2019; 7(5): 105-5.
[http://dx.doi.org/10.21037/atm.2019.01.42] [PMID: 31019955]

[91] Coënon L, Villalba M. From CD16a Biology to Antibody-Dependent Cell-Mediated Cytotoxicity Improvement. Front Immunol 2022; 13: 913215.
[http://dx.doi.org/10.3389/fimmu.2022.913215] [PMID: 35720368]

[92] Kavian N, Hachim A, Poon LLM, Valkenburg SA. Vaccination with ADCC activating HA peptide epitopes provides partial protection from influenza infection. Vaccine 2020; 38(37): 5885-90.
[http://dx.doi.org/10.1016/j.vaccine.2020.07.008] [PMID: 32718818]

[93] Zahavi D, AlDeghaither D, O'Connell A, Weiner LM. Enhancing antibody-dependent cell-mediated cytotoxicity: a strategy for improving antibody-based immunotherapy. Antib Ther 2018; 1(1): 7-12.
[http://dx.doi.org/10.1093/abt/tby002] [PMID: 33928217]

[94] Moujalled D, Strasser A, Liddell JR. Molecular mechanisms of cell death in neurological diseases. Cell Death Differ 2021; 28(7): 2029-44.
[http://dx.doi.org/10.1038/s41418-021-00814-y] [PMID: 34099897]

[95] Khalid N, Azimpouran M. Necrosis Pathology. [Updated 2023 Mar 6]. In: StatPearls [Internet]. Treasure Island (FL): StatPearls Publishing; 2025 Jan-. Available from: https://www.ncbi.nlm.nih.gov/books/NBK557627/

[96] Adigun R, Basit H, Murray J. Cell Liquefactive Necrosis. In: StatPearls [Internet]. Treasure Island (FL): StatPearls Publishing; 2025 Jan-. Available from: https://www.ncbi.nlm.nih.gov/books/NBK430935/

[97] Ye K, Chen Z, Xu Y. The double-edged functions of necroptosis. Cell Death Dis 2023; 14(2): 163.
[http://dx.doi.org/10.1038/s41419-023-05691-6] [PMID: 36849530]

[98] Ros U, Pedrera L, Garcia-Saez AJ. Partners in crime: The interplay of proteins and membranes in regulated necrosis. Int J Mol Sci. MDPI AG 2020.

[99] Yee PP, Li W. Tumor necrosis: A synergistic consequence of metabolic stress and inflammation. BioEssays 2021; 43(7): 2100029
[http://dx.doi.org/10.1002/bies.202100029] [PMID: 33998010]

[100] Conrad M, Angeli JPF, Vandenabeele P, Stockwell BR. Regulated necrosis: Disease relevance and therapeutic opportunities. Nat Rev Drug Discov. Nature Publishing Group 2016; pp. 348-66.

[101] Galluzzi L, Baehrecke EH, Ballabio A, *et al.* Molecular definitions of autophagy and related processes. EMBO J 2017; 36(13): 1811-36.
[http://dx.doi.org/10.15252/embj.201796697] [PMID: 28596378]

[102] Ren K, Pei J, Guo Y, *et al.* Regulated necrosis pathways: a potential target for ischemic stroke. Burns Trauma 2023; 11tkad016.
[http://dx.doi.org/10.1093/burnst/tkad016] [PMID: 38026442]

[103] Halestrap AP. What is the mitochondrial permeability transition pore? J Mol Cell Cardiol 2009; 46(6): 821-31.
[http://dx.doi.org/10.1016/j.yjmcc.2009.02.021] [PMID: 19265700]

[104] Robichaux DJ, Harata M, Murphy E, Karch J. Mitochondrial permeability transition pore-dependent necrosis. J Mol Cell Cardiol 2023; 174: 47-55.
[http://dx.doi.org/10.1016/j.yjmcc.2022.11.003] [PMID: 36410526]

[105] Kist M, Vucic D. Cell death pathways: intricate connections and disease implications. EMBO J 2021; 40(5): e106700.
[http://dx.doi.org/10.15252/embj.2020106700] [PMID: 33439509]

[106] Tower J. Programmed cell death in aging. Ageing Res Rev. Elsevier Ireland Ltd 2015; pp. 90-100.

CHAPTER 14

Stem Cells: Breakthroughs in Medicine and Therapeutics

Vanshikaa Karthikeyan[1], Janani Balaji[1], SriSri SatishKartik[1], Dannie Macrin[2,*] and K.N. Aruljothi[1]

[1] *Department of Genetic Engineering, SRM Institute of Science and Technology, Kattankulathur, Chengalpattu, Tamil Nadu, India*

[2] *Department of Computational Biology, Institute of Bioinformatics, Saveetha School of Engineering, SIMATS Saveetha Institute of Medical and Technical Sciences, Chennai, India*

Abstract: Stem cells are usually referred to as undifferentiated cell masses, which can differentiate into many cell types. Stem cells are present in all organisms in various evolved forms. These cells can self-renew. Additionally, stem cells can be broadly classified based on their origin and differentiation potential. Thus, the therapeutic potential of these cells can be harnessed and utilized to treat various disorders, including blood disorders, tissue and organ regeneration, and others, which can be classified under the division of stem cell therapy. Although stem cells possess the capacity for use in disease treatment, their implementation also has its own risks. This chapter covers the basics of stem cells, including the associated cell types, applications, epigenetics, and recent trends.

Keywords: Blood disorders, Differentiation potential, Epigenetics, Recent trends, Stem cell therapy, Stem cells, Tissue regeneration.

INTRODUCTION

Stem cells are defined as undifferentiated masses of cells present in all living organisms. Furthermore, stem cells represent the building blocks of all tissues and organs, with major properties including:

i. Extensive proliferation
ii. An ability to arise from a single cell (clonal).
iii. An ability to differentiate to form different cell types (potency).

---

* **Corresponding author Dannie Macrin:** Department of Computational Biology, Institute of Bioinformatics, Saveetha School of Engineering, SIMATS Saveetha Institute of Medical and Technical Sciences, Chennai, India; E-mail: danniem.sse@saveetha.com

K.N. Aruljothi, Prakash Gangadaran, Krishnan Anand, Satish Ramalingam, K. Kumaran & Kruthika Prakash (Ed.)
All rights reserved-© 2026 Bentham Science Publishers

However, these properties vary among the different cell types. For instance, Embryonic Stem Cells (ESCs) undergo extensive proliferation, whereas Adult Stem Cells (ASCs) primarily differentiate into tissue-specific cells.

Thus, stem cells are currently employed in cell therapy and drug development processes. Additionally, stem cells have improved understanding of pathogenesis. Nonetheless, despite these significant advances in stem cell biology, certain ethical issues remain that limit the utilization of these cells [1].

STEM CELL FUNCTIONS

Stem cells are predominantly used due to their ability to self-renew and differentiate into various types of cells. In adult organisms, stem cells can either actively replenish tissues to repair damage or remain dormant, as is the case with neural stem cells in the mammalian brain. There are two types of stem cell division: Symmetric and asymmetric. Symmetric division occurs when the cells reproduce to produce more identical cells, while asymmetric division occurs when the cell differentiates to form a specific cell type. In the asymmetric model, the stem cell forms an exact copy and one differentiated cell, which helps maintain the homeostasis of the stem cell pool at the individual level. A major disadvantage of asymmetric division is that the stem cell pool cannot be replenished at the time of injury. However, this limitation can be resolved by the symmetric division, which maintains the homeostasis of the stem cell pool at the population level. Meanwhile, two types of symmetric division exist: the first is a proliferation division, which leads to the formation of two stem cells; the second is a differentiation division, which leads to the formation of two differentiated cells. Proliferation and differentiation of these cells occur due to the signals from the surrounding tissues as well as the stem cell [2, 3].

TYPES OF STEM CELLS

Stem cells are classified based on the following factors:

i. Differentiation potential
ii. Origin

Stem Cells Based on Differentiation Potential

The ability of stem cells to differentiate is based on their origin and associated derivation; thus, stem cells can be divided into five major categories.

Totipotent/Omnipotent Stem Cells

These are the cells that have the potential to differentiate into the various cell types found in the entire organism. Totipotent cells have the highest potential to differentiate. For example, the zygote is a totipotent cell that can differentiate into all cell types, including the three germ layers and extraembryonic tissues, such as the placenta. Meanwhile, totipotent cells found in an embryo differentiate into the inner cell mass (ICM) and the extra-embryonic cell lineage, such as the trophectoderm. These totipotent embryos exhibit a unique transcriptome, characterized by dramatic changes in epigenetic and chromatin features, including the de novo assembly of nucleosomes, demethylation of DNA, chromatin remodeling, and histone modifications.

Pluripotent Stem Cells

These stem cells play a crucial role in forming the three primary germ layers, which, in turn, help develop all tissues and organs: ectoderm, mesoderm, and endoderm. These stem cells are only short-lived during the early stages of embryo development; however, pluripotent stem cells can develop into either fetal cells or adult cells, but not into a complete organism. However, these cell lines can be immortalized and grown *in vitro*. These cells can then differentiate into all cell types through various mechanisms, similar to those employed by ESCs.

Multipotent Stem Cells

Meanwhile, Multipotent Stem Cells (MSCs) have a more restricted differentiation potential, transitioning to specific cell types within a particular lineage. For example, Hematopoietic Stem Cells (HSCs) can differentiate into different types of blood cells. These cells are considered ASCs due to their limited capacity for differentiation and multipotency. MSCs are found in various organs, including the periosteum, adipose tissue, placenta, dermis, and umbilical cord blood. MSCs can be identified by their adherence and the ability to differentiate into cartilage, fat, and bone, as well as the expression of markers, such as CD90, CD105, and CD73. Currently, these cells are being used in tissue healing, development, and defense.

Oligopotent Stem Cells

These cells possess the ability for self-renewal and can differentiate into various types of cells. Moreover, these stem cells can be viewed as progenitor cells for ASCs and can differentiate into specific cell types; for example, myeloid stem cells can differentiate into white blood cells but not red blood cells. Furthermore, these progenitor-oligo cells have also been observed to differentiate into more mature cells, such as alveolar epithelial cells and bronchiolar epithelium.

Unipotent Stem Cells

These cells possess the ability to undergo repeated divisions, although with limited differentiation capacity; specifically, unipotent stem cells can only differentiate into a single cell type. Nonetheless, these cells exhibit fluid plasticity and play a major role in repairing cells affected by sudden trauma or structural damage. Moreover, these cells can proliferate at a faster rate and are ideal for reprogramming into induced pluripotent stem cells (iPSCs). For example, epidermal stem cells in the skin only differentiate into keratinocytes [4 - 7].

Stem Cells Based on Origin

Stem cells are divided into four major categories (Fig. **1**):

Embryonic Stem Cells

ESCs are pluripotent and are extracted from the ICM of the blastocyst, a structure that forms during the early stages of embryonic development. Meanwhile, MSCs differentiate into the three germ layers. These cells can also be preserved in an undifferentiated state for an extended period. Indeed, ESCs without any genetic defects are preserved and can subsequently be used for several experiments. Cryopreservation in liquid nitrogen at -196 °C or in nitrogen vapor at -150 °C represents the most common method of maintaining ESCs. Another rapid freezing process used for preserving ESCs is called vitrification; however, this method can only be conducted on a small scale. The Open-Pulled Straw (OPS) method represents another important method of ESC preservation, where 1–20 µL clumps of ESCs are stored in special straws.

Types of Embryonic Stem Cells

i. Fetal stem cells: These stem cells help in the development of fetal cells and tissues before birth. Fetal stem cells are pluripotent, and the different types include HSCs, mesenchymal stem cells, and endothelial stem cells. Fetal stem cells are also replicated at a much higher rate than ASCs. The pre-immune state of fetal stem cells is crucial in the context of mismatched transplants.
ii. Umbilical cord stem cells: The blood present in the umbilical cord has cells that are genetically similar to the cells of a newborn child. These cells are multipotent. These stem cells can be divided into HSCs and Mesenchymal Stem Cells (MSCs). HSCs can be used to treat blood, bone marrow, and immune system disorders. MSCs are of mesodermal origin. These cells can differentiate into adipocytes, chondrocytes, or osteocytes, and are characterized by their ability to adhere to plastic surfaces.

Adult Stem Cells

ASCs, such as mesenchymal stem cells from placental tissue, secrete anti-inflammatory factors that facilitate tissue repair. Additionally, ASCs have a more limited differentiation potential compared to ESCs. The primary function of ASCs is to aid in the *in vivo* restoration of organs. These cells also secrete specific molecular mediators that facilitate the repair of organs.

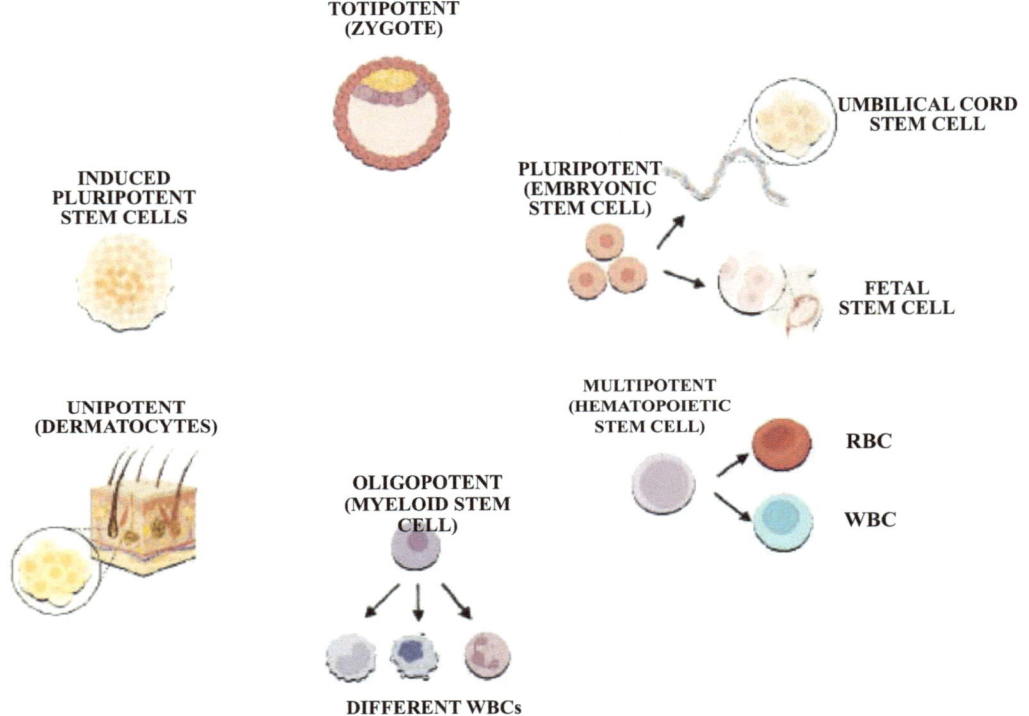

Fig (1). Types of stem cells.

Tissue-Resident Stem Cells

These stem cells differentiate to form tissue-specific cells; most injury-related renewals are dependent on these cells. These cells are mostly dormant; however, activation signals are produced at the time of injury and repair. These cells reside in specialized microenvironments known as stem cell niches, which regulate their self-renewal and differentiation in response to signals such as tissue damage.

Induced-Pluripotent Stem Cells

These iPSCs are generated by reprogramming adult somatic cells through the introduction of specific genes that restore them to a pluripotent, embryonic-like state. Similar to ESCs, iPSCs differentiate into the three germ layers *in vitro*. Thus, these cells are used in drug development and regenerative medicine techniques. The advantages of these iPSCs include their human origin, ease of accessibility and expansion, ability to produce various cell types, and potential for development into personalized medicine using patient-specific iPSCs [8 - 13].

Perinatal Stem Cells

Perinatal stem cells, including those derived from placental blood or tissue, refer to stem cells collected around the time of birth from sources such as the umbilical cord and amniotic fluid. Perinatal stem cells include various types, including HSCs derived from umbilical cord blood, Mesenchymal Stromal Cells (MSCs) from Wharton's jelly in the umbilical cord, and epithelial stem cells and trophoblasts from placental tissues.

Types of Perinatal Stem Cells

i. HSCs: These cells are derived from umbilical cord blood. There are approximately 10 times fewer functional HSCs in cord blood than in a bone marrow graft. These HSCs can be separated from the cord blood by Fluorescent-Activated Cell Sorting (FACS) and immunomagnetic precipitation assays. The capacity for HSCs to proliferate in cord blood is higher than that of HSCs in adult bone marrow and peripheral blood cells.

ii. Mesenchymal Stromal Cells (MSCs): These cells are derived from umbilical cord blood and tissue, which is collectively known as Wharton's jelly. These cells are multipotent and can be differentiated into adipocytes, chondrocytes, and osteocytes. MSCs are isolated from the umbilical cord by explant and enzymatic techniques. Cells isolated using the explant technique exhibit high yields and viability, a reduced risk of contamination, and successful differentiation into the mesodermal lineage. However, isolation performed through the explant method takes longer than through the enzymatic method [11, 14].

STEM CELL NICHE

ASCs, which are present in various tissues, play a crucial role in maintaining homeostasis and responding to environmental stimuli. To perform these functions, stem cells rely on their surrounding environment, known as the stem cell niche, which provides necessary signals for regulation. This microenvironment comprises cellular and molecular factors that regulate the behavior of stem cells. In multicellular organisms, stem cells are influenced by the associated niche, which provides the cells with properties that can be altered according to the requirements of the organism. Stem cell niches vary in structure; these niches can consist of a single cell, as seen in *Caenorhabditis elegans*, or a group of interacting cells, such as those in *Drosophila melanogaster*. Furthermore, stem cell niches are highly specific, requiring precise organization to balance stem cell self-renewal and differentiation [15].

Types of Niche

There are three major types of niches present for stem cells:

Simple Niches

Some niches, such as those in gonadal, epithelial, and digestive tissues, are relatively simple and function through common regulatory mechanisms. These niches feature adherens junctions that closely associate stem cells with neighboring permanent cells, facilitating the reception of intercellular signals. The interaction with the extracellular matrix also plays a major role. When these stem cells are tightly associated with normal cells, they can obtain many critical intercellular signals. This niche must also make sure that the daughter cells differentiate appropriately upon leaving the niche.

Complex Niches

Some niches—such as the subventricular zone (SVZ), which is home to neural stem cells—are more complex and involve close interactions with neuroblasts, endothelial cells, and other surrounding cell types. This is due to the complexity of the nervous system, which requires these stem cells to be controlled more than the different tissues. This also occurs when multiple stem cells are present in a niche, since these cells coordinate and interact with each other to produce specific cells needed by the surrounding tissue, which are regulated by niche signals.

Storage Niches

Storage niches contain quiescent stem cells, which are cells that remain inactive in a niche until they are required to respond to specific extracellular signals, after which the cells divide and migrate. The bulge region of the hair follicle in mice is a well-established example of a storage niche, housing quiescent epithelial stem cells. During the hair cycle, epithelial stem cells in the follicle are active, contributing to hair growth, while the bulge stem cells remain dormant. However, upon injury, the bulge stem cells are activated to facilitate hair regeneration [16].

EVOLUTION OF STEM CELLS

The first unicellular organisms are often considered analogs to stem cells due to their ability to self-replicate and differentiate into various forms. The transition between unicellular organisms and hydra could be more structured. In *Hydra*, a single epithelial cell is capable of performing many physiological functions, demonstrating a remarkable level of cellular plasticity. Similarly, in plants, single cells from adult tissues, such as those of carrot or tobacco, have the capacity to regenerate into a complete organism. The *piwi* gene in *Drosophila*, which regulates germline stem cells, and the *ZWILLE* gene in *Arabidopsis*, which controls shoot meristematic stem cells, share significant homology.

SIMILARITIES AND DIFFERENCES BETWEEN PLANT AND ANIMAL STEM CELLS

Similarities

Both plants and animals exhibit cellular, tissue, structural, organ, and systemic levels of regeneration, though the mechanisms may differ. In both plants and animals, injury often triggers regenerative processes to restore damaged tissues. Plant callus and animal blastema are specialized, undifferentiated structures that regenerate new tissues.

Differences

The regenerative capacity of plants and animals varies significantly. Generally, regeneration in animals is quite slow and varies considerably across different body parts. For instance, the skin and other tissues regenerate quickly, whereas the regeneration of organs such as the heart and stomach is a slow process. Neural cells in animals have limited regenerative capacity, and severe brain damage is often irreparable, although stem cell therapy offers a potential treatment. Meanwhile, in plants, the ability to regenerate is generally stronger than in animals, although this ability varies between species. For example, species such as *Taxus chinensis* and *Ginkgo biloba* have weaker regenerative abilities compared to lower plants such as *Ficus virens* and *Laminaria japonica*.

The distribution of these stem cells in plants and animals also varies to a greater extent. In plants, the root and shoot apical meristems (RAMs and SAMs) contain pluripotent stem cells, and these stem cells are present in the meristem for a long time. Animal tissues lack the aggregation of stem cells, as seen in plants;

however, stem cells are widely distributed in tissues and organs in small numbers [17].

INTRODUCTION TO STEM CELL THERAPEUTICS

Stem cells have made significant contributions to the medical field in the modern world. Indeed, stem cells possess a wide range of properties, including extensive proliferation, clonal properties, and differential potencies, which are applied in both biological and clinical fields. These stem cells are mainly used to repair damaged tissues and for organ regeneration. Cell therapy has shown promise in treating various degenerative disorders and other conditions, including diabetes mellitus, cirrhosis, liver diseases, and heart failure. Notably, the liver is the largest internal organ in the body, performing numerous functions, including detoxification, metabolism, and supporting digestion, as well as processing vitamins. Thus, liver homeostasis disruption leads to cirrhosis, non-alcoholic fatty liver disease (NAFLD), alcoholic liver disease, autoimmune liver disease (ALD), and liver failure. Stem cell therapy is one of the most effective methods for treating liver failure, as it can lead to liver regeneration and recovery. However, concerns regarding the genetic stability of certain stem cells remain, as these cells could potentially lead to tumor formation.

A randomized trial demonstrated that liver function improved and infections decreased in patients with hepatitis B virus-related acute-on-chronic liver failure (ACLF) who received allogeneic bone marrow-derived MSCs (BM-MSCs) by peripheral infusion. HSCs were also used in this therapy, where HSCs from peripheral blood were isolated and used in a Phase 1 clinical trial for patients with chronic liver failure. Interestingly, this treatment improved the serum albumin and bilirubin levels in these patients. Nonetheless, ethical concerns, particularly surrounding the use of ESCs, cloning, and informed consent, continue to be a major issue in stem cell research and therapy. Thus, the National Academy of Sciences (NAS) in the United States established a committee in 2003 to develop guidelines for institutions and individuals researching human ESCs. Such committees and various policies can minimize the risk of potential harm, establish informed consent procedures, and reduce the likelihood of therapeutic misconception, among other benefits [1, 18, 19].

Types of Stem Cells used in Therapeutics

The major stem cells that are used in therapeutics are as follows:

Embryonic Stem Cells

ESCs are extremely pluripotent and can be easily isolated and cultured. Furthermore, ESCs exhibit high productivity in culture and can also integrate easily into fetal tissue. However, the main disadvantages of ESCs are that these cells can be rejected by the immune system, differentiate inadequately, form tumors, or become contaminated.

Adult Stem Cells

ASCs are multipotent, have a reduced capacity for differentiation, and can sometimes be stimulated by drugs. The disadvantages of these cells include difficulty in isolation, poor differentiation when cultured, and handling challenges (Table 1) [20].

Table 1. Advantages, disadvantages, and therapeutic potential of embryonic and adult stem cells.

Attributes	Embryonic stem cells	Adult stem cells
Advantages	Pluripotent, easy to isolate, high productivity in cultures, and easy integration into fetal tissue.	Multipotent, can be stimulated by drugs.
Disadvantages	It can be rejected by the immune system, can differentiate inadequately, and can form tumors.	It has a reduced differentiation capacity, is difficult to isolate, and is difficult to handle.
Therapeutic potential	Treatment of blood disorders.	Tissue repair, organ restoration.

STEPS IN STEM CELL THERAPY

Four major steps exist in stem cell therapy.

Determination of Stem Cell Source

There are various types of stem cells available due to the increased demand for regenerative medical techniques in the market. However, certain risks are associated with these stem cells, primarily due to the methods used in cell manufacturing. Therefore, the strengths and weaknesses of all cells must be analyzed to produce the best therapeutic effects for various diseases. This process helps in preparing disease-targeting stem cells by selecting the appropriate stem cell for the specific disease.

Specification of Cell Dosage

The effectiveness of stem cell dosage in stem cell therapy must be determined through *in vitro* or *in vivo* practices. It is also crucial that the stem cell dosages are both safe and highly effective. Indeed, the standard method for determining stem cell dosage cannot be applied when stem cell therapy is used in high-risk areas, such as the Central Nervous System (CNS). However, most processes fail when a higher dosage is required for higher efficacy. Hence, each dosage for a separate stem cell must be determined individually before treatment. An increased dosage of CD34 cells has been observed to stimulate multilineage hematopoiesis at the early stage and reconstitution at the post-transplant stage. However, in Acute Myeloblastic Leukemia (AML), a high dose of HSCs was found to restore function and recover hematological and immunological functions; however, these effects were not unconditional. In the above study, higher doses of HSCs were shown to yield poorer results than lower doses.

Administrative Methods

The administration of stem cells can be divided into two types: systemic and local administration. Numerous targeted injections, including intramuscular and intracardiac injections, are employed in local transmission. Vascular routes, including intravenous and intra-arterial pathways, are used in systemic transmission. Intravenous administration can be used for myocardial repair in patients with diffuse myocardial disease by introducing MSCs. Once introduced, the MSCs are delivered into the infected myocardium through the systemic circulation and receive signals from injured tissues.

Stem Cell Manipulation for Effective Treatment

All stem cell therapies possess individual pros and cons. Therefore, we must be one hundred percent sure of the safety of these therapies. The risk factors associated with these therapies vary due to differences in stem cell manufacturing techniques, clinical experience, *etc*. Thus, clinicians and researchers must closely monitor the distribution of the administrative and target organ sites, as well as whether distribution across the entire body is required. The organ to which stem cells are distributed must be analyzed fully, including the administrative sites. CRISPR-based editing techniques have been utilized in the genetic engineering of hPSCs; a process that can be achieved by making the cell membrane of the hPSCs permeable and then delivering the CRISPR-Cas9 tools. Nanoparticles, extracellular vesicles, and viruses can be used for this carrier-dependent delivery, while physical and mechanical delivery methods can be used for carrier-independent delivery [21 - 23].

MAJOR BREAKTHROUGH IN STEM CELL THERAPY

A notable breakthrough in stem cell therapy occurred in 2006, when iPSCs were found to possess the ability to reprogram and regain their pluripotency with the help of four major factors: OCT4, SOX2, KLF4, and MYC; these are collectively known as the "Yamanaka factors." Prof. Shinya Yamanaka received the Nobel Prize in 2012 for this path-breaking discovery. Meanwhile, several methods exist for de-differentiating somatic cells to reclaim their pluripotency. Firstly, the differentiated cells (somatic cells) are incubated with extracts of pluripotent stem cells; it has been reported that pluripotency markers, such as OCT3/4, are re-expressed.

Another approach can also be employed, in which mouse embryonic fibroblasts and fibroblasts from the tail tip are cultivated using the knock-in technique, which involves inserting a neomycin reporter gene into the *Fbx* gene locus of these cells, a locus that is highly useful for maintaining pluripotency in these mouse cells. Additionally, 24 pluripotency-inducing factors have been introduced into these cells retrovirally, which were later narrowed down to four major iPSC genes: *Klf4*, *Sox2*, *c-Myc*, and *Oct3/4*. Interestingly, the *neomycin* gene can serve as a selection marker for the ESC-like colonies, which are pluripotent, from which iPSCs are selected.

Consequently, human iPSC technology has evolved in response to these discoveries, leading to its application in drug screening and disease modeling. Additionally, these approaches are gaining popularity due to the advantages of utilizing human iPSCs over traditional screening technologies, such as a reduced immune response when using autologous cells. Disease modeling in iPSCs involves isolating cells that contain the mutation(s) associated with the disease. This method is used in various neurodegenerative diseases, wherein cells containing the mutation are first isolated and then differentiated into the pre-disposed cell type. These cells are then used to identify the cause of the disease and the underlying pathological mechanisms. One of the major diseases for which iPSCs are used is Parkinson's disease (PD). The major features of PD are the selective loss of A9-type dopaminergic neurons, which project from the substantia nigra pars compacta (SNPc) region of the brain in the dorsal striatum. The metabolism of α-synuclein also plays a crucial role in the development of PD. Therefore, PD can be studied by creating three-dimensional (3D) models of midbrain tissue, also known as brain organoids.

Notably, iPSCs have also helped in shaping the current regenerative medicine research. For example, iPSCs are now being used for the ex vivo expansion of various blood components, which occurs during red blood cell (RBC) depletion in case of any injury or disease [24 - 26].

MAJOR TYPES OF STEM CELL THERAPY

Somatic Cell Nuclear Transfer

Somatic cell nuclear transfer (SCNT) is a crucial method for generating biologically identical pluripotent stem cells, which are essential for patient treatment. In this method, first, the nucleus is removed from an unfertilized oocyte—a process called enucleation. The nucleus from the donor somatic cell is then transferred into the enucleated cell. Following this, condensed metaphase chromosomes are formed when the nuclear membrane of the donor cell nucleus begins to degrade. This phenomenon is known as premature chromosomal condensation (PCC) and is triggered by the M-phase-promoting factors (MPFs) in the ooplasm. Oocyte activation is then induced by sperm-borne phospholipase C zeta 1 (PLCZ1). After activation, the genome in the donor cell begins to form the nuclear membrane. The donor somatic cells can then represent the healthy cells in the patient. Next, the cell undergoes complete reprogramming, during which the cell acts and behaves similarly to a somatic cell (Fig. 2). This method is highly advantageous as it generates genetically identical cells, thereby preventing these cells from being rejected by the immune system. However, this method has been completed in mice and primates, whereas many complications remain in human SCNT, primarily due to the low efficiency of blastocyst production. Nonetheless, SCNT has great therapeutic potential and can be used to treat degenerative diseases caused by mutations in mitochondrial DNA (mtDNA). Here, one of the major concerns was obtaining human oocytes, which was overcome by the egg-sharing scheme. The selection of the donor cell type and the method of preparation for SCNT are also crucial in realizing the therapeutic potential of this method for such diseases. Indeed, nuclear transfer-derived ESCs have shown low levels of mtDNA mutations and are suitable for therapeutic use in this disease. Although SCNT is successful in more than 20 mammalian species, the rate at which cloned animals reach term remains quite low. For example, about 70% of SCNT embryo development in mice is arrested before the blastocyst stage, and only 1–2% of the embryos reach term after being transferred to surrogate mothers. Although SCNT blastocysts and ntESCs were developed in Rhesus monkeys by Mitalipov's group more than 10 years ago, the rate of blastocyst formation remained at only about 16% and most of the subsequent attempts to clone a monkey have, until recently, been unsuccessful. One of the biggest challenges in the SCNT method is the epigenetic reprogramming, which is mainly responsible for the low birth rate in clones. Another major disadvantage is abnormal placental development, which can lead to multiple pathologies. Such cases are highly inevitable in SCNT pregnancies and represent the primary process through which the majority of the embryos are lost during the first trimester.

Nevertheless, many milestones have been achieved in SCNT research—the first was the successful cloning of Dolly the sheep in 1997. However, more recently, genetically engineered (GE) livestock has been produced, following trials that have been ongoing for over two decades. Meanwhile, precise insertions and deletions can be introduced for gene knock-ins and knockouts using engineered nucleases. SCNT represents the primary method for creating GE livestock with 70% knock-ins and 50% knock-outs. This method also eliminates mosaicism, thereby conveniently reducing the time and cost needed for GE animals [27 - 31].

Implementation of Stem Cell Therapy in Cancer Treatment

Stem cell therapy is used to treat various diseases, while cancer represents a major disease, with up to 90% of deaths related to cancer caused by metastasis. For cancer patients who have received an early diagnosis, understanding the molecular basis of metastasis is crucial to avoiding this progression. According to current theories, the colonization and physical transfer of a cancer cell from the primary tumor to the microenvironment of a distant tissue comprises two main events in the metastatic cascade.

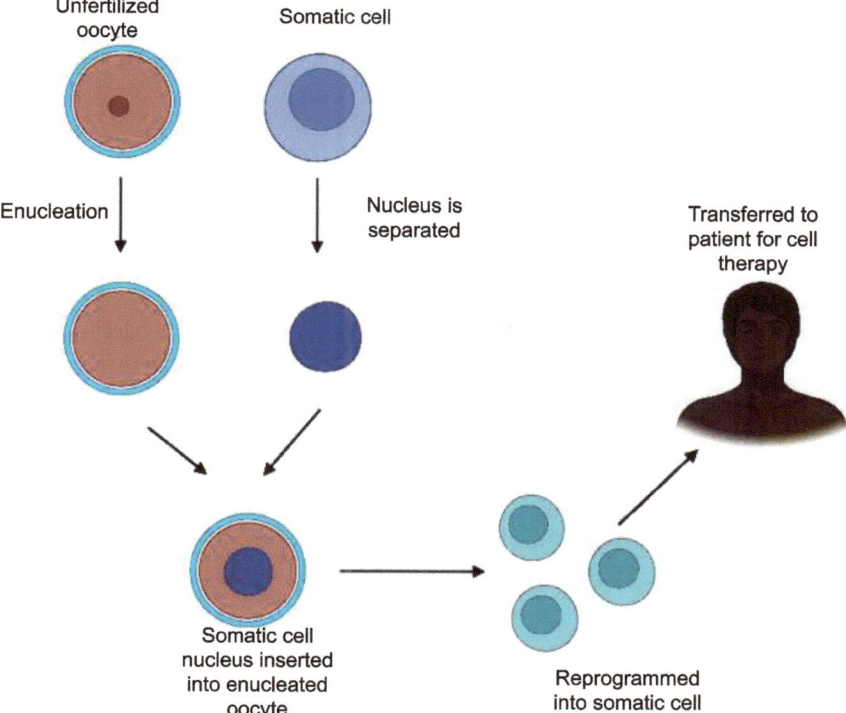

Fig. (2). A schematic of somatic cell nuclear transfer.

However, a major challenge in cancer treatment is the recurrence of these cancer cells, which may lead to the formation of chemotherapy-resistant cancer stem cells (CSCs). In such cases, therapeutic methods are involved that target the CSC-associated signaling pathways, which include the Wnt pathway, the Notch pathway, and the Hedgehog (Hh) pathway. The Wnt pathway plays a key role in maintaining the stemness of cancerous tumors; many CSC markers, such as CD24 and OCT-4, are upregulated through this pathway. Notch signaling hyperactivation has led to the maintenance of CSCs in various cancer types, such as colon, ovarian, pancreatic, and hepatic. Meanwhile, high expression of the Notch pathway is associated with a poor prognosis in hepatocellular carcinoma (HCC). This pathway also contributes to the initiation of tumors by $CD133^+$ CSCs. Aberrant Hh signaling is required to maintain the stemness and proliferation capacity of CSCs. This pathway is activated in many myeloma CSCs, with the *SMO* gene being overexpressed and high *Gli1* transcription.

Stem cells that promote tissue regeneration are used as a therapeutic method in the treatment of hematological cancers, such as those affecting the blood. Apart from targeting CSCs, normal stem cells, such as HSCs, are also used in bone marrow transplants and as drug-delivering agents to treat cancers. Autologous HSC transplantation (autoHCT) is used to treat various aggressive forms of myeloma and lymphoma. In this treatment, stem cells are reinfused into the patient after undergoing high-dose chemotherapy. Similarly, alloHCT is performed for patients with high-risk acute leukemia. AlloHCT, along with tyrosine kinase inhibitors, has been shown to decrease recurrence rates and also aids in the subsequent maintenance therapy of alloHCT. Stem cells can also be induced with prodrug-activating enzymes and interleukins, which act as anticancer agents, producing anti-tumor responses against various cancers [13]. Mesenchymal stem cells (MSCs) are now being genetically engineered to deliver enzymatic prodrugs to treat multiple cancers. For this purpose, human adipose tissue-derived MSCs (AT-MSCs) are being genetically engineered with a fused transgene encoding yeast cytosine deaminase (CD): uracil phosphoribosyltransferase (CDy::UPRT). This process has displayed high cytotoxic effects against breast, gastric, and glioma cancer cells *in vitro*. Similarly, genetically engineered MSCs producing CD are used to treat glioblastoma multiforme (GBM), the most aggressive and common brain tumor. These CD-MSCs exhibit anticancer effects by converting the non-toxic prodrug 5-fluorocytosine (5-FC) into the anticancer drug 5-fluorouracil (5-FU).

While stem cells have immense therapeutic potential, CSCs present a significant obstacle in cancer treatment due to their role in tumor initiation and resistance to therapy. Tumor initiation can occur through transformed differentiated cells or transformed tissue resident stem cells. This transformation can occur during tissue regeneration or as a result of an accelerated response to toxins, infections, diseases, or other factors. Meanwhile, oncogenes are overexpressed during this transformation, while tumor suppressor genes are underexpressed, leading to uncontrolled growth. Thus, since stem cells can already undergo extensive proliferation, only a few genomic changes are required to convert normal stem cells into CSCs. For example, low mutagenic changes that promote cancerous tumors from tissue-resident stem cell populations have been identified in more than 10% of gastric cancers. The transformation of such a tissue-resident stem cell population occurs due to uncontrolled, niche-independent proliferation resulting from certain genomic changes. CSCs controlling tumor initiation divide asymmetrically to maintain the pool of CSCs. This leads to the production of more CSCs, resulting in increased proliferative capacity. Tumor initiation also occurs through oncogenes, which are formed from proto-oncogenes, *via* chromosomal translocation, where a proto-oncogene is translocated from a location where it cannot be transcribed to an area where it can be transcribed. Tumor initiation also occurs through point mutations and gene amplification. Indeed, examples of chromosome translocation are seen in the MYC family of proto-oncogenes. The transactivation of the C-MYC transcription factor, resulting from C-MYC deletion, leads to the negative regulation of the *Adenomatous Polyposis Coli (APC)* gene. The subsequent knockdown of the *APC* gene can lead to carcinogenic activity of C-MYC. Human pancreatic cancer represents a type of cancer that possesses a point mutation, specifically a single missense mutation in the *KRAS* gene, which is responsible for transforming protein 21. Here, the codon 12 is modified to produce either aspartate or valine instead of glycine, leading to protein activation. An example of gene amplification is the increased production of the *c-MYC* gene in neuroblastoma [32 - 38].

Risks Associated with Stem Cell Therapy

Risk is defined as the probability of danger, and a risk factor is defined as a cause of harm. Thus, the risk factors associated with stem cell therapy include the source of the stem cells, the manipulation the stem cells have previously undergone, and the site of injection. Meanwhile, to assess the risks associated with stem cells, one must consider both the important and theoretical risks. Additionally, there are two types of risk factors: Intrinsic and extrinsic. The intrinsic properties of risks are based on a particular cell type, such as the origin of cells; for example, diseased or healthy donor, allogenic or autologous. In contrast, extrinsic properties consider the procurement, culturing, and storage of cells. The final risk factor lies in clinical characteristics, such as surgical procedures, mode of administration, and comorbidities. The formation of a tumor poses a significant risk in stem cell therapy since stem cells often share properties with cancer cells, including a long lifespan, resistance to apoptosis, and increased replication capacity. Moreover, both stem cells and cancer cells have similar growth regulation and control mechanisms, which makes both cell types a potential source of tumor formation. Stem cell administration can lead to modifications in the immunity of the host cells, potentially affecting the immune system. Thus, immune modulation is a major concern in this aspect. For example, adult MSCs have been shown to play a major role in immunomodulation by suppressing T cell proliferation, cytokine secretion, and cytotoxicity. These MSCs affect the production and secretion of antibodies, and also block the maturation and activation of antigen-presenting dendritic cells. Although MSCs help stimulate B cell viability, these cells may also lead to cell cycle arrest, thereby inhibiting the proliferation of these B cells. Another critical issue is the procurement of pure stem cells since contamination with other stem cells can cause numerous undesirable effects [39, 40].

RESEARCH PROSPECTS OF STEM CELL THERAPY

While using stem cells in clinical practices, it is important to consider the developmental role of stem cells and the cues for which the cells were designed, *e.g.*, ESCs can be used to treat various neurodegenerative diseases, where these cells are directed by specific cell cultures to grow into multiple neuronal subtypes. Thus, ESCs can be used in cell replacement therapy. Similarly, somatic stem cells can be used for bone marrow transplantation. Various studies have noted that stem cell therapies can be used to treat spinal cord injury (SCI). For this, stem cell therapy can be divided into two major categories. The first category comprises non-neural stem cells, including bone marrow MSCs, umbilical cord MSCs, and adipose tissue-derived MSCs. These cells are administered *via* intravenous or intrathecal methods, where grafting to injured areas is facilitated by a homing mechanism. Due to this, neural cell differentiation is limited, and replenishment of nervous lesions is also restricted. In the second category, neural stem progenitor cells (NSPCs) or iPSCs can be directly transferred and transplanted into the injured spinal cord; however, the transplantation procedure is invasive and requires surgery. Engrafting decides the fate of these transplanted cells, replacing the lost neural cells and restoring their function. The above-discussed strategy is called loading therapy.

The development of these stem cells depends on fundamental developmental principles and basic pathophysiological principles. An understanding of these two interactions will make cell therapy more tractable for various diseases.

According to a recent study, future directions in stem cell research may involve developing cardiomyocytes (CMs) using human pluripotent stem cells (hPSCs). Notably, previous disease modeling efforts were limited to transgenic animal models and heterologous expression studies. In contrast, creating CMs using hPSCs is advantageous, as this method can lead to unlimited proliferation and the generation of cells specific to the patient for disease modeling. Meanwhile, a major challenge primarily noted for this method is that different heart diseases target different regions and different CM subtypes in the heart. Therefore, chamber-specific hPSC–CMs are necessary for modeling such diseases. Most hPSC–CM differentiation results in the formation of ventricular CMs. However, variations have been noted between laboratories and cell lines, which is considered a significant challenge [41 - 43].

PLANT STEM CELLS

These stem cells are located in the meristem of plants, which plays a crucial role in the development of plant organs. These meristems contribute to the plasticity, growth, and development of plants. Regeneration of plant cells depends on the action of plant hormones, such as auxins and cytokinins, through signaling pathways. Target of rapamycin (TOR) plays a role in both plant and animal regeneration. In plants, TOR plays a role in promoting root and stem growth, as well as callus formation. Moreover, TOR facilitates the integration of nutrients, energy, hormones, and environmental signals [44].

Plant Tissue Culture

Plant tissue culture is a general term for the cultivation of any component of a plant, including cells, tissues, and organs, in a nutritive medium under aseptic circumstances. Since Haberlandt established this technique, it has been widely used in plant research at both basic and applied levels. This technique is now being used to culture disease-free plants, plants with rare genotypes, plant-derived metabolites, and other valuable products.

Plant tissue culture can also be used in agricultural biotechnology through methods such as micropropagation. Here, the plant is first prepared for *in vitro* culture, then these plants are inoculated in a suitable medium containing all the plant growth regulators and nutrients required for growth (Fig. **3**). Next, the plant is multiplied by sub-culturing, which helps in the development of shoots. Then, the shoots produced in the previous stage are inoculated in the same or a different medium for root production. Lastly, plants grown in such *in vitro* conditions cannot be directly transferred to the field, as they have not been exposed to the harsh field conditions. Therefore, these plants are acclimatized by reducing the humidity, altering the light requirements, and transferring the plants to pots containing a mixture of clay, sand, and peat, among other materials. Then, the plants are transferred to the greenhouse, and finally, to the field [44, 45].

Steps in Plant Tissue Culture

Pre-propagation

Initially, the plant of interest is chosen; the plants must be free from insects and diseases. If necessary, some pretreatments, such as the application of fungicides or pesticides, can be used.

In vitro Cultivation

Small plant fragments (explants) must be removed from the process, or seeds must be used and their surfaces sterilized. This sterilization can be achieved using various chemicals, such as sodium hypochlorite, ethanol, sterile distilled water, and detergent.

Culturing of Explants

The explants are then placed in the appropriate culture medium and cultured for a brief period. This leads to the multiplication of shoots or roots. Then, the contaminated explants are identified and removed, and the surviving explants are processed to the next stage.

Micropropagation

This step can vary depending on the desired culture. In the case of micropropagation, the propagated shoots are transferred to root-promoting culture media, and the callus is multiplied through callogenesis. Then, the callus and root cultures are cultured using bioreactors to scale up the associated cultivation.

Hardening

After being micropropagated, the plants are hardened to create individual plants that can perform photosynthesis. The hardening process is accomplished gradually by allowing the plants to acclimatize to the ex vitro environment, whereby the plants are taken from low to high light intensity and from high to low humidity [44].

Challenges in Plant Tissue Culture

Despite the many advantages of plant tissue culture, this technique also presents challenges. For example, economically viable plants and those on the verge of extinction are not currently being cultivated using plant tissue culture. Therefore, new plant tissue culture techniques must be developed to grow and protect these plants. Moreover, certain plants show poor growth during plant tissue culture. Thus, this technique must be further improved to incorporate all variations of plants [45].

INTRODUCTION TO EPIGENETICS IN STEM CELLS

Epigenetics refers to the effects that alter gene expression without modifying the underlying DNA sequence. These can be categorized into DNA methylation, histone modification, and activities of small interfering RNAs. These modifications can help us understand gene expression patterns and the mechanisms involved in gene modulation during development. As epigenetic markers can be replicated during cell division, analyzing these markers can help elucidate the mechanisms through which pluripotency is achieved at a molecular level.

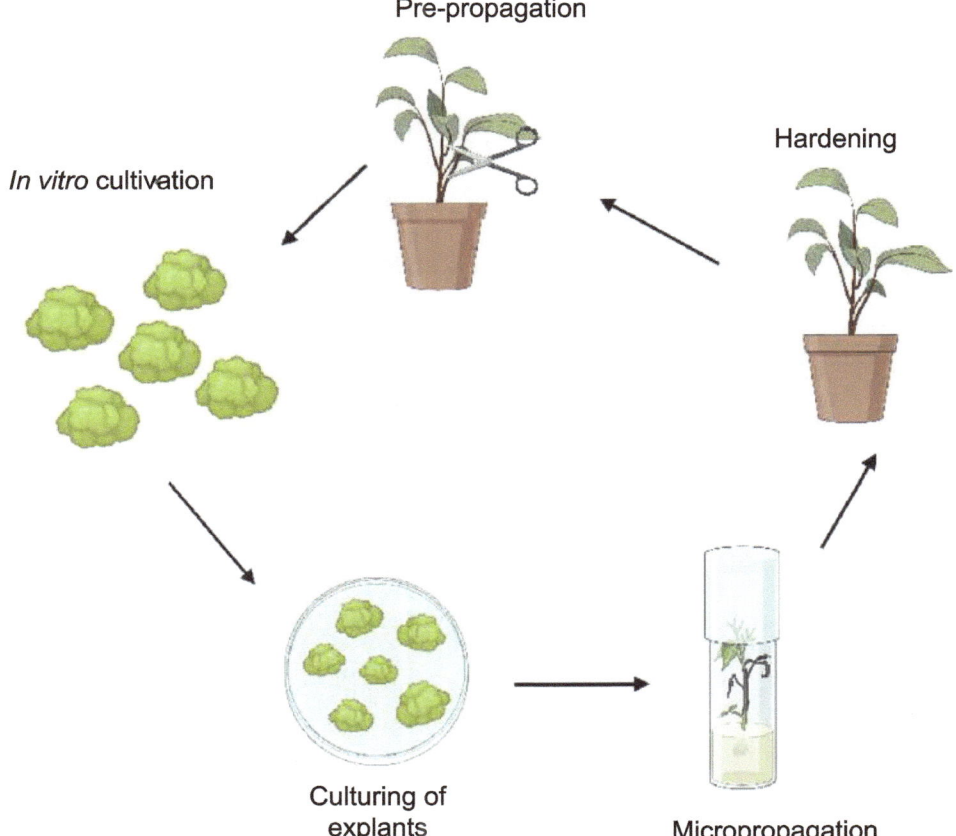

Fig. (3). A schematic of plant tissue culture.

Some epigenetic signals regulate the division of stem cells. For example, liver stem cells and neural stem cells are derived from the same precursor, but form either mature liver cells or mature neural cells during differentiation. Here, cellular memory plays a major role in the differentiation of these stem cells. However, the epigenetic and memory mechanisms have ceased to prevent proliferation permanently.

The epigenetics of stem cells dates back to the 1900s when Conrad Waddington proposed the concept of an epigenetic landscape. Indeed, Waddington's landscape consisted of crests and troughs, while the landscape of the surface represented the genes that have undergone epigenetic modifications.

Epigenetics plays a key role in dental pulp stem cells (DPSCs), which are isolated from human dental pulp. These cells are unique owing to their ability to differentiate into diverse cell types, including neural progenitors, odontoblasts, osteoblasts, melanocytes, and adipocytes. DPSCs can be used for craniomaxillofacial regeneration and dental tissue engineering. Meanwhile, DPSCs exhibit low DNA methylation levels both *in vivo* and *in vitro*, as well as high acetylation levels of histones. These levels weaken the chromatin–DNA interaction, which allows gene expression. Similarly, lncRNA H19 suppresses DNMT3b activity, which, in turn, decreases the methylation level of DLX3 in DPSCs, leading to odontogenic differentiation [46 - 48].

KEY EPIGENETIC MECHANISMS IN STEM CELLS

There are three major types of epigenetic mechanisms in stem cells:

DNA Methylation

DNA methylation involves the covalent modification of cytosine at the C5 nucleotide in the CpG island. Almost 70% of the CpG dinucleotides in mammals are methylated, predominantly by DNA methyltransferases (DNMTs). DNA methylation plays a major role in the epigenetic reprogramming of ESCs. The pluripotent properties of these stem cells are influenced by genes such as *OCT4*, *SOX2*, and *NANOG*. Studies have shown that pluripotency can be induced in fibroblasts derived from ESCs or ASCs through the exogenous expression of genes such as *KLF4*, *OCT4*, and *SOX2*. Meanwhile, iPSCs that are similar to ESCs have less methylation in the *OCT4* promoter and have high efficiency of germline incorporation [49].

Histone Modification

Histone modifications involve post-translational histone modifications that occur within the first 30 amino acids of the N-terminal histone domains, also known as the histone tails. There are various types of modifications, including methylation, acetylation, and phosphorylation, among others. These modifications play a major role in mesenchymal stem cells (MSCs), which are known to influence major processes such as adipogenesis, which is controlled by transcription factors including PPARG and CEBPA. The disruption of MSC differentiation leads to the abnormal accumulation of adipose tissue. Properly differentiated adipogenic MSCs have a unique histone modification. Trimethylation of histone H3 marks the activation of genes for transcription or the repression of genes for silencing transcription. Cell stemness and gene differentiation lead to bivalent histone methylation, which activates H3K4 trimethylation and represses H3K27 trimethylation in DPSCs [48, 50].

Chromatin Remodeling

Two major chromatin states exist in stem cells: Euchromatin and heterochromatin. Euchromatin can be transcribed easily and is less condensed, whereas heterochromatin is highly condensed and cannot be easily transcribed. ESCs regulate their chromatin through specialized transcription factors. Chromatin in ESCs is packaged and controlled by a high euchromatin–heterochromatin ratio, special nucleosome remodelers, transcription factors, and bivalent gene promoters. Chromatin remodeling also involves the action of chromatin remodeling proteins. Chromatin remodeling proteins are ATP-dependent complexes that contain a catalytic ATPase subunit in addition to proteins that regulate protein–protein and chromatin–protein interactions. Moreover, chromatin remodeling proteins play a crucial role in the pluripotency and function of ESCs, and this also significantly contributes to stemness and early development. Proteins such as chromodomain☐helicase DNA☐binding (CHD) proteins, especially CHD1, have been seen to regulate pluripotency and open chromatin in mouse ESCs, in association with H3K4me3. CHD7 has also been noted to regulate neural crest migration and neural crest gene expression in human ESCs during embryogenesis through an association with the PBAF chromatin remodeling complex [49, 51, 52].

Retinitis pigmentosa is an eye disease associated with retinal damage. Patients often experience retinal dystrophy, which can lead to a loss of visual field, blurred vision, or poor vision in low-light conditions. This inherited disorder is also associated with early-onset night blindness, for which there is no treatment for restoration. This disorder is caused by an inactivating mutation in the *rhodopsin*

(*RHO*) gene, which codes for the GPCR of the same name. Rhodopsin is necessary for low-light vision, as this protein is found in the rods of the retina. When activated by photons, the rhodopsin proteins initiate a chain of chemical reactions that are converted into electrical signals, which the brain interprets as vision.

EPIGENETIC MECHANISMS IN MESENCHYMAL STEM CELLS (HUNTINGTON'S DISEASE)

Huntington's disease is an autosomal dominant and neurodegenerative disorder that is caused by over-expansion of a CAG repeat in the *Huntington* gene, which results in the production of an abnormal huntingtin protein. Another observed cause is the downregulation of trophic factors. MSCs can be engineered to overexpress these factors, especially those derived from the bone marrow, potentially leading to a cure for this disease.

Bone marrow-derived MSCs can differentiate into three lineages: osteogenic, adipogenic, and chondrogenic. However, epigenetic modifications are crucial for converting MSCs into neuronal cells. This is achieved through the expansion of MSCs in culture or passaging, during which some of the genes undergo methylation and acetylation.

When these BM-MSCs are introduced into the brains of mice with HD, cells with a higher number of passages in the brain are observed to be more effective in reducing behavioral problems in mouse models with HD [47].

EPIGENETIC CELL REPROGRAMMING IN IPSCS

Cell-type identity and function are known to be determined by epigenetic processes. Therefore, the underlying epigenome must be rewired to reprogram one cell type into another. Furthermore, somatic cells can be converted into iPSCs and instructed to develop into specific cell types. Regenerative medicine uses cells that have been reprogrammed to repair organs and tissues that have been damaged both physically and functionally. SCNT, cell fusion, ectopic expression of specific transcription factors, and microRNA production are some of the reprogramming techniques employed. Cellular reprogramming clearly results in the original cell replacing its original characteristics with newly gained traits (Fig. 4).

Some methods that are used for reprogramming cells in iPSCs are as follows:

Ectopic Expression of Transcription Factors

The common methods for reprogramming cells involve using viruses, such as adenovirus, lentivirus, or retrovirus, to deliver one or more transcription factors into primary cells. These transcription factors can alter the epigenetic state of gene regulatory regions, thereby influencing the expression of genes. The transcription factors Oct4, Sox2, and Klf4 are known to maintain the pluripotency and differentiation potential of ESCs. During reprogramming, these factors facilitate the Polycomb repressive complex 2 (PRC2) proteins in repressing lineage-specific genes in differentiated cells, thereby converting these cells into iPSCs.

CRISPR–Cas9-Based Genome Editing for Reprogramming

CRISPR–Cas9 is used for the cellular reprogramming of cells by targeting specific genes and epigenetic regulators. Moreover, CRISPR–Cas9 activates the promoters of the *Oct4*, *Sox2*, *Klf4*, and *Myc* genes, thereby converting human fibroblast cells into iPSC cells. The reprogramming efficiency is increased by targeting genes involved in the activation of the Alu-motif embryonic genome.

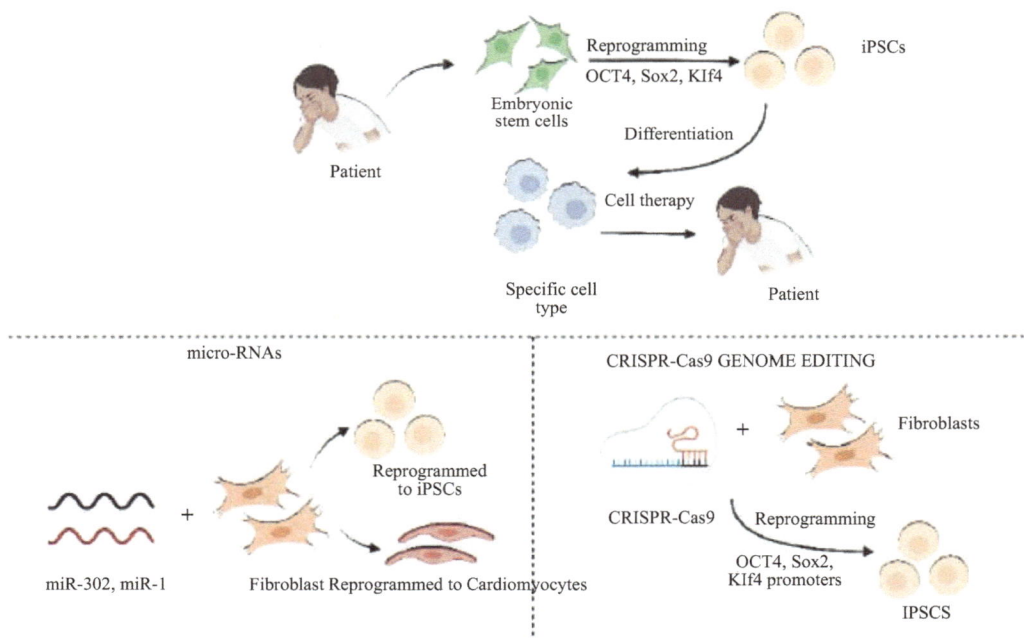

Fig. (4). Epigenetic cell reprogramming in iPSCs.

CRISPR–Cas9 gene editing can be used to modify DNA methylation, regulating and facilitating cell reprogramming. DNA methyltransferase Dnmt3a or DNA demethylase Tet1 is fused with Cas9 to target and alter the regulatory elements of genes that require epigenetic reprogramming. Thus, the created iPSCs can be used for either *in vitro* drug screening to find treatments for diseases or for cell replacement therapy. It has been observed that patient-specific reprogrammed cells help avoid cell rejection and provide a limitless supply of cells for treatment. This approach also promotes the concept of transdifferentiation, which enables the direct reprogramming of cells within the affected tissue or organ, thereby eliminating the need to derive pluripotent cells from outside the body. The epigenetic reprogramming of cells has been successfully evaluated for its safety and effectiveness in treating diseases [53].

EPIGENETIC THERAPY TARGETING BONE MARROW STEM CELLS

As individuals age, the bone marrow mesenchymal stem cells (BMSCs) gradually lose their ability to renew and differentiate properly, resulting in impaired bone tissue regeneration and disruption of bone tissue homeostasis. Moreover, changes in epigenetic modifications are a key reason for BMSC dysfunction during epigenetic therapies. Therefore, targeting these abnormal modifications could help reduce stem cell aging and restore their function.

Mesenchymal stem cells (MSCs) originate from the embryonic mesoderm, which can self-renew and differentiate into multiple cell types. The epigenetic modifications are reversible; thus, the epigenetic changes during BMSC aging provide potential targets for treatment.

DNA methylation is catalyzed and maintained by the DNA methyltransferase (DNMT) family. The inhibition of DNMT1, DNMT3A, and DNMT3B in BMSCs can prevent aging, promote angiogenesis, and inhibit cancer progression. The inhibition of DNMT3A reduces the methylation of CpG islands upstream of the *Sod2* gene, leading to the expression of a dominant antioxidant enzyme for anti-aging.

Methylation is reduced at most CpG sites in aging stem cells, although a few cells express increased methylation. The heterochromatin region predominantly exhibits reduced methylation, leading to loose chromosomes and genomic instability during the aging process. Moreover, increasing TET enzymes in BMSCs may provide new therapeutic strategies to prevent bone loss and promote the recovery of age-related bone diseases [54].

RECENT TRENDS

Stem Cells in Neurodegenerative Diseases

Stem cells can differentiate and replace damaged or lost cells, thereby supporting the growth and repair of neural tissue. Many neurological disorders are marked by widespread neuron death and the limited ability of the brain to regenerate. The combination of additional drugs enhances the success of stem cell therapy for treating neurological conditions with immunomodulatory properties, paracrine effects, and the ability to differentiate into various cell types. Neurological stem cells (NSCs) are utilized for neuroregeneration, as these cells can differentiate into neurons and astrocytes, and promote remyelination and axonal growth. Stem cells can release paracrine factors that support neuroregeneration, such as neurotrophic factors, cytokines, and growth factors. Given the ability of NSCs to migrate to tumor sites, this approach could be highly effective for the targeted delivery of chemotherapy drugs [55].

Therapeutic Potential of Dental Stem Cells

ASCs are found in nearly all adult tissues, including bone marrow, dental tissues, and skin. Dental stem cells (DSCs), which are obtained from premolars and wisdom teeth, are often extracted for orthodontic purposes. Human DSCs are utilized for the restoration of structural defects and the treatment of various diseases. These multipotent cells are present in numerous easily accessible dental tissues. For example, DPSCs are best obtained from the third molar; these cells exhibit a strong ability to proliferate, self-renew, and differentiate into odontoblast-like cells and osteoblasts, thereby forming dentin and bone. DPSCs can generate bone-like tissues *in vivo* and regulate cell differentiation, adhesion, developmental maturation, and the production of cytoskeletal elements. Thus, these characteristics increase the potential of DSCs for pulp, dentin, and periodontal regeneration [56].

Cytokines Regulate the Fates of Hematopoietic Stem Cells

HSCs are a rare cell type capable of regenerating the entire blood and immune systems after transplantation, which makes these cells a potential cure for various blood disorders. HSCs are obtained from umbilical cord blood, which is a convenient source for transplantation; however, umbilical cord blood often contains a limited number of HSCs, necessitating effective engraftment and long-term blood cell regeneration. Hematopoietic stem and progenitor cells (HSPCs) are regulated by external signals, such as cytokines, to adapt to the ever-changing needs of the body for blood cell production. Meanwhile, various recombinant cytokines are often added to human HSC cultures, along with serum albumin, to promote the rapid induction of intracellular signaling pathways. These specific cytokine combinations then direct particular cell fates by activating overlapping signaling pathways.

The functional heterogeneity present in human HSPCs *in vitro* and *in vivo* helps to measure signaling dynamics in individual human HSPCs over time upon delivery of combined cytokines with known concentrations [57].

Bioengineered Scaffolds to Deliver Stem Cells for Wound Healing

It has been demonstrated that microRNAs (miRNAs), either alone or in conjunction with other transcription factors, can reprogram cells. Interestingly, miRNAs are readily transported into cells to initiate reprogramming due to their small size. For instance, certain microRNAs, such as miR-1, miR-133, miR-208, and miR-499, promote the conversion of fibroblasts into CMs. Meanwhile, miR-1 and miR-133 are known to inhibit CM proliferation and the transition from the G1 to S phase in the cell cycle. Furthermore, miR-208 facilitates cardiac lineage determination by promoting the expression of cardiac transcription factors. Additionally, miR-499 controls cell division in the latter phases of heart development. Furthermore, miR-9 and miR-124 may also be used to reprogramme fibroblasts into neurons. These molecules affect the differentiation process into neuronal lineages by modulating the SWI/SNF-like BAF chromatin-remodeling complexes in neural progenitor cells [58].

ASSESSING THE RISKS AND FUTURE POTENTIAL IN STEM CELL REPROGRAMMING

Numerous risk factors can influence the risks associated with stem cell treatment, including the type of stem cells employed, their history of acquisition and culture, the degree of modification, and the injection location. The dangers connected to various stem cell-based medications may also vary significantly due to the wide range of associated risk factors. First, risk factors related to the inherent cellular characteristics of a specific cell type or class of stem cells; second, extrinsic risk factors brought about by the acquisition, handling, culture, or storage of the cells; last, risk factors related to the clinical features (such as immunosuppression, surgical procedures, site and mode of administration, or co-morbidities). The third is the possibility of adventitious agents and the spread of human pathogens. Lastly, there can be other risk variables that provide undiscovered dangers to patients.

Pluripotent stem cells, including ESCs and iPSCs, have not yet been utilized in therapeutic settings. The hazards associated with products containing these cells, such as the possibility of cancer development, are considered, although not intolerable, given the traits expressed by these cells of unrestricted self-renewal and rapid multiplication. However, the overwhelming majority of small-scale clinical studies using mesenchymal stem/stromal cells (MSCs) in regenerative medicine applications have not shown significant health issues, indicating that (MSCs) treatments are potentially safe. In contrast, serious adverse effects have been documented in some clinical studies; nevertheless, this highlights the need for further research, especially in the areas of long-term safety and biological processes.

It appears illogical to increase the dose and frequency of transplanted iPSCs to attain the desired therapeutic potential, since this might raise the risk of pulmonary embolism and increase expenses. Therefore, the primary goal is to maximize the therapeutic advantages and potency of MSCs. Numerous methods have been developed to maximize stem cells, which are broadly classified into two groups: genetic alteration and non-genetic modification (pre-activation). Ultimately, iPSCs will generate or overexpress functional genes that allow these cells to enhance paracrine actions, promote migration and homing, and resist apoptosis and a hostile microenvironment. Although several studies suggest that gene-transfected iPSCs have greater therapeutic potential than wild-type iPSCs, safety concerns remain the primary obstacle to the future clinical therapeutic use of genetically modified iPSCs. Viral integration into the host genome poses a carcinogenic risk, and viral expression systems may trigger immunological and inflammatory responses in the host. Furthermore, it is essential to thoroughly elucidate the therapeutic potential and long-term functional benefits of genetically modified iPSCs. As a result, pre-activation—a collection of very effective non-genetic alteration techniques—is a practical and alternative strategy to enhance IPSC treatment results [59].

CONCLUSION

Stem cells hold immense potential due to their unique ability to differentiate into various cell types and self-renew, making these cell types invaluable in regenerative medicine, tissue engineering, and the treatment of a wide range of diseases. From ESCs with high pluripotency to ASCs with more restricted differentiation capabilities, each cell type presents unique advantages and challenges. The employment of these cells in therapeutic applications, such as in blood disorders and tissue regeneration, is promising yet accompanied by significant ethical, technical, and safety concerns. Recent advancements in stem cell research, including epigenetics, continue to open new avenues for understanding diseases and developing novel treatments. However, a balance must be struck between advancing science and addressing the ethical implications to ensure responsible and safe applications in clinical settings.

CONSENT FOR PUBLICATION

All visual representations were created by the authors involved in this publication and were not sourced from any other publications. No further consent is required.

CONFLICT OF INTEREST

This book chapter received no specific grant from any funding agency in the public, commercial, or not-for-profit sector. The authors have no conflict of interest related to this publication.

ACKNOWLEDGEMENTS

We would like to acknowledge Ms. Kruthika P, Mr. Kumaran K, Ms. Raksa Arun, Ms. Srisri SatishKartik, Ms. Sanjana Dhayalan, and Mr. Sasitharan T for their contribution in proofreading and editing works.

REFERENCES

[1] Kolios G, Moodley Y. Introduction to stem cells and regenerative medicine. Respiration 2013; 85(1): 3-10.
[http://dx.doi.org/10.1159/000345615] [PMID: 23257690]

[2] Alison MR, Poulsom R, Forbes S, Wright NA. An introduction to stem cells. J Pathol 2002; 197(4): 419-23.
[http://dx.doi.org/10.1002/path.1187] [PMID: 12115858]

[3] Shahriyari L, Komarova NL. Symmetric vs. asymmetric stem cell divisions: an adaptation against cancer? PLoS One 2013; 8(10): e76195.
[http://dx.doi.org/10.1371/journal.pone.0076195] [PMID: 24204602]

[4] Lu F, Zhang Y. Cell totipotency: molecular features, induction, and maintenance. Natl Sci Rev 2015; 2(2): 217-25.
[http://dx.doi.org/10.1093/nsr/nwv009] [PMID: 26114010]

[5] De Miguel MP, Fuentes-Julián S, Alcaina Y. Pluripotent stem cells: origin, maintenance and induction. Stem Cell Rev Rep 2010; 6(4): 633-49.
[http://dx.doi.org/10.1007/s12015-010-9170-1] [PMID: 20669057]

[6] Sobhani A, Khanlarkhani N, Baazm M, et al. Multipotent Stem Cell and Current Application. Acta Med Iran 2017; 55(1): 6-23.
[PMID: 28188938]

[7] Tatullo M, Codispoti B, Paduano F, Nuzzolese M, Makeeva I. Strategic Tools in Regenerative and Translational Dentistry. Int J Mol Sci 2019; 20(8): 1879.
[http://dx.doi.org/10.3390/ijms20081879] [PMID: 30995738]

[8] Zakrzewski W, Dobrzyński M, Szymonowicz M, Rybak Z. Stem cells: past, present, and future. Stem Cell Res Ther 2019; 10(1): 68.
[http://dx.doi.org/10.1186/s13287-019-1165-5] [PMID: 30808416]

[9] Coopman K. Large scale compatible methods for the preservation of human embryonic stem cells: Current perspectives. Biotechnol Prog 2011; 27(6): 1511-21.
[http://dx.doi.org/10.1002/btpr.680] [PMID: 22235484]

[10] O'Donoghue K, Fisk NM. Fetal stem cells. Best Pract Res Clin Obstet Gynaecol 2004; 18(6): 853-75.
[http://dx.doi.org/10.1016/j.bpobgyn.2004.06.010] [PMID: 15582543]

[11] Alatyyat SM, Alasmari HM, Aleid OA, Abdel-maksoud MS, Elsherbiny N. Umbilical cord stem cells: Background, processing and applications. Tissue Cell 2020; 65: 101351.
[http://dx.doi.org/10.1016/j.tice.2020.101351] [PMID: 32746993]

[12] Biswas A, Hutchins R. Embryonic stem cells. Stem Cells Dev 2007; 16(2): 213-22.

[http://dx.doi.org/10.1089/scd.2006.0081] [PMID: 17521233]

[13] Shi Y, Inoue H, Wu JC, Yamanaka S. Induced pluripotent stem cell technology: a decade of progress. Nat Rev Drug Discov 2017; 16(2): 115-30.
[http://dx.doi.org/10.1038/nrd.2016.245] [PMID: 27980341]

[14] Torre P, Flores AI. Current Status and Future Prospects of Perinatal Stem Cells. Genes (Basel) 2020; 12(1): 6.
[http://dx.doi.org/10.3390/genes12010006] [PMID: 33374593]

[15] Walker MR, Patel KK, Stappenbeck TS. The stem cell niche. J Pathol 2009; 217(2): 169-80.
[http://dx.doi.org/10.1002/path.2474] [PMID: 19089901]

[16] Ohlstein B, Kai T, Decotto E, Spradling A. The stem cell niche: theme and variations. Curr Opin Cell Biol 2004; 16(6): 693-9.
[http://dx.doi.org/10.1016/j.ceb.2004.09.003] [PMID: 15530783]

[17] Liu L, Qiu L, Zhu Y, et al. Comparisons between Plant and Animal Stem Cells Regarding Regeneration Potential and Application. Int J Mol Sci 2023; 24(5): 4392.
[http://dx.doi.org/10.3390/ijms24054392] [PMID: 36901821]

[18] Hoang DM, Pham PT, Bach TQ, et al. Stem cell-based therapy for human diseases. Signal Transduct Target Ther 2022; 7(1): 272.
[http://dx.doi.org/10.1038/s41392-022-01134-4] [PMID: 35933430]

[19] King NMP, Perrin J. Ethical issues in stem cell research and therapy. Stem Cell Res Ther 2014; 5(4): 85.
[http://dx.doi.org/10.1186/scrt474] [PMID: 25157428]

[20] Bagno LL, Salerno AG, Balkan W, Hare JM. Mechanism of Action of Mesenchymal Stem Cells (MSCs): impact of delivery method. Expert Opin Biol Ther 2022; 22(4): 449-63.
[http://dx.doi.org/10.1080/14712598.2022.2016695] [PMID: 34882517]

[21] Liras A. Future research and therapeutic applications of human stem cells: general, regulatory, and bioethical aspects. J Transl Med 2010; 8(1): 131.
[http://dx.doi.org/10.1186/1479-5876-8-131] [PMID: 21143967]

[22] Lotfi M, Morshedi Rad D, Mashhadi SS, et al. Recent Advances in CRISPR/Cas9 Delivery Approaches for Therapeutic Gene Editing of Stem Cells. Stem Cell Rev Rep 2023; 19(8): 2576-96.
[http://dx.doi.org/10.1007/s12015-023-10585-3] [PMID: 37723364]

[23] Mousaei Ghasroldasht M, Seok J, Park HS, Liakath Ali FB, Al-Hendy A. Stem Cell Therapy: From Idea to Clinical Practice. Int J Mol Sci 2022; 23(5): 2850.
[http://dx.doi.org/10.3390/ijms23052850] [PMID: 35269990]

[24] Lewitzky M, Yamanaka S. Reprogramming somatic cells towards pluripotency by defined factors. Curr Opin Biotechnol 2007; 18(5): 467-73.
[http://dx.doi.org/10.1016/j.copbio.2007.09.007] [PMID: 18024106]

[25] Valadez-Barba V, Cota-Coronado A, Hernández-Pérez OR, et al. iPSC for modeling neurodegenerative disorders. Regen Ther 2020; 15: 332-9.
[http://dx.doi.org/10.1016/j.reth.2020.11.006] [PMID: 33426236]

[26] Jiang Z, Han Y, Cao X. Induced pluripotent stem cell (iPSCs) and their application in immunotherapy. Cell Mol Immunol 2014; 11(1): 17-24.
[http://dx.doi.org/10.1038/cmi.2013.62] [PMID: 24336163]

[27] Hemmat S, Lieberman D, Most S. An introduction to stem cell biology. Facial Plast Surg 2010; 26(5): 343-9.
[http://dx.doi.org/10.1055/s-0030-1265015] [PMID: 20853224]

[28] Greggains GD, Lister LM, Tuppen HAL, et al. Therapeutic potential of somatic cell nuclear transfer for degenerative disease caused by mitochondrial DNA mutations. Sci Rep 2014; 4(1): 3844.

[http://dx.doi.org/10.1038/srep03844] [PMID: 24457623]

[29] Matoba S, Zhang Y. Somatic Cell Nuclear Transfer Reprogramming: Mechanisms and Applications. Cell Stem Cell 2018; 23(4): 471-85.
[http://dx.doi.org/10.1016/j.stem.2018.06.018] [PMID: 30033121]

[30] Malin K, Witkowska-Piłaszewicz O, Papis K. The many problems of somatic cell nuclear transfer in reproductive cloning of mammals. Theriogenology 2022; 189: 246-54.
[http://dx.doi.org/10.1016/j.theriogenology.2022.06.030] [PMID: 35809358]

[31] Polejaeva IA. 25th ANNIVERSARY OF CLONING BY SOMATIC CELL NUCLEAR TRANSFER: Generation of genetically engineered livestock using somatic cell nuclear transfer. Reproduction 2021; 162(1): F11-22.
[PMID: 34042607]

[32] Kontomanolis EN, Koutras A, Syllaios A, et al. Role of Oncogenes and Tumor-suppressor Genes in Carcinogenesis: A Review. Anticancer Res 2020; 40(11): 6009-15.
[http://dx.doi.org/10.21873/anticanres.14622] [PMID: 33109539]

[33] Walcher L, Kistenmacher AK, Suo H, et al. Cancer Stem Cells—Origins and Biomarkers: Perspectives for Targeted Personalized Therapies. Front Immunol 2020; 11: 1280.
[http://dx.doi.org/10.3389/fimmu.2020.01280] [PMID: 32849491]

[34] Chang D-Y, Jung J-H, Kim AA, et al. Combined effects of mesenchymal stem cells carrying cytosine deaminase gene with 5-fluorocytosine and temozolomide in orthotopic glioma model. Am J Cancer Res 2020; 10(5): 1429-41.
[PMID: 32509389]

[35] Ho YK, Woo JY, Tu GXE, Deng LW, Too HP. A highly efficient non-viral process for programming mesenchymal stem cells for gene directed enzyme prodrug cancer therapy. Sci Rep 2020; 10(1): 14257.
[http://dx.doi.org/10.1038/s41598-020-71224-2] [PMID: 32868813]

[36] Yin W, Wang J, Jiang L, James Kang Y. Cancer and stem cells. Exp Biol Med (Maywood) 2021; 246(16): 1791-801.
[http://dx.doi.org/10.1177/15353702211005390] [PMID: 33820469]

[37] Bair SM, Brandstadter JD, Ayers EC, Stadtmauer EA. Hematopoietic stem cell transplantation for blood cancers in the era of precision medicine and immunotherapy. Cancer 2020; 126(9): 1837-55.
[http://dx.doi.org/10.1002/cncr.32659] [PMID: 32073653]

[38] Manni W, Min W. Signaling pathways in the regulation of cancer stem cells and associated targeted therapy. Beijing: MedComm 2022; p. 3.

[39] Gao F, Chiu SM, Motan D A L, et al. Mesenchymal stem cells and immunomodulation: current status and future prospects. Cell Death Dis 2016; 7(1): e2062-2.
[http://dx.doi.org/10.1038/cddis.2015.327] [PMID: 26794657]

[40] Herberts CA, Kwa MSG, Hermsen HPH. Risk factors in the development of stem cell therapy. J Transl Med 2011; 9(1): 29.
[http://dx.doi.org/10.1186/1479-5876-9-29] [PMID: 21418664]

[41] Shinozaki M, Nagoshi N, Nakamura M, Okano H. Mechanisms of Stem Cell Therapy in Spinal Cord Injuries. Cells 2021; 10(10): 2676.
[http://dx.doi.org/10.3390/cells10102676] [PMID: 34685655]

[42] Daley GQ, Goodell MA, Snyder EY. Realistic prospects for stem cell therapeutics. Hematology (Am Soc Hematol Educ Program) 2003; 2003(1): 398-418.
[http://dx.doi.org/10.1182/asheducation-2003.1.398] [PMID: 14633792]

[43] Reilly L, Munawar S, Zhang J, Crone WC, Eckhardt LL. Challenges and innovation: Disease modeling using human-induced pluripotent stem cell-derived cardiomyocytes. Front Cardiovasc Med 2022; 9: 966094.

[http://dx.doi.org/10.3389/fcvm.2022.966094] [PMID: 36035948]

[44] Espinosa-Leal CA, Puente-Garza CA, García-Lara S. *In vitro* plant tissue culture: means for production of biological active compounds. Planta 2018; 248(1): 1-18.
[http://dx.doi.org/10.1007/s00425-018-2910-1] [PMID: 29736623]

[45] Aslam A, Bibi A, Bibi S, *et al.* Plant Tissue Culture and Crop Improvement Climate-Resilient Agriculture. Cham: Springer International Publishing 2023; Vol. 2: pp. 841-62.

[46] Spivakov M, Fisher AG. Epigenetic signatures of stem-cell identity. Nat Rev Genet 2007; 8(4): 263-71.
[http://dx.doi.org/10.1038/nrg2046] [PMID: 17363975]

[47] Srinageshwar B, Maiti P, Dunbar G, Rossignol J. Role of Epigenetics in Stem Cell Proliferation and Differentiation: Implications for Treating Neurodegenerative Diseases. Int J Mol Sci 2016; 17(2): 199.
[http://dx.doi.org/10.3390/ijms17020199] [PMID: 26848657]

[48] Hussain A, Tebyaniyan H, Khayatan D. The Role of Epigenetic in Dental and Oral Regenerative Medicine by Different Types of Dental Stem Cells: A Comprehensive Overview. Stem Cells Int 2022; 2022: 1-15.
[http://dx.doi.org/10.1155/2022/5304860] [PMID: 35721599]

[49] Li X, Zhao X. Epigenetic regulation of mammalian stem cells. Stem Cells Dev 2008; 17(6): 1043-52.
[http://dx.doi.org/10.1089/scd.2008.0036] [PMID: 18393635]

[50] Ren J, Huang D, Li R, Wang W, Zhou C. Control of mesenchymal stem cell biology by histone modifications. Cell Biosci 2020; 10(1): 11.
[http://dx.doi.org/10.1186/s13578-020-0378-8] [PMID: 32025282]

[51] Klein DC, Hainer SJ. Chromatin regulation and dynamics in stem cells. Curr Top Dev Biol 2020; 138: 1-71.

[52] Harikumar A, Meshorer E. Chromatin remodeling and bivalent histone modifications in embryonic stem cells. EMBO Rep 2015; 16(12): 1609-19.
[http://dx.doi.org/10.15252/embr.201541011] [PMID: 26553936]

[53] Basu A, Tiwari VK. Epigenetic reprogramming of cell identity: lessons from development for regenerative medicine. Clin Epigenetics 2021; 13(1): 144.
[http://dx.doi.org/10.1186/s13148-021-01131-4] [PMID: 34301318]

[54] Wang X, Yu F, Ye L. Epigenetic control of mesenchymal stem cells orchestrates bone regeneration. Front Endocrinol (Lausanne) 2023; 14: 1126787.
[http://dx.doi.org/10.3389/fendo.2023.1126787] [PMID: 36950693]

[55] Mattei V, Delle Monache S. Mesenchymal Stem Cells and Their Role in Neurodegenerative Diseases. Cells 2024; 13(9): 779.
[PMID: 38727315]

[56] Zhou Y, Xu T, Wang C, Han P, Ivanovski S. Clinical usage of dental stem cells and their derived extracellular vesicles. 2023; pp. 297-326.

[57] Wang W, Zhang Y, Dettinger P, *et al.* Cytokine combinations for human blood stem cell expansion induce cell-type- and cytokine-specific signaling dynamics. Blood 2021; 138(10): 847-57.
[PMID: 33988686]

[58] Hazrati R, Davaran S, Omidi Y. Bioactive functional scaffolds for stem cells delivery in wound healing and skin regeneration. React Funct Polym 2022; 174: 105233.

[59] Drela K, Stanaszek L, Nowakowski A, Kuczynska Z, Lukomska B. Experimental Strategies of Mesenchymal Stem Cell Propagation: Adverse Events and Potential Risk of Functional Changes. Stem Cells Int 2019; 2019: 7012692.
[PMID: 30956673]

CHAPTER 15

Cancer Stem Cells: Catalysts of Cancer Progression

Vedika Kartha[1], Saloni Semwal[1], Lakshmi Sai Varshini Yedavalli[1], Disha Kamath[1], S. Pooja[1], Dannie Macrin[2], Satish Ramalingam[1,*] and K.N. Aruljothi[1]

[1] *Department of Genetic Engineering, SRM Institute of Science and Technology, Kattankulathur, Chengalpattu, Tamil Nadu, India*

[2] *Department of Computational Biology, Institute of Bioinformatics, Saveetha School of Engineering, SIMATS Saveetha Institute of Medical and Technical Sciences, Chennai, India*

Abstract: Within tumors, a small number of cells exist alongside the majority of more rapidly dividing cells that constitute most of the tumor. This slowly growing subset of cells is referred to as cancer stem cells (CSCs), and they can regenerate themselves, allowing for the continuous growth of the tumor. CSCs were first reported in acute myeloid leukemia by Bonnet and Dick in 1997 and have been identified in other forms of cancer, including the brain, breast, lung, and liver tumors. These cells also show a striking resemblance to normal stem cells, for instance, in their ability to give rise to both the progenitor and malignant cells via asymmetric mitosis. These cells act on their own and play a significant role in the progression of tumors, inducing metastatic foci and bearing chemoresistance and radiotherapy resistance to standard treatments. This resistance can be attributed to several mechanisms, including active DNA repair, non-cycling state, and efflux of the drugs. They are components of the immune system that allow them to interact with other immune cells, thus escaping any immune responses while at the same time aiding in the recurrence of the tumors. Many crucial signaling pathways in the body, such as Wnt/β-catenin, Notch, and Hedgehog, control neural stem cells' actions. New therapies have been integrated against cancer aimed at cancer stem cells, such as specific marker blocking, miRNA/LncRNA therapy, and immunotherapy. There is also great potential in new strategies, such as nanotechnology-based targeting of cancer stem cells to reduce the chances of tumor recurrence. Management techniques seek to eradicate the cancer stem cell population to mitigate the chances of recurrence and enhance treatment success.

---

[*] **Corresponding author Satish Ramalingam:** Department of Genetic Engineering, SRM Institute of Science and Technology, Kattankulathur, Chengalpattu, Tamil Nadu, India; E-mail: satishr@srmist.edu.in

K.N. Aruljothi, Prakash Gangadaran, Krishnan Anand, Satish Ramalingam, K. Kumaran & Kruthika Prakash (Ed.)
All rights reserved-© 2026 Bentham Science Publishers

Keywords: ALDH, Asymmetric division, Cancer heterogeneity, Cancer stem cells, Chemoresistance, Differentiation, DNA repair, Epigenetics, Epithelial-mesenchymal transition, Hedgehog pathway, Immunotherapy, Markers, Metastasis, Microenvironment, Notch pathway, Proliferation, Quiescence, Self-renewal, Tumor growth, Tumor microenvironment.

INTRODUCTION TO CANCER STEM CELLS

Cancer stem cells are small subunits of cells formed within a solid tumor that are capable of self-renewal, leading to continuous tumor growth. They are similar to normal stem cells and can divide to form undifferentiated masses of cells. The mitosis of a cancer stem cell is asymmetric and can give rise to both progenitor and malignant cells. The identification and characterization of these cells were first done by Bonnet and Dick (1997) in acute myeloid leukemia (AML) in CD34+ CD38- sets; later, cancer stem cells were also spotted in other types of tumors in the brain, breasts, lung, and liver. These stem cells, which are originally undifferentiated cells, would differentiate for the renewal of cells or tumorigenesis.

History and Evolution of Cancer Stem Cells

1997: Scientists Bonnet and John Dick identified masses of cells that would divide uncontrollably in leukemia when those cells were put in mice. They would then provide evidence that supported the hypothesis of the formation of cancer stem cells.

2000: Cancer stem cells were studied in a broader spectrum, and their existence was attributed to solid tumors that would be present in the brain, breast, and colon cancer. The solid tumors were identified when the cell samples were tested with molecular/surface markers and functional assays.

2010: Later, around 2010, researchers began looking in depth for the functional understanding of cancer stem cells. They focused on differentiating cancer stem cells from normal stem cells and their interactions with other cells, the pathways associated with the cells, and how to treat cancers that are represented by cancer stem cells.

Current Concerns and Challenges

With ongoing research on cancer stem cells, researchers have developed target therapies that work under the mechanism of targeting the specific cancer stem cell and working to eliminate them. However, cancer stem cells are very complex structures and are heterogeneous in nature. Different cancers have different types of cancer stem cells, and to use target technology to destroy them is extremely complicated; hence, newer and more effective approaches have been developed to destroy these cells in cancerous tumors.

CHARACTERISTICS OF CANCER STEM CELLS

Cancer stem cells can be caused by differentiation and self-renewal in multipotent, tissue-specific, mature, or progenitor stem cells. In self-renewal, stem cells can continue to exist throughout an organism's lifetime, whereas in differentiation, progenitors and mature cells are passed on for regeneration purposes or tissue genesis. Differentiation is typically a one-way process, as after division, the cell loses the self-renewal property, but in the case of self-renewal in stem cells, the progenitor cells end up dividing to form mature cells and do not lose the self-renewal characteristic (Figs. **1** and **2**) [1, 2].

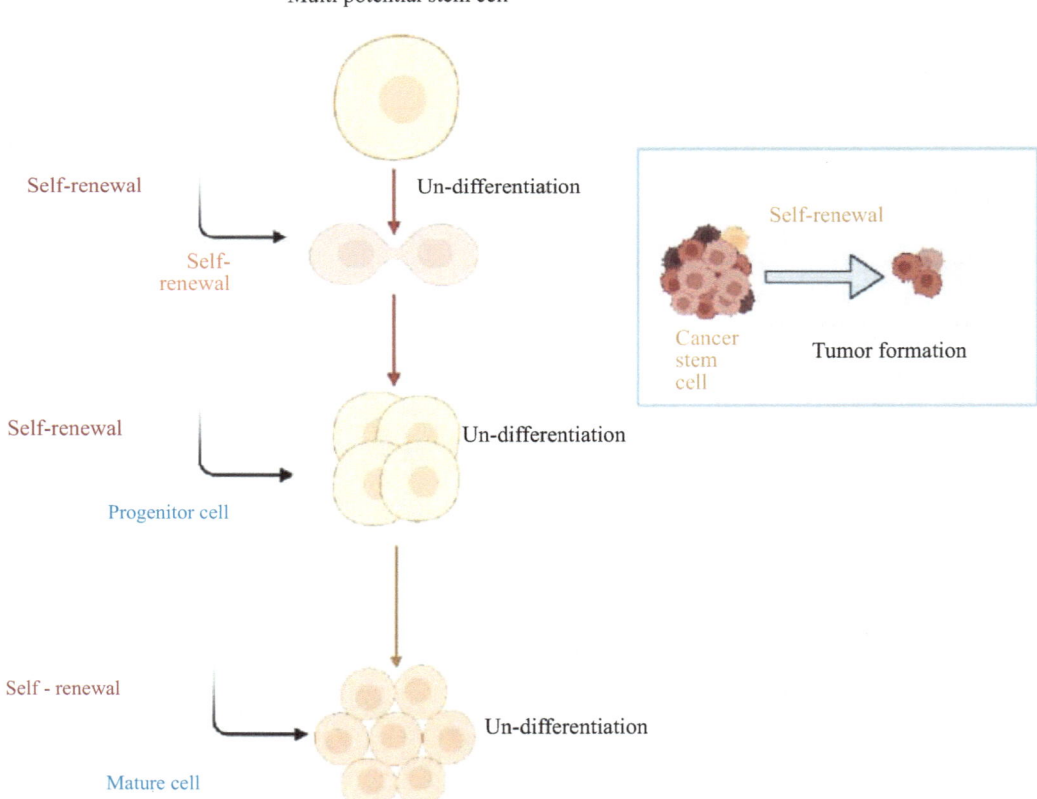

Fig. (1). Differentiation of multi-potent stem cells.

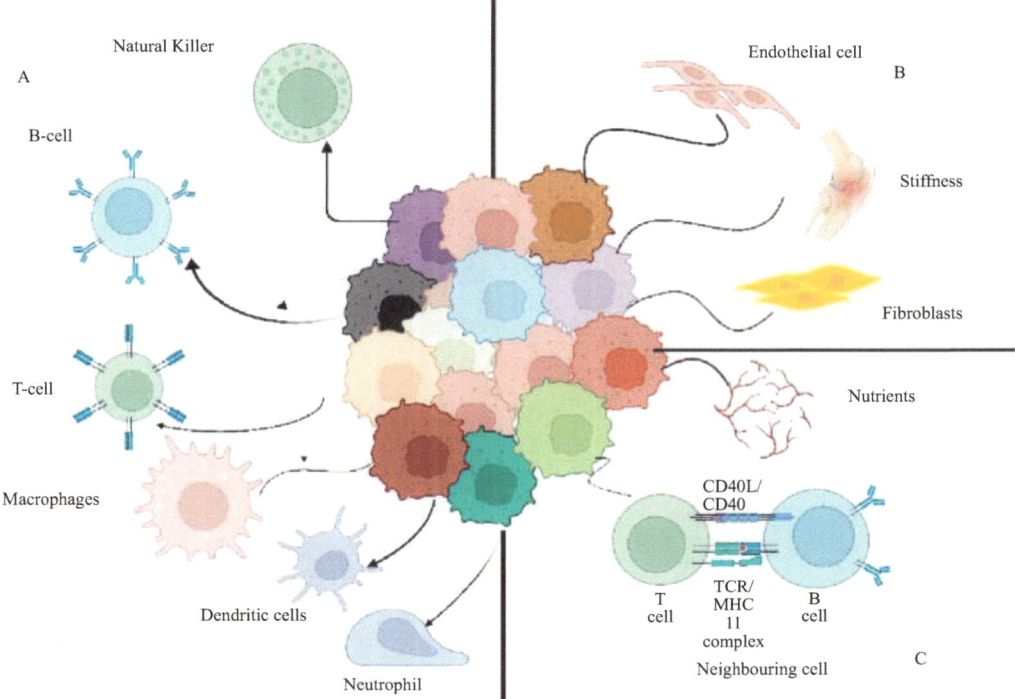

Fig. (2). Different types of cancer stem cells.

There are various surface markers that can detect stem cells by targeting specific antigens that are present in cells. These types of markers contain different forms of expression patterns and sublocalization, especially in cancer stem cells, in comparison to normal stem cells. Cancer stem cells with surface markers such as CD133, CD444, and nestin are found in tumors and are usually produced through the self-renewal or differentiation process. For easy identification of the cancer stem cells, certain techniques like functional assays, filtration methods, and xenotransplantation are used.

Interaction between Immune and Cancer Stem Cells

Various methods have been introduced to treat cancer, such as chemotherapy, open surgery, therapeutic methods, etc., but the possibility of recurrence is still high. Within the tumors themselves, the ability of tumor cells to spread and proliferate is caused by certain clusters of cells that exist in tumor cells known as cancer stem cells. Cancer stem cells also manage to control immune cells, which contain T cells, B cells, NK cells, and macrophages, and for the maintenance of the tumor microenvironment, certain inhibitors (Immune Checkpoint Inhibitors) stimulate the release of inhibitors programmed cell death ligand (PD-L1), lymphocyte activation gene-3 (LAG3) as such. Cancer cells can eventually take over and evade immune cells through a process known as immunoediting, which enables tumor progression. Cancer stem cells have not only been presented through immune targets but also EVs with several other factors such as interleukins (ILs), MMP, VEGF, that take part in the cross-talk between cancer stem cells and tumor microenvironment [3, 4].

Intrinsic Features of Cancer Stem Cells

Cancer stem cells usually undergo asymmetric division, in which:

1. Some cells keep the organelle compartments mainly for maintaining self-renewal
2. Other types proliferate and create new forms of tumor cells.

Cancer stem cells also have certain genetic/epigenetic alterations that help with further progression, survival rate, and division of tumor cells. These genetic/epigenetic features are what build up the intrinsic heterogeneity level of cancer stem cells. Intrinsic factors also contribute to the plasticity of cancer stem cells, and retainment of cell plasticity plays a significant role in the growth and survival of the cells. Genomic changes such as DNA mutations, chromosomal amplifications, and deletions control the maintenance of the aggressive cell state of cancer, but compared to these genomic changes, epigenetic reprogramming manages adaptation and resistance of the cell to its particular environment.

BIOLOGICAL PROPERTIES OF CANCER STEM CELLS

Marker Expression

Cancer stem cells are a special cell population present within the tumor mass that possesses the capability of self-renewal and induction of the progression of cancer. The identification of such cells is done with the aid of specific markers present on the cell surface. These markers act as identifiers and facilitate understanding of cancer stem cells' roles in growth and treatment resistance within cancers.

CD44

CD44 can be expressed on many cell types, including cancer stem cells; it participates in cell-cell adhesion and the migration of cells. In a range of cancers, including breast cancer and head and neck cancer, cancer stem cells have been identified expressing the marker CD44. This marker, importantly, pertains to the ability of cancer stem cells to adhere to the ECM, while maintaining their proliferative capabilities and expansion potential. Furthermore, CD44 also plays a part in conferring resistance to cancer stem cells against various forms of cancer therapy. Thus, by interacting with the ECM, the cell surface protein CD44 enables cancer stem cells to attach to their surroundings and maintain many of their stem cell-like traits. Such interaction supports tumor growth and has implications for therapy resistance.

CD133

Another cell surface protein having broad associations with cancer stem cells is prominin-1, generally known as CD133. The expression of CD133 has been utilized to identify cancer stem cells within various cancers of the brain, gliomas, colon, and prostate. Cells that express CD133 are typically more tumorigenic-capable of initiating a tumor and sustaining it. This marker becomes important for studies on the properties of cancer stem cells and why they could resist conventional therapies. It has served as a specific marker for cancer stem cells in brain and colon cancers. Being more tumorigenic, CD133+ cells generally outweigh the CD133- ones. In the diagram, one can better visualize the distribution and importance of CD133 in cancer stem cells' capability to both initiate and sustain tumors.

ALDH (Aldehyde Dehydrogenase)

ALDH refers to a class of enzymes involved in the intracellular processing of aldehydes. High activity levels of ALDH are one of the distinctive features of cancer stem cells. The ALDH-mediated detoxification is involved in the development of resistance capacity against chemotherapy drugs by cancer stem cells. The more cellular activity ALDH possesses, the more these cells are able to maintain their properties as stem cells and resist treatment. Thus, ALDH is considered a very important marker point in understanding drug resistance by cancer stem cells and their roles in recurrence. The activity of ALDH within the cancer stem cells means describing the mode in which such cells detoxify harmful substances and develop resistance to chemotherapy. High activities of ALDH are associated with the potential of cancer stem cells to evade treatment and their contribution to recurrence.

Microenvironment

The niche, wherein the cancer stem cells reside, sculpturally molds its characteristics into what is commonly referred to as the tumor microenvironment. It comprises a myriad of components responsible for the maintenance of cancer stem cells per se and also for their resistance to therapies.

Extracellular Matrix

The ECM is made up of a three-dimensional network of proteins and molecules surrounding cells and extracellular structures, providing their architectural integrity. These maintain cell functions and tissue architecture and thus, by maintaining the three-dimensional structure of the ECM, mediate the adhesion of cancer stem cells, modulating their growth and sensitivity to treatments. Generally, the general proteins in the tumor ECM, which include fibronectin, laminin, and collagen, interact directly with cancer stem cells, influencing their survival and resistance properties to therapies. These include the roles of interaction with cancer stem cells, which allow for their adhesion and, by that means, modulate their growth and response to therapy. All these may influence the ways through which cancer stem cells may respond to the treatments for their role in tumor progression.

Immune Cell

Immune cells are part of the body's defense machinery that is engineered to recognize and destroy such abnormal cells, including tumor cells. Through communication, these cancer stem cells protect the immune cells from destruction. For example, cancer stem cells can secrete some factors in the microenvironment that dampen the immune response and, as such, escape the immunity to give less effective treatments. Cancer stem cells can influence immune cells to protect themselves from being destroyed, complicating the effectiveness of immunotherapy

Signaling Molecules

Signaling molecules are chemical means through which cells communicate with one another. They include growth factors and cytokines.

On the other hand, the signaling molecules were significantly involved in the regulation of cancer stem cells. In this aspect, Wnt, Notch, and Hedgehog pathways have been implicated in maintaining properties of cancer stem cells and tumorigenesis. These pathways can also impact responsiveness to therapies and the re-population of tumors by cancer stem cells following treatments. Signaling pathways such as Wnt, Notch, and Hedgehog drive the properties of cancer stem cells. These pathways are driving the behavior of cancer stem cells toward tumor growth and responses to therapies [5, 6].

TYPES OF CELLS SIGNALLING PATHWAYS RESPONSIBLE FOR CANCER STEM CELLS

There are some pathways that are involved in regulating cancer stem cells, such as Wnt/β-catenin, Notch, and Hedgehog (Hh) pathways, which play a major role in the growth, self-renewal, and differentiation of the cancer stem cells. In the case of the Wnt/β-catenin pathway, the regulation of transcription factor β-catenin is controlled by this pathway, where canonical Wnt/β-catenin controls factors like cell differentiation, cell proliferation, etc., whereas the non-canonical pathway of Wnt/β-catenin protein focuses on self-renewal characteristics, cell polarity, migration, and expression of molecular markers associated with the cancer stem cell. Genes coding for the Wnt/β-catenin pathway can give rise to several mutations in cancers, including lymphoma, leukemia, breast cancer, gastric cancer, and colorectal cancer (CRC) [7 - 9].

Notch Pathway

The notch pathway consists of transmembrane proteins - notch ligands (DLL1, DLL3, DLL4, JAG1, and JAG2) and notch receptors (Notch1-4). When a notch ligand on one cell binds to a receptor of another cell, it initiates proteolytic cleavage of the receptor, which is controlled by enzymatic pathways of metalloproteinases (ADAMs) and γ-secretase. This interaction secretes a notch intracellular domain (NICD) into the cytoplasm, which in turn activates the transcription of target genes through the CBF1 region. The notch pathway also deals with the self-renewal and cell differentiation of cancer stem cells in breast, ovaries, and hepatocellular carcinoma (HCC) [10].

Hedgehog Pathway

Hedgehog pathway signaling deals with tumor growth and self-renewal of cancer stem cells, and in the case of colorectal cancer, the cells express the Indian hedgehog (IHH) gene, which contributes to the maintenance of colorectal cancer-initiating cells. By regulating cellular proliferation, differentiation, and migration, the Hedgehog pathway is mostly responsible for the development of many organs during embryogenesis, including the nervous system, skeleton, limbs, lung, heart, and gut. The Hedgehog pathway is mostly dormant in most postnatal tissues and is usually segregated within the skin cells, keratin in the hair and skin, and tissues in the teeth. Compared to the other pathways, it has been studied that the hedgehog pathways help control the population formation of stem and progenitor cells [11 - 13].

CANCER HETEROGENEITY

The concept of cancer heterogeneity, or the structural and functional diversity of cells within one tumor, has contributed to the deeper understanding of tumor biology and the ways of targeting them. The tumoral heterogeneity can stem from cancer stem cells, which are a unique group of small tumor subpopulations with special qualities that initiate, sustain, and defend treatment against cancer.

Cancer heterogeneity can be defined as the variation of cell makeup within any one tumor, in terms of genotype, epigenotype, and phenotype. This variability can occur at the following levels:

- Intertumoral Heterogeneity: This refers to differences in the same type of tumors seen in different patients. For example, there are two patients with breast cancer, each of whose tumors will have different traits and will require a different way of treatment.
- Intratumoral Heterogeneity: It refers to the differences within a single tumor, and it includes genetic makeup differences of individual tumor cells. It will lead to the emergence of subpopulations with different properties.
- Metastatic Heterogeneity: It defines the differences between the primary tumor and its metastases. Metastases will change and orient genetically and physiologically to that new environment.

Key Role in Creating Heterogeneity: The ability of cancer stem cells and self-renew and differentiate into different populations of tumor cells is crucial for creating heterogeneous cellular variability in the tumor. This organization can also be compared to that of tissues, where the embryonic or adult stem cell can give rise to any of its differentiated cell lineages. Further, cancer stem cells are well-defined by other molecular patterns and functions, such as communication, and their dysregulation is frequently linked to diseases like cancer, underscoring their critical role in maintaining normal cellular function.

- Enhanced Resistance: This is a well-known distinguishing functional characteristic of the cancer stem cell population, where this generation is known to be more resistant to standard treatment regimens such as chemotherapy and radiotherapy.
- Tumor Initiation: Cancer stem cells are highly efficient at initiating tumor growth due to their self-renewal properties.

THE ROLE OF CANCER STEM CELLS IN TUMOR DEVELOPMENT AND METASTASIS

Cancer stem cells represent a small proportion of tumor cells that express characteristic properties of usual stem cells, including self-renewal and multilineage differentiation potential. They are considered to play an important role in tumor progression, metastasis, and recurrence. Unlike the bulk of the tumor cells, cancer stem cells usually reside in a quiescent state, allowing them to evade conventional therapies that mainly target actively dividing cells. Therefore, such cells are associated with the recurrence of tumors and the development of tumor growth to distant organs. They also play an active role in metastasis.

Mechanisms through which Cancer Stem Cells Mediate Tumor Progression and Metastasis

- Self-renewal and Differentiation: Cancer stem cells provide the tumor with heterogeneity through the generation of tumor cells of different lineages. Their self-maintaining properties help them to sustain the tumor continuously, while their differentiating properties encourage cell diversity in the tumor. By this principle, cancer stem cells are capable of initiating and propagating tumors even when the bulk of the tumor cells has been treated.
- Tumor Initiation and Growth: Cancer stem cells have very high tumor-generating capacity compared to the usual tumor cells. Indeed, studies have demonstrated that a few cancer stem cells can give rise to whole tumors, explaining their crucial role in the initiation and maintenance of tumor growth. These cells have typically different signaling pathways, such as Wnt, Notch, and Hedgehog, that help in their survival and proliferation and make them resistant to therapeutic interventions.
- Intratumoral and Intertumoral Heterogeneity: Cancer stem cells give rise to diverse tumor cells with genetic and phenotypic variations, leading to intratumoral and intertumoral heterogeneity. Cancer stem cells drive this heterogeneity, particularly in metastatic lesions, which can vary significantly from the primary tumor, resulting in differing therapeutic vulnerabilities.

Cancer Stem Cells and Metastasis

- Metastasis is the leading cause of cancer-related mortality, and cancer stem cells play a central role in this process. A key link between cancer stem cells and metastasis is their ability to undergo epithelial-mesenchymal transition (EMT), a cellular transformation that imparts increased aggressiveness, invasiveness, migratory potential, and survival capabilities.
- Epithelial-Mesenchymal Transition: EMT represents an important process in metastasis, wherein cancer stem cells lose their epithelial features, such as cell

adhesion and cell polarity, and gain mesenchymal features, such as motility and invasiveness. This transition is driven by the downregulation of epithelial markers like E-cadherin and the upregulation of mesenchymal markers like vimentin and N-cadherin. Acquiring mesenchymal properties allows cancer stem cells to detach from the primary tumor, invade local tissues, enter the bloodstream, and migrate through the blood flow to distant sites.
- Migration and Invasion: EMT provides migratory and invasive capacities to the cancer stem cells, enabling them to disrupt the basement membrane and extracellular matrix-essential barriers to metastatic dissemination. After reaching the circulatory system, the survival and immune-evading capabilities of the cancer stem cells are increased, thereby allowing them to travel to organs that are far away. At these metastatic sites, the cancer stem cells have the ability to return to an epithelial phenotype through mesenchymal-to-epithelial transition, which facilitates the colonization and expansion of secondary tumors (Fig. 3).

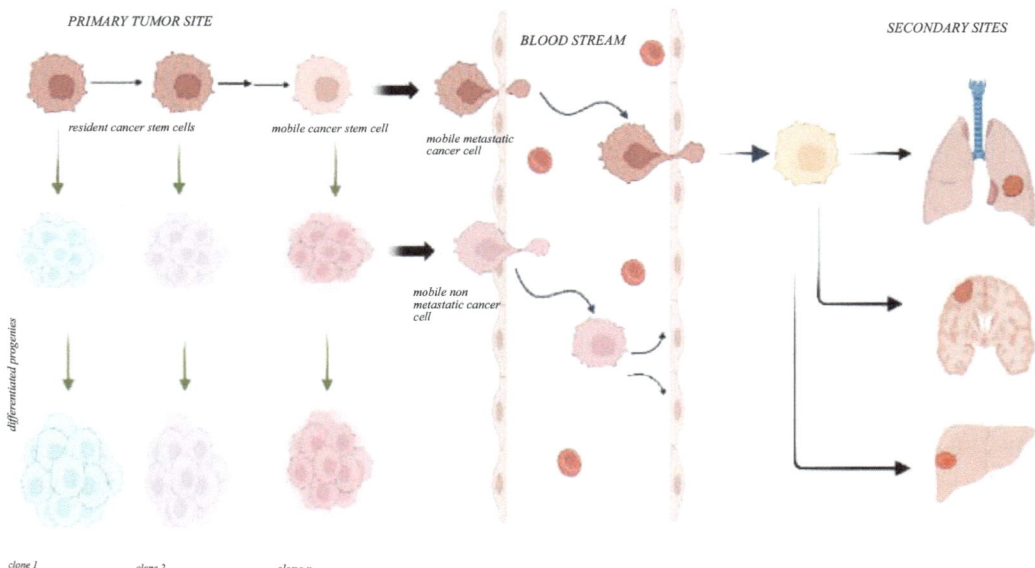

Fig. (3). Metastasis of cancer.

Interaction with the Tumor Microenvironment (TME)

Interaction with the tumor microenvironment in metastasis is a highly dynamic and complex ecosystem. This has been associated with an important role in the metastasis of cancer stem cells. It offers a niche that supports better survival, stemness, and metastatic potential of cancer stem cells.

- Cancer Stem Cell Niches in Metastasis: Cancer stem cells enjoy specialized niches within the TME, a location that is particularly safe from threats such as immune attacks and therapeutic agents. These niches are often hypoxic and rich in stromal cells that further promote cancer stem cells' stemness and metastatic capabilities. The interaction between cancer stem cells and their microenvironment is vital for the preparation of cancer stem cells toward metastatic dissemination, thus enabling them to acquire additional survival traits that facilitate metastasis.
- Immune Evasion and Metastasis: Cancer stem cells evade immune surveillance both in the primary tumor and during metastasis by manipulating the immune microenvironment. These cells secrete immunosuppressive factors such as TGF-β, IL-10, and VEGF, which recruit immune cells to dampen anti-tumor immune responses and provide a more permissive environment for metastasis. This immune evasion not only promotes the survival of the primary tumor but also highly facilitates the cancer stem cells' survival in the circulation and during metastatic colonization.

Cancer Stem Cells and Angiogenesis in Tumor Progression

Cancer stem cells promote tumor angiogenesis, that is, the formation of new blood vessels, providing a blood supply carrying oxygen and nutrition for tumor growth and an efficient route for metastatic spread of the tumor.

- Secretion of Pro-Angiogenic Factors: Cancer stem cells produce vascular endothelial growth factor (VEGF) that facilitates the formation of new blood vessels. The new formed vessels support not only the growth of the primary tumor but also the metastasis of cancer stem cells.
- Vasculogenic Mimicry: The cancer stem cells form vessel-like structures that provide pathways for nutrient delivery and metastatic spread. This non-endothelial vascular network of vessels supports the aggressive nature of metastatic cancer stem cells by facilitating their transport to distant organs.

Cancer Stem Cells in Metastatic Niche Formation

Once the cancer stem cells arrive at the distant organ, creating a supportive metastatic niche is very important for successful colonization and tumor outgrowth. Cancer stem cells can interact with the local microenvironment to optimize it for themselves by creating a supportive niche for survival and growth of the tumor.

- Microenvironment Re-Engineering: Cancer stem cells secrete matrix metalloproteinases (MMPs), among other enzymes, to degrade the tissue

barriers, which can facilitate their invasion and colonization into distant tissues. Besides that, they also recruit fibroblasts, immune cells, and endothelial cells, forming a niche in their support for tumor growth.
- Dormancy and Reactivation: Cancer stem cells may arrive at a metastatic site, enter a state of dormancy, and consequently avoid elimination by the immune system and also evade chemotherapy. Under favorable conditions, dormant cancer stem cells can reactivate into new metastatic tumors, often after many years, by responding to changes in their microenvironment.

Therapeutic Implications of Targeting Cancer Stem Cells in Metastasis

Cancer stem cells are causing significant challenges to cancer therapies, particularly in the aspect of metastasis. The traditional therapies often fail in eradicating the cancer stem cells, leading to tumor recurrence and metastasis.

- Resistance to Therapeutic Interventions: Due to their quiescent nature, enhanced DNA repair mechanisms, and activation of drug efflux pumps, cancer stem cells show high resistance to chemotherapy and radiotherapy. This resistance brings forth the need for novel therapies that specifically target cancer stem cells.

Example: Targeting cancer stem cell pathways, including the main pathways for cancer stem cells signaling, Wnt, Notch, and Hedgehog, are targeted; therapies are being explored to prevent metastasis driven by cancer stem cells. Apart from this, strategies are aimed at disrupting the cancer stem cells' niche or re-sensitizing cancer stem cells to traditional therapies [14].

- Combination Therapies: This can be achieved by the use of conventional therapies in combination with treatments targeted against cancer stem cells. This approach would efficiently kill not only the bulk tumor cells but also the cancer stem cells, leading to a decrease in metastasis and improvement in patient outcomes [15, 16].

METHODS OF CANCER STEM CELL IDENTIFICATION AND ISOLATION

Cancer stem cells form a small population of tumor cells that have the ability of self-renewal, differentiation, and the initiation of tumor development, thus known as tumor-initiating cells (TIC). Cancer stem cells have also been found to play crucial roles in progression and therapy resistance. Therefore, one very important step in cancer research and therapeutic development is the identification and isolation of cancer stem cells from tumors or cell lines, so that they can help in the research and therapeutic design and development. There are four major methods of identification and isolation of cancer stem cells. Each method utilizes specific biological properties: cell surface markers, SP analysis, activity assays using ALDH, and spheroid formation assays.

Isolation using Cell Surface Markers

The cell surface marker is one of the most commonly used ways of identifying and isolating cancer stem cells. The key strategy for cell surface markers depends on the variable expression of surface proteins that are unique to cancer stem cell populations. The technique mainly used in this strategy includes FACS and MACS.

Fluorescence-Activated Cell Sorting (FACS) uses fluorescently labeled antibodies for binding to specific cell surface markers. Cells undergo a flow cytometer in which lasers can detect the fluorescent signal, thereby allowing the separation of cancer stem cells depending on their size, granularity, and expression of surface markers. In this way, FACS will be used in the identification and isolation of cancer stem cells. FACS offers high specificity.

Magnetic-Activated Cell Sorting (MACS) involves magnetic beads along with antibodies that bind to particular cell surface markers. The magnetically labeled cells are retained within a column that is within a magnetic field, thus enabling their isolation from the unlabeled cells. Cells bound to the beads are retained, while the ones that are not are washed away. It has a very minimal impact on the viability of the cell, and it can be used on a large scale, too.

Cancer stem cells have been identified and isolated in several tumors using specific markers like CD24, CD44, CD133, and EpCAM. For example, it was documented that in human breast cancer, the expression of stem cell characteristics was presented by cells with phenotypes CD44+/CD24−/low [8, 17].

Limitations

A large number of markers are shared with normal stem cells, and both the tumor microenvironment and the culture conditions may affect their expression. Furthermore, sorting is very time-consuming and also affects surface markers. The ongoing developments in antibody technology and improvements in sorting techniques continue to improve the overall efficiency of cancer stem cell isolation. For instance, the development of specific antibodies and the use of multi-parametric FACS amplifies the accuracy of cancer stem cells harvesting while reducing risks related to contamination from non-cancer stem cells lineage.

Side Population (SP) Assay

This side population assay is based on the ability of stem cells to efflux fluorescent dyes such as Hoechst 33342 through ATP-binding cassette transporters. The technique uses high levels of such transporters in cancer stem cells, which contribute to resistance against cytotoxic agents and chemotherapy.

In the SP test, cancer stem cells are characterized as a small, distinct population of cells, excluding Hoechst dye and displaying low fluorescence. SP cells are defined as those with the ability to efflux Hoechst 33342 via the activity of ABC (ATP-binding cassette) transporters such as ABCG2, which has been described as one of the major determinants of the SP phenotype. The FACS-sorted cells show that SP is a small population of low-fluorescent cells. SP cells were also reported to exhibit higher tumorigenic potential and to be resistant to radiation and chemotherapy.

Limitations

SP assays need optimization in staining conditions in order to minimize toxicity, and due to a lack of standardization, they may give variable results. The recent studies and research have successfully exploited the advanced flow cytometry techniques and novel fluorescent dyes to enhance the sensitivity and specificity of the SP assay. These improvements help in addressing issues like dye toxicity and variability, potentially enhancing the reliability of the SP assay.

Aldehyde Dehydrogenase (ALDH) Assay

Another marker applied for the identification of cancer stem cells is represented by aldehyde dehydrogenase activity. ALDH, especially ALDH1, takes part in the oxidation of intracellular aldehydes; its activity has been previously associated with stemness and chemotherapy resistance. The ALDEFLUOR assay is usually applied for the detection and isolation of cells with high activity of ALDH.

ALDEFLUOR Assay: This is a technique in which cells are incubated with a fluorescent ALDH substrate-BAAA-that is converted by ALDH into a fluorescent product, accumulating in the cells. Flow cytometry measures the level of fluorescence, and cells with high ALDH activity are considered cancer stem cells.

Limitations

The ALDEFLUOR assay is relatively simple and highly specific, yet its power can be affected by substances like the inhibitor DEAB used in control samples to inhibit ALDH activity. Furthermore, not all cancer stem cells show high ALDH activity, and, therefore, the assay may underestimate the complete population of cancer stem cells. Researchers are working on the development of more sensitive substrates and optimizing protocols that minimize the impact of inhibitors

Spheroid Formation Assay

Another defining feature of cancer stem cells is the capacity to form spheroids or 3D clusters under non-adherent, serum-free conditions. This method, called the spheroid or sphere formation assay, has been one of the widely used methods to analyze the self-renewal and differentiation potential of cancer stem cells.

Tumor Spheres: Under serum-free, non-adherent conditions, only cells with properties similar to those of stems will survive and proliferate through spheroid formation, while the others die. Spheroid-formation assays have since found widespread application in the field of breast cancer research, in which it was documented that cells with the CD44+/CD24− phenotype and high activity of ALDH could form mammospheres.

Limitations

However, along with their broad utilization, there are significant shortcomings to spheroid assays, including variation in the capability of disparate cell lines to form spheroids and the fact that non-cancer stem cells are able to form spheres. Furthermore, sphere formation tends to select mainly for actively proliferating cells and might fail to detect quiescent cancer stem cells. Researchers and scientists are working on more standardized culture conditions and additional assays to evaluate the functional characteristics of spheroids. More advancements in 3D culture technologies and bioprinting technologies are being used to mimic the in vivo conditions and to improve the spheroid formation assay.

Combination Approaches

Due to the individual limitations of these methods, the combined use of such techniques often results in the identification and isolation of cancer stem cells. Combining SP analysis with ALDH assays or the identification of surface markers can better define the cancer stem cell population. This will enable the isolation of a subset of cancer stem cells expressing multiple stem cell markers or functional properties and therefore give an extended view of the cancer stem cells' biology.

MECHANISM OF DRUG RESISTANCE AND METASTASIS

Cancer stem cell-mediated therapy resistance deals with controlling the stemness and dormancy of these cells. They do this by involving drug resistance mechanisms, which are used to improve anti-tumor strategies and overall improvement of cancer therapy in patients [18].

ATP-binding Cassette Transporter (ABC Transporter)

They are transmembrane proteins that transport drugs using energy derived from ATP hydrolysis. They contain ABCG2, ABCB1, and ABCC1, where ABCG2 transports drugs like doxorubicin and methotrexate, and ABCB1 is present in chemoresistant tumors. Other ABC transporter proteins, like multidrug-resistant protein-1, maintain Hh signaling; hence, these ABC transporters can be used as markers to identify and isolate cancer stem cells.

Apoptosis Avoidance

Members of the GTP family proteins known as Rho proteins have an effector Rho Rho-associated protein kinase (ROCK), which plays a role in the progression of cancer stem cells. In melanoma and breast cancer stem cells, survivin an anti-apoptotic protein's expression is inhibited by ROCK and this in return increases the pancreatic wall lining to be more permeable to drug absorption.

Numb Protein

These proteins are stemness genes that take part in cell growth and migration. Specifically, in prostate cancer, NOTCH or Hh pathways are used to enrich a castration-resistant prostate cancer cell subpopulation. Self-renewal and tumorigenesis are promoted by phosphorylation of NUMB protein by NANOG, which results in p53 proteolysis.

Musashi (MSI) is a member of the RNA protein family. Musashi1(MSI1) and Musashi2 (MSI2) maintain activation and proliferation of cancer stem cells. MSI1 regulates CD44+ production by forcing the formation of anti-apoptotic stress granules, which promotes resistance to tumors. MSI2 regulates the expression of the self-renewing gene Lin28A, which also promotes resistance to tumors [19].

STRATEGIES TO TARGET CANCER STEM CELLS

Target Cancer Stem Cells Markers

Stemness can be induced by integrin $\alpha v\beta 3$, and in breast, lungs, and pancreatic cancer, stemness is induced by the KRas/RalB/NF-κB pathway. In nasopharyngeal carcinoma, FOXM1 is associated with stem cell markers like NANOG and SOX2. FOXM1 affects cancer cells by providing self-renewal features and increasing tumor spheres in them. They also stimulate the overexpression of CD133 and PRX3 to regulate the stemness of colon cancer stem cells [20 - 22].

Target miRNA/LncRNA to Cancer Stem Cells

LNC00284/LNCRNA-NRAD1 in triple-negative breast cancer (TNBC) can also be targeted with cancer stem cells. Antisense oligonucleotides, when paired with NRAD1, reduce tumor growth and cell survival rate. miR-328-3p (miR-328) is used for the upregulation of cancer stem cells in ovaries; however, it is noted that intense expression by targeting DNA damage binding protein 2(DDB2) can inhibit ovarian cancer stem cells. Targeting miR-328 may be used to eliminate cancer stem cells and can help prevent tumor metastasis [23].

Cancer Stem Cells and Resistance to Therapies

Cancer stem cells also have a high resistance to conventional modes of therapy against cancers, such as chemotherapy and radiation. Their resistance is amongst the most frequent causes of recurrence in cancer and the major cause of treatment failure. Generally, mechanisms of cancer stem cell resistance may be explained by three main categories: high efficiency in DNA repair, quiescence, and active drug efflux.

Superior DNA Repair

More importantly, cancer stem cells are much more proficient in restoring DNA damage compared to non-stem cancer cells and thus recover from inflicted damage due to chemotherapy and radiation. Most key proteins thought to be involved in DNA repair processes, including BRCA1, BRCA2, and RAD51, were found to be expressed at much higher levels in a large number of cancer stem cells. These proteins are very actively engaged in repairing DNA breaks, a fact which may contribute to the survival of cancer stem cells by their continued proliferation in spite of treatment-induced DNA damage. This may include overexpression of enzymes like PARP, poly (ADP-ribose) polymerase, involved in single-strand DNA break repair. All these enhanced activities help the cancer stem cells to resist the treatments, which induce DNA damage.

Quiescence

Generally speaking, quiescence is a resting phase in which the cycling of cancer stem cells is not active. The quiescent state thus acts somewhat like a shield against therapies directed at the rapidly cycling cells. Cancer stem cells may enter into a resting phase, more specifically known as the G0 phase of the cell cycle, in which they are less sensitive to therapies targeting actively dividing cells in a preferential manner, such as chemotherapy and radiation. Thus, dormant cancer stem cells outlast treatments and regrow afterwards.

Compared to the quiescent status, cancer stem cells also show lower metabolic activities and reduced oxidative stress, and thus are less sensitive to therapeutic agents targeting high metabolic rates or ROS production. Often, cancer stem cells express high levels of drug efflux pumps that, by active removal of therapeutic agents from the cells, reduce effective treatment.

Some of the ABC transporters, which are often overexpressed through the cancer stem cells, include but are not limited to, P-glycoprotein or simply P-gp, Multidrug Resistance Protein 1 or MRP1, and breast cancer resistance protein (BCRP). These transporters utilize ATP in order to pump out the drugs from the cells, therefore reducing their intracellular concentration and hence their effectiveness. Generally speaking, high expression of these efflux pumps is consistent with the very stem properties of cancer stem cells. Thus, such an effective removal of drugs can enable cancer stem cells to survive the treatment and continue driving the growth of a tumor.

CANCER STEM CELLS IN DIFFERENT CANCERS

Cancer Stem Cells in Breast Cancer

The cells begin to proliferate without regulation, resulting in a type of cancer referred to as breast cancer, which is based in the tissue of the breast. It is most common in women over forty years and often presents itself as a lump or mass within the breast. It is regarded as the second most common cancer in women, with skin cancer being at the top. The increasing incidence of breast cancer and its serious negative effect on women's health has made organizations like the American Cancer Society take serious notice.

- Tumor Initiation and Maintenance: Most of the time, cancer stem cells are responsible for tumor initiation and maintenance. Due to their special self-renewal ability, they can maintain tumor growth for a longer period.
- Metastasis: Cancer stem cells are able to detach from the primary tumor and migrate to other parts of the body, hence leading to metastasis and secondary tumors [24]
- Resistance to Treatment: The main problem with cancer stem cells is that they are resistant to conventional therapies, including radiation and chemotherapy.

This resistance plays a role in recurrence after initial therapy.
- Recurrence of Cancer: Even when the treatment is effective, the cancer can relapse. Their persistiveness makes them an essential subject for further research and corrective measures.

Markers: The researchers depend on specific surface markers to identify the cancer stem cells. They are essential to understand how the cells work and also to develop specific treatments [25 - 28].

Functional Tests: These are done in the lab to determine whether cancer stem cells are capable of forming tumors, as well as the efficiency of new medicines against them.

Genetic and Protein Analysis: By analyzing the genes and proteins expressed by the cancer stem cells, scientists garner more information about how these cells behave and find novel targets for them.

New Approaches to Treatment

- Targeted Therapy: New treatments specifically focus on the cancer stem cells. The drugs aim to block those mechanisms that enhance the survival and growth of cancer stem cells.
- Immunotherapy: It enhances the ability of the immune system to recognize and destroy the cancer stem cells. Several new drugs and vaccines are under study that will enhance the immune response against these cells [29, 30]
- Combined modality treatment: It aims at the eradication of the principal tumor and residual cancer stem cells responsible for recurrence by the addition of cancer stem cells-targeting agents to the more conventional treatments like radiation and chemotherapy.
- Epigenetic Therapy: This approach attempts to change the chemical modifications in DNA that determine the behavior of cancer stem cells and perhaps diminish their tumorigenicity [31, 32].
- Nanoparticles: Utilizing nanoparticles creatively, drugs could be delivered right at the site of cancer stem cells. This will enhance the precision of treatment while reducing side effects [33 - 35].

Follow-up and Preventive Care

1. Early Detection: Better techniques for the early identification of cancer stem cells are bound to translate into an exact diagnosis and hence better prognosis of the cancerous ailment. Today, personalized medicine, which is a treatment that is tailored to the particular malignancy a patient possesses, including cancer stem cells, is an increasingly common treatment administered to patients

[36].
2. Cancer Stem Cells Level Monitoring: Monitoring levels of cancer stem cells at the time of and after treatment allows for therapeutic adjustments and aids in the early detection of recurrence signs and symptoms.

Cancer Stem Cells in Skin Cancer

Skin cancer is a result of an uncontrollable growth of skin cells on the surface of the skin. It is characterized by new growth or the change of an existing mole or sore. The following are some main skin cancer types:

Basal cell carcinoma: It is the most frequent skin cancer; it originates from basal cells on the outer layer of the skin. Though it is the most frequent skin cancer, it is not that serious and seldom spreads, compared to other types of skin cancers.

SCC or Squamous Cell Carcinoma: This kind of cancer originates from the top layers of the squamous cells in the skin. SCC has a higher risk of metastasis than BCC, which means the cancer spreads to other parts of the body. It may also be more aggressive.

Melanoma: Melanoma is the most dangerous form because melanocytes are the cells responsible for skin pigmentation in melanoma. Melanoma is a critical threat to health because this type of cancer has the potential to migrate from the site of origin to other organs. This ability of metastasis is what makes it so dangerous [37].

Importance of Cancer Stem Cells in Skin Cancer

- Tumor Formation and Maintenance: Cancer stem cells are critical in the genesis and progression of tumors. The self-renewal properties can be responsible for the advancement of the tumor over a certain period.
- Cancer Metastasis: Cancer stem cells are capable of migrating from the original tumor site and lodging in other parts of the body, resulting in metastatic disease. This is particularly of concern since melanoma has the potential to metastasize to other organs [24].
- Resistance to Treatments: Cancer stem cells are typically resistant to conventional medications, such as chemotherapy and radiation. Because of this resistance, skin cancers are relatively difficult to treat and often recur.
- Recurrence: Even though a tumor has disappeared, cancer stem cells can still survive and cause the recurrence of cancer. Since recurrence is most often observed as a result of the persistence of these cells, targeting them becomes of great significance in avoiding recurrence in skin cancer [38].

Methods for Identifying and Investigating Cancer Stem Cells

- Surface Markers: The researchers look for certain markers on the surface of cancer stem cells to identify the surface markers. These surface markers are important to understand the behavior of cancer stem cells for targeted therapies.
- Laboratory Tests: A number of lab tests are conducted to study the role that cancer stem cells play in malignancy development and their sensitivity to novel therapeutic agents. These studies further advance the therapeutic approaches.
- Genetic and Protein Analysis: Through the analysis of the genes and proteins involved in cancer stem cells, there will be the design of specific therapies for cancer stem cells and further clarification of their behavior [39, 40].

Novel Strategies for the Management of Cancer Stem Cells

- Targeted Therapy: New treatments are under study that will selectively attack cancer stem cells without putting healthy cells at risk. This form of targeting may well prove to be far more effective and efficient in treatment [41].
- Immunotherapy: It is another approach that enhances the potential of the immune system in recognizing and eliminating cancer stem cells. New vaccines and therapies are under research for the enhancement of immune responses against these cells [29, 30].
- Combination therapy: Drugs acting on cancer stem cells can be combined with traditional radiation and chemotherapy to target not only the primary tumor but also the cancer stem cells responsible for recurrence [42].
- Epigenetic Treatment: Epigenetic treatment aims to transform the chemical changes in DNA responsible for changing the behavior of cancer stem cells. With these modifications, it is foreseen that the capabilities of the cancer stem cells in driving tumor growth will be reduced [40, 43].
- Nanoparticles: In developing small particles targeting malignant cells with drugs, the side effects of cancer treatments on the body will be minimal when directly targeted to cancer stem cells [44].

Continuing and Preventive Care

- Early Detection: Efforts continue to make improvements in the early detection of cancer stem cells, hence presenting quick and successful treatments for skin cancer.
- Personalized Medicine: The treatments are now tailored according to the specific characteristics of an individual's cancer, such as the presence of cancer stem cells themselves. The idea behind this personal approach is to find the best possible treatment for a patient [45].
- Monitoring Routinely: Monitoring the cancer stem cells levels before, during,

and after treatment helps medical professionals to have an early warning for the recurrence of cancer and assists them in managing medicines for better results for the patients [46 - 49].

Cancer Stem Cells in Novel Therapies

Targeted Therapy: Currently, new drugs that target only cancer stem cells are under consideration. Due to the fact that this treatment is supposed to kill only cancer stem cells while leaving the healthy cells, the therapy has proven to be more effective.

- Immunotherapy: It allows the immune system to identify and destroy cancer stem cells. Researchers are investigating the use of new methods to make the immune system better protect against them [30, 50].
- Combination Therapy: The goal of this therapy is to provide treatment not only to the subpopulation of cancer stem cells that may lead to recurrence but also to the original tumor. These drugs, which specifically target the cancer stem cells, may be used in combination with other more conventional modes of treatment, such as chemotherapy and radiotherapy [51].
    - Epigenetic treatment: It targets the reversal of these chemical changes to DNA that control the behavior of cancer stem cells. It is envisaged that changing such modifications can downplay the efficiency of cancer stem cells to act as drivers of tumor growth.
- Early detection: Cancer stem cells can be identified even in the early stage of colorectal cancer, with the advancement of identification techniques. This may help open ways for more effective treatment and care.
- Personalized medicine: Attention is now focused on the distinct traits of cancer stem cells associated with a patient's cancer. A personalized plan can make an individual reach their full potential.

It monitors the amount of cancer stem cells both pre- and post-treatment so that therapeutic measures can be performed if needed and check recurrences of the cancer well in advance [32, 46].

Colorectal cancer is a type of cancer that occurs in the epithelial cells in either the colon or the rectum of patients. This can be related to the benign form of polyps, although it can also give rise to malignant tumors. This is the type of cancer that affects the colon, which is the largest section of the large intestine. Rectal cancer, on the other hand, originates in the rectum, which is the last section of the large intestine before the anus [52, 53].

Cancer Stem Cells Play an Important Role in Colorectal Cancer

Tumor Initiation: Cancer stem cells play a part in the development or initiation of colorectal malignancies. Like a catalyst, these cells play a role in tumor initiation and progression.

Metastasis: Cancer stem cells have the potential to proliferate from the primary site of the tumor and give rise to new tumors in other parts of the body. It has a tendency for high-risk conditions to develop into secondary cancers.

Resistance to Treatment: One of the most disconcerting features of cancer stem cells is that they hardly respond to conventional treatments like chemotherapy and radiation. Such resistance may favor the survival of cancer stem cells and recurrence of cancer [24].

Recurrence of Cancer: Despite apparently successful treatments, cancer stem cells are able to retain their capability to persist in the body and to eventually cause recurrence of cancer. For these reasons, the targeting of cancer stem cells is important in the long-term management of cancers [48].

Methods for Identification and Study of Cancer Stem Cells

- Surface markers: Cancer stem cells are identified by researchers through the detection of specific signs on their surface. These markers are crucial to our capability of isolating and studying cancer stem cells, as well as understanding their role in disease progression.
- Laboratory Models: These include cell cultures, animal studies, and many other types of laboratory models that help in understanding cancer stem cells. These further explain how cancer stem cells take part in the development of malignancies and their responses to various treatments [54].
- Genetic and Protein Analysis: The genes and proteins associated with cancer stem cells will lead to a better understanding of their activities and behaviors. This information will add to the development of specific therapeutic interventions against cancer stem cells [40, 50, 54 - 56].

Novel Therapies Targeting Cancer Stem Cells

- Targeted therapy: New drugs that selectively target only cancer stem cells are being developed. These therapies aim at eradicating cancer stem cells while sparing healthy cells, which may lead to an improvement in therapeutic outcomes due to fewer adverse side effects.
- Immunotherapy: This is a technique aimed at increasing the sensitivity and efficiency of the immune system to better identify and destroy cancer stem cells.

New vaccines and other pharmaceuticals are under study that can enhance the immune response against such cells [16, 30, 57, 58].
- Combination Therapy: Drugs that target cancer stem cells are combined with other traditional modes of treatment, including chemotherapy and radiation, to formulate an overall treatment approach. Such a strategy launches an attack on the main tumor and the cancer stem cells responsible for recurrence.
- Epigenetic Therapy: This type of treatment modifies the chemical structure of DNA to control the functions of cancer stem cells. Researchers hope that, if they can reverse these modifications, it may reduce the ability of cancer stem cells to form malignancies [32].
- Nanoparticles: Nanoparticle formulation allows drugs to be delivered directly to cancer stem cells. This technique aims at making the treatment more precise so that healthy cells are minimally devastated [32, 42, 44, 59].

Current Preventive Treatment

Detection: Efforts are being made to detect cancer stem cells early in the hope of increasing the early detection rate in CRC. Early detection can enable more constructive outcomes other than providing affordable costs of treatment [60].

Personalized Medicine: Treatment designs are becoming increasingly individualistic for each patient's particular cancer, considering whether cancer stem cells are or are not present in that patient. Such tailored therapy would include treatment optimized for each patient [56].

Follow-up on a Routine Basis: At intervals before, during, and following treatment, the level of cancer stem cells is routinely assessed to identify early signs of recurrence. These facilitate modification of therapy, if necessary. In malignancy, management has to be done well in advance.

Cancer Stem Cells in Cervical Cancer

Introduction and Importance

Usually, cervical cancer develops from the epithelial cells of the cervix as a result of a persistent infection with high-risk strains of human papillomavirus. In spite of improvements in diagnosing and curing methods, cervical cancer remains a significant health issue, particularly in underdeveloped areas. The significance of cancer stem cells in cervical cancer research has grown due to their important role in tumor development, advancement, and resistance to treatment. Cancer stem cells are a unique proportion of cells found within tumors, possessing characteristics such as self-renewal ability and differentiation, which could play a role in tumor growth and return. This ability to drive disease persistence identifies them as a focal point for the development of more effective treatment strategies [61].

Characteristics of Cancer Stem Cells in Cervical Cancer

Cancer stem cells in cervical cancer express some markers, including CD133 and CD49f; these markers help in the identification of these cells. CD133 is associated with stem-like properties and is one of the key markers in the identification of cancer stem cells in a variety of cancers, including cervical cancer. CD49f, or integrin alpha-6, mediates cell adhesion and signaling and points further to stem cell features. Cancer stem cells may exhibit self-renewal by asymmetric cell division, maintaining the stem cell pool by producing differentiated cells, contributing to tumor heterogeneity. This characteristic of cancer stem cells represents a requirement for continuous tumor growth and practically presents a complication for treatment. Approaches like flow cytometry and immunohistochemistry are employed in the identification of these cells for understanding their highly needed function within the tumor microenvironment [62 - 64].

Mechanisms of Resistance to Therapy

Cancer stem cells in cervical cancer are considered resistant to conventional therapies, including chemotherapy and radiotherapy, because of their well-developed DNA damage response and repair mechanisms. Key pathways include the ATM and ATR kinases, which activate repair processes through proteins such as BRCA1 and PARP-1. These pathways are critical for the repair of DNA double-strand breaks and single-strand breaks, respectively, and are often hyperactivated in cancer stem cells, enabling their survival post-therapy. The high activity of ALDH in cancer stem cells further confers added resistance due to protection against oxidative stress and DNA damage. Cancer stem cells also express quiescence, which is a state of dormancy that shields them from the action of cytotoxic therapies targeting rapidly dividing cells. This quiescent state allows the cancer stem cells to escape treatment and then re-enter the cell cycle for tumor recurrence. The other contributors to this resistance include TGF-β signaling and EMT. TGF-β signaling can promote cancer stem cells' survival and drug resistance, while EMT contributes to a more invasive phenotype, further complicating the effectiveness of therapy.

Other reasons for the resistance of tumors to therapy involve hypoxic or low oxygen conditions within the tumor environment. Through the action of low oxygen levels, hypoxic environments activate HIFs that control genes in charge of angiogenesis and cell survival. More precisely, HIF1α and HIF2α increase cancer stem cells' properties and decrease their sensitivity to therapy by the activation of pathways like PI3K/Akt. In fact, higher oxygenation of tumors is related to increased success with radiotherapy; however, cancer stem cells are more often found within hypoxic niches where they are more resistant to therapy [65].

Current Therapeutic Methods and Strategies

Targeting cancer stem cells directly in cervical cancer involves a few strategies. Development of monoclonal antibodies against some cancer stem cell markers, like CD133 and CD49f, that selectively eliminate cancer stem cells and thus reduce tumor recurrence, is one strategy. Small molecule inhibitors of critical pathways that mediate cancer stem cells' functions, such as Notch, Wnt, and Hedgehog, may also be explored to disrupt cancer stem cells' functions and improve treatment responses.

Other indirect targeted mechanisms involve the modification of the tumor microenvironment, including the modulation of the cancer stem cell niche targeting CAFs and TAMs. These components support the survival and resistance to therapy of cancer stem cells; therefore, there may be other targets that could disrupt the supportive niche. Other promising approaches to selectively attack the cancer stem cells and improve therapeutic outcomes include hypoxia-activated prodrugs, which target hypoxic tumor regions.

Clinical trials are studying epigenetic therapies, including DNA methyltransferase inhibitors and histone deacetylase inhibitors, as a strategy to alter chromatin structure and change gene expression in cancer stem cells. Preclinical studies have shown that these agents can effectively reverse abnormal epigenetic modifications associated with the properties of cancer stem cells and enhance the efficacy of standard treatments. Combination therapies, using conventional chemotherapy and radiotherapy in combination with cancer stem cell-targeting agents, are also under development to circumvent resistance and realize better patient outcomes.

Targeted Mechanisms against Cancer Stem Cells

Targeting the Molecular Signaling Pathways

Disruption of key signaling pathways in cancer may lead to the formation of cancer stem cells, which drive tumorigenesis. Thus, this new approach focuses on these abnormal pathways for cancer therapy.

Targeting Cancer Stem Cell Markers

The surface markers employed for the isolation and identification of cancer stem cells also serve as ideal targets. Targeting cytotoxic drugs to specific cancer stem cells using these markers can prove to be an effective approach.

Targeting the Cancer Stem Cell Niche And The Quiescent State

Cancer stem cells depend on a specialized microenvironment known as the cancer stem cell niche. This niche delivers appropriate signals and pathways that regulate the self-renewal and normal homeostatic processes. Cancer stem cells are frequently in a dormant or quiescent state, which makes them less susceptible to these traditional therapies. To overcome this resistance, dormant cancer stem cells are induced to re-enter the cell cycle [66].

Manipulation of miRNA Expression

miRNAs regulate various cellular processes, including self-renewal, differentiation, and cell division. They can function as both tumor suppressors and oncogenes, thus serving as an ideal target for cancer treatment.

Induction of Cancer Stem Cell Apoptosis

Altering the apoptotic machinery to induce cell death in cancer stem cells proves to be a potential strategy to eliminate cancer stem cells.

Induction of Cancer Stem Cell Differentiation

Differentiation therapy compels cancer stem cells to undergo terminal differentiation, thereby losing self-renewal capabilities. Differentiation therapy can reduce their tumor-initiating capacity and potentially decrease their ability to sustain the tumor.

Cancer Stem Cells Targeting Therapy

Employing oncolytic adenoviruses presents an antitumor strategy for eliminating cancer stem cells.

Therapeutic Implications of Cancer Stem Cells in Cancer Therapy

Even with notable progress in diagnosing, treating, and surgically removing tumors, recurrence and metastasis continue to present significant challenges in cancer treatment. This drawback is due to the failure to evaluate cancer stem cells in tumors. As per the present cancer stem cells model, tumors are composed of a small proportion of cancer stem cells surrounded by a bulk of differentiated tumor cells. Consequently, because of the fundamental differences between these two cell populations, it is possible that traditional therapies primarily target differentiated cells, conferring cancer stem cells with an advantage. Henceforth, integrating traditional therapies with targeted cancer stem cells-specific factors could address the whole tumor and provide a promising approach for long-term treatment [16]

CONCLUSION

Cancer stem cells (CSCs) are crucial drivers of tumor growth, metastasis, and recurrence due to their ability to self-renew and resist standard therapies. Their identification has advanced cancer research, highlighting the need for targeted therapies. Despite their complexity and heterogeneity, promising new approaches like immunotherapy and nanotechnology are being developed to target CSCs. Eliminating these cells is key to improving cancer treatment outcomes and reducing the likelihood of relapse.

ACKNOWLEDGEMENT

We would like to acknowledge Ms. Kruthika P, Mr. Kumaran K, Ms. Raksa Arun, Ms. Srisri SatishKartik, Ms. Sanjana Dhayalan, and Mr. Sasitharan T for their contribution in proofreading and editing works.

REFERENCES

[1] Nguyen LV, Vanner R, Dirks P, Eaves CJ. Cancer stem cells: an evolving concept. Nat Rev Cancer 2012; 12(2): 133-43.
[http://dx.doi.org/10.1038/nrc3184] [PMID: 22237392]

[2] Abbaszadegan MR, Bagheri V, Razavi MS, Momtazi AA, Sahebkar A, Gholamin M. Isolation, identification, and characterization of cancer stem cells: A review. J Cell Physiol 2017; 232(8): 2008-18.
[http://dx.doi.org/10.1002/jcp.25759] [PMID: 28019667]

[3] Li S, Li Q. Cancer stem cells, lymphangiogenesis, and lymphatic metastasis. Cancer Lett 2015; 357(2): 438-47.
[http://dx.doi.org/10.1016/j.canlet.2014.12.013] [PMID: 25497008]

[4] Baisiwala S, Budhiraja S, Goel C, Nandoliya KR, Saathoff MR, Ahmed AU. Spelling Out CICs: A Multi-Organ Examination of the Contributions of Cancer Initiating Cells' Role in Tumor Progression. Stem Cell Rev Rep 2022; 18(1): 228-40.
[http://dx.doi.org/10.1007/s12015-021-10195-x] [PMID: 34244971]

[5] Karamboulas C, Ailles L. Developmental signaling pathways in cancer stem cells of solid tumors. Biochim Biophys Acta, Gen Subj 2013; 1830(2): 2481-95.
[http://dx.doi.org/10.1016/j.bbagen.2012.11.008] [PMID: 23196196]

[6] Liang L, Kaufmann AM. The Significance of Cancer Stem Cells and Epithelial–Mesenchymal Transition in Metastasis and Anti-Cancer Therapy. Int J Mol Sci 2023; 24(3): 2555.

[7] Kahn M. Can we safely target the WNT pathway? Nat Rev Drug Discov 2014; 13(7): 513-32.
[http://dx.doi.org/10.1038/nrd4233] [PMID: 24981364]

[8] Takebe N, Miele L, Harris PJ, *et al.* Targeting Notch, Hedgehog, and Wnt pathways in cancer stem cells: clinical update. Nat Rev Clin Oncol 2015; 12(8): 445-64.
[http://dx.doi.org/10.1038/nrclinonc.2015.61] [PMID: 25850553]

[9] Clevers H, Nusse R. Wnt/β-catenin signaling and disease. Cell 2012; 149(6): 1192-205.
[http://dx.doi.org/10.1016/j.cell.2012.05.012] [PMID: 22682243]

[10] Guo Y, Zhang K, Cheng C, *et al.* Numb$^{-/low}$ Enriches a Castration-Resistant Prostate Cancer Cell Subpopulation Associated with Enhanced Notch and Hedgehog Signaling. Clin Cancer Res 2017; 23(21): 6744-56.
[http://dx.doi.org/10.1158/1078-0432.CCR-17-0913] [PMID: 28751447]

[11] Merchant AA, Matsui W. Targeting Hedgehog--a cancer stem cell pathway. Clin Cancer Res 2010; 16(12): 3130-40.
[http://dx.doi.org/10.1158/1078-0432.CCR-09-2846] [PMID: 20530699]

[12] Petrova R, Joyner AL. Roles for Hedgehog signaling in adult organ homeostasis and repair. Development 2014; 141(18): 3445-57.
[http://dx.doi.org/10.1242/dev.083691] [PMID: 25183867]

[13] Sari IN, Phi LTH, Jun N, Wijaya YT, Lee S, Kwon HY. Hedgehog Signaling in Cancer: A Prospective Therapeutic Target for Eradicating Cancer Stem Cells. Cells 2018; 7(11): 208.
[http://dx.doi.org/10.3390/cells7110208] [PMID: 30423843]

[14] Herreros-Pomares A. Identification, Culture and Targeting of Cancer Stem Cells. Life (Basel) 2022; 12(2): 184.
[http://dx.doi.org/10.3390/life12020184] [PMID: 35207472]

[15] Yang L, Shi P, Zhao G, *et al.* Targeting cancer stem cell pathways for cancer therapy. Signal Transduct Target Ther 2020; 5(1): 8.
[http://dx.doi.org/10.1038/s41392-020-0110-5] [PMID: 32296030]

[16] Zeng Z, Fu M, Hu Y, Wei Y, Wei X, Luo M. Regulation and signaling pathways in cancer stem cells: implications for targeted therapy for cancer. Mol Cancer 2023; 22(1).
[http://dx.doi.org/10.1186/s12943-023-01877-w]

[17] Lin S, Xu Y, Gan Z, *et al.* Monitoring cancer stem cells: insights into clinical oncology. OncoTargets Ther 2016; 9: 731-40.
[PMID: 26929644]

[18] Moitra K. Overcoming Multidrug Resistance in Cancer Stem Cells. BioMed Res Int 2015; 2015: 1-8.
[http://dx.doi.org/10.1155/2015/635745] [PMID: 26649310]

[19] Siddique HR, Feldman DE, Chen CL, Punj V, Tokumitsu H, Machida K. NUMB phosphorylation destabilizes p53 and promotes self+renewal of tumor+initiating cells by a NANOG+dependent mechanism in liver cancer. Hepatology 2015; 62(5): 1466-79.
[http://dx.doi.org/10.1002/hep.27987] [PMID: 26174965]

[20] Luo W, Gao F, Li S, Liu L. FoxM1 Promotes Cell Proliferation, Invasion, and Stem Cell Properties in Nasopharyngeal Carcinoma. Front Oncol 2018; 8: 483.
[http://dx.doi.org/10.3389/fonc.2018.00483] [PMID: 30416986]

[21] Song IS, Jeong YJ, Jeong SH, *et al.* FOXM1-Induced PRX3 Regulates Stemness and Survival of Colon Cancer Cells via Maintenance of Mitochondrial Function. Gastroenterology 2015; 149(4): 1006-1016.e9.
[http://dx.doi.org/10.1053/j.gastro.2015.06.007] [PMID: 26091938]

[22] Qiu G, Ma D, Li F, Sun D, Zeng Z. lnc-PKD2-2-3, identified by long non-coding RNA expression

profiling, is associated with pejorative tumor features and poor prognosis, enhances cancer stemness and may serve as cancer stem-cell marker in cholangiocarcinoma. Int J Oncol 2019; 55(1): 45-58.
[http://dx.doi.org/10.3892/ijo.2019.4798] [PMID: 31059014]

[23] Zhang X, Powell K, Li L. Breast Cancer Stem Cells: Biomarkers, Identification and Isolation Methods, Regulating Mechanisms, Cellular Origin, and Beyond. Cancers (Basel) 2020; 12(12): 3765.
[http://dx.doi.org/10.3390/cancers12123765] [PMID: 33327542]

[24] Castaneda M, den Hollander P, Kuburich NA, Rosen JM, Mani SA. Mechanisms of cancer metastasis. Semin Cancer Biol 2022; 87: 17-31.
[http://dx.doi.org/10.1016/j.semcancer.2022.10.006] [PMID: 36354098]

[25] Velasco-Velázquez MA, Popov VM, Lisanti MP, Pestell RG. The role of breast cancer stem cells in metastasis and therapeutic implications. Am J Pathol 2011; 179(1): 2-11.
[http://dx.doi.org/10.1016/j.ajpath.2011.03.005] [PMID: 21640330]

[26] Al-Hajj M, Wicha MS, Benito-Hernandez A, Morrison SJ, Clarke MF. Prospective identification of tumorigenic breast cancer cells. Proc Natl Acad Sci USA 2003; 100(7): 3983-8.
[http://dx.doi.org/10.1073/pnas.0530291100] [PMID: 12629218]

[27] Walther W, Schlag PM. Current status of gene therapy for cancer. Curr Opin Oncol 2013; 25(6): 659-64.
[http://dx.doi.org/10.1097/CCO.0000000000000004] [PMID: 24100345]

[28] Holland JD, Klaus A, Garratt AN, Birchmeier W. Wnt signaling in stem and cancer stem cells. Curr Opin Cell Biol 2013; 25(2): 254-64.
[http://dx.doi.org/10.1016/j.ceb.2013.01.004] [PMID: 23347562]

[29] Köseer AS, Di Gaetano S, Arndt C, Bachmann M, Dubrovska A. Immunotargeting of Cancer Stem Cells. Cancers (Basel) 2023; 15(5): 1608.
[http://dx.doi.org/10.3390/cancers15051608] [PMID: 36900399]

[30] Ruiu R, Tarone L, Rolih V, et al. Cancer stem cell immunology and immunotherapy: Harnessing the immune system against cancer's source. 2019. p. 119–88.

[31] Wang Y, Cardenas H, Fang F, et al. Epigenetic targeting of ovarian cancer stem cells. Cancer Res 2014; 74(17): 4922-36.
[http://dx.doi.org/10.1158/0008-5472.CAN-14-1022] [PMID: 25035395]

[32] French R, Pauklin S. Epigenetic regulation of cancer stem cell formation and maintenance. Int J Cancer 2021; 148(12): 2884-97.
[http://dx.doi.org/10.1002/ijc.33398] [PMID: 33197277]

[33] Ertas YN, Abedi Dorcheh K, Akbari A, Jabbari E. Nanoparticles for Targeted Drug Delivery to Cancer Stem Cells: A Review of Recent Advances. Nanomaterials (Basel) 2021; 11(7): 1755.
[http://dx.doi.org/10.3390/nano11071755] [PMID: 34361141]

[34] Krzyszczyk P, Acevedo A, Davidoff EJ, et al. The growing role of precision and personalized medicine for cancer treatment. Technology (Singap) 2018; 6(03n04): 79-100.
[http://dx.doi.org/10.1142/S2339547818300020] [PMID: 30713991]

[35] Walcher L, Kistenmacher A, Suo H, et al. Cancer stem cells—origins and biomarkers: perspectives for targeted personalized therapies. Front Immunol 2020; 11.
[http://dx.doi.org/10.3389/fimmu.2020.01280]

[36] Crosby D, Bhatia S, Brindle KM, Coussens LM, Dive C, Emberton M, et al. Early detection of cancer. Science 1979; 2022: 375.
[PMID: 35298272]

[37] Schlaak M, Schmidt P, Bangard C, Kurschat P, Mauch C, Abken H. Regression of metastatic melanoma by targeting cancer stem cells. Oncotarget 2012; 3(1): 22-30.
[http://dx.doi.org/10.18632/oncotarget.437] [PMID: 22289880]

[38] Sagar J, Chaib B, Sales K, Winslet M, Seifalian A. Role of stem cells in cancer therapy and cancer stem cells: a review. Cancer Cell Int 2007; 7(1): 9.
[http://dx.doi.org/10.1186/1475-2867-7-9] [PMID: 17547749]

[39] Loescher LJ, Janda M, Soyer HP, Shea K, Curiel-Lewandrowski C. Advances in skin cancer early detection and diagnosis. Semin Oncol Nurs 2013; 29(3): 170-81.
[http://dx.doi.org/10.1016/j.soncn.2013.06.003] [PMID: 23958215]

[40] Bisht S, Nigam M, Kunjwal SS, Sergey P, Mishra AP, Sharifi-Rad J. Cancer Stem Cells: From an Insight into the Basics to Recent Advances and Therapeutic Targeting. Stem Cells Int 2022; 2022: 1-28.
[http://dx.doi.org/10.1155/2022/9653244] [PMID: 35800881]

[41] Borlongan MC, Wang H. Profiling and targeting cancer stem cell signaling pathways for cancer therapeutics. Front Cell Dev Biol 2023; 11
[http://dx.doi.org/10.3389/fcell.2023.1125174]

[42] Chu X, Tian W, Ning J, et al. Cancer stem cells: advances in knowledge and implications for cancer therapy. Signal Transduct Target Ther 2024; 9(1).
[http://dx.doi.org/10.1038/s41392-024-01851-y]

[43] Galassi C, Manic G, Esteller M, Galluzzi L, Vitale I. Epigenetic regulation of cancer stemness. Signal Transduct Target Ther 2025; 10(1).
[http://dx.doi.org/10.1038/s41392-025-02340-6]

[44] Rahimkhoei V, Akbari A, Jassim AY, Hussein UA, Salavati-Niasari M. Recent advances in targeting cancer stem cells by using nanomaterials. Int J Pharm 2025; 125381.
[http://dx.doi.org/10.1016/j.ijpharm.2025.125381]

[45] Helgadottir H, Rocha Trocoli Drakensjö I, Girnita A. Personalized Medicine in Malignant Melanoma: Towards Patient Tailored Treatment. Front Oncol 2018; 8: 202.
[http://dx.doi.org/10.3389/fonc.2018.00202] [PMID: 29946532]

[46] Santos SND, Witney TH. Molecular imaging of cancer stem cells and their role in therapy resistance. J Nucl Med 2025; 66(1): 14-9.
[http://dx.doi.org/10.2967/jnumed.124.267657]

[47] Phi LTH, Sari IN, Yang Y, et al. Cancer stem cells (CSCs) in drug resistance and their therapeutic implications in cancer treatment. Stem Cells Int 2018; 2018: 1-16.
[http://dx.doi.org/10.1155/2018/5416923]

[48] Islam F, Gopalan V, Smith RA, Lam AKY. Translational potential of cancer stem cells: A review of the detection of cancer stem cells and their roles in cancer recurrence and cancer treatment. Exp Cell Res 2015; 335(1): 135-47.
[http://dx.doi.org/10.1016/j.yexcr.2015.04.018] [PMID: 25967525]

[49] Morand du Puch CB, Vanderstraete M, Giraud S, Lautrette C, Christou N, Mathonnet M. Benefits of functional assays in personalized cancer medicine: more than just a proof-of-concept. Theranostics 2021; 11(19): 9538-56.
[http://dx.doi.org/10.7150/thno.55954] [PMID: 34646385]

[50] Jung HJ. Chemical Proteomic Approaches Targeting Cancer Stem Cells: A Review of Current Literature. Cancer Genomics Proteomics 2017; 14(5): 315-27.
[PMID: 28870999]

[51] Dzobo K, Senthebane DA, Ganz C, Thomford NE, Wonkam A, Dandara C. Advances in therapeutic targeting of cancer stem cells within the tumor microenvironment: an updated review. Cells 2020 Aug; 9(8): 1896.
[http://dx.doi.org/10.3390/cells9081896] [PMID: 32823711] [PMCID: PMC7464860]

[52] Chiou GY, Yang TW, Huang CC, et al. Musashi-1 promotes a cancer stem cell lineage and chemoresistance in colorectal cancer cells. Sci Rep 2017; 7(1): 2172.

[http://dx.doi.org/10.1038/s41598-017-02057-9] [PMID: 28526879]

[53] O'Brien CA, Pollett A, Gallinger S, Dick JE. A human colon cancer cell capable of initiating tumour growth in immunodeficient mice. Nature 2007; 445(7123): 106-10.
[http://dx.doi.org/10.1038/nature05372] [PMID: 17122772]

[54] Hervieu C, Christou N, Battu S, Mathonnet M. The Role of Cancer Stem Cells in Colorectal Cancer: From the Basics to Novel Clinical Trials. Cancers (Basel) 2021; 13(5): 1092.
[http://dx.doi.org/10.3390/cancers13051092] [PMID: 33806312]

[55] Ally A, Balasundaram M, Carlsen R, et al. Comprehensive and Integrative Genomic Characterization of Hepatocellular Carcinoma. Cell 2017; 169(7): 1327-1341.e23.
[http://dx.doi.org/10.1016/j.cell.2017.05.046]

[56] Vaseghi Maghvan P, Jeibouei S, Akbari ME, et al. Personalized medicine in colorectal cancer. Gastroenterol Hepatol Bed Bench 2020; 13 (Suppl. 1): S18-28.
[PMID: 33585000]

[57] Yi S, Wei M, Zhao L. Targeted immunotherapy to cancer stem cells: a novel strategy of anticancer immunotherapy. Crit Rev Oncol Hematol 2024; 196: 104313.
[http://dx.doi.org/10.1016/j.critrevonc.2024.104313]

[58] Martínez-Pérez J, Torrado C, Domínguez-Cejudo MA, Valladares-Ayerbes M. Targeted Treatment against Cancer Stem Cells in Colorectal Cancer. Int J Mol Sci 2024; 25(11): 6220.
[http://dx.doi.org/10.3390/ijms25116220] [PMID: 38892410]

[59] Yeh YC, Huang TH, Yang SC, Chen CC, Fang JY. Nano-Based Drug Delivery or Targeting to Eradicate Bacteria for Infection Mitigation: A Review of Recent Advances. Front Chem 2020; 8: 286.
[http://dx.doi.org/10.3389/fchem.2020.00286] [PMID: 32391321]

[60] Li D. Recent advances in colorectal cancer screening. Chronic Dis Transl Med 2018; 4(3): 139-47.
[http://dx.doi.org/10.1016/j.cdtm.2018.08.004] [PMID: 30276360]

[61] Di Fiore R, Suleiman S, Drago-Ferrante R, et al. Cancer Stem Cells and Their Possible Implications in Cervical Cancer: A Short Review. Int J Mol Sci 2022; 23(9): 5167.
[http://dx.doi.org/10.3390/ijms23095167] [PMID: 35563557]

[62] Kapoor-Narula U, Lenka N. Cancer stem cells and tumor heterogeneity: Deciphering the role in tumor progression and metastasis. Cytokine 2022; 157: 155968.
[http://dx.doi.org/10.1016/j.cyto.2022.155968] [PMID: 35872504]

[63] Zhu L, Jiang M, Wang H, et al. A narrative review of tumor heterogeneity and challenges to tumor drug therapy. Ann Transl Med 2021; 9(16): 1351-1.
[http://dx.doi.org/10.21037/atm-21-1948] [PMID: 34532488]

[64] Proietto M, Crippa M, Damiani C, et al. Tumor heterogeneity: preclinical models, emerging technologies, and future applications. Front Oncol 2023; 13: 1164535.
[http://dx.doi.org/10.3389/fonc.2023.1164535] [PMID: 37188201]

[65] Makena MR, Ranjan A, Thirumala V, Reddy AP. Cancer stem cells: Road to therapeutic resistance and strategies to overcome resistance. Biochim Biophys Acta Mol Basis Dis 2020; 1866(4): 165339.
[http://dx.doi.org/10.1016/j.bbadis.2018.11.015] [PMID: 30481586]

[66] Keyvani V, Riahi E, Yousefi M, et al. Gynecologic Cancer, Cancer Stem Cells, and Possible Targeted Therapies. Front Pharmacol 2022; 13: 823572.
[http://dx.doi.org/10.3389/fphar.2022.823572] [PMID: 35250573]

SUBJECT INDEX

A

Abiogenesis 1, 2
Acidic keratins 183, 191
Actin 184, 185, 186, 187, 188, 189, 190
   bundles 185, 186
   filaments 184, 185, 186, 187, 188, 189, 190
   networks 185
Active transport 138, 140, 151
Adult stem cells (ASCs) 290, 293, 298
Alcoholic liver disease 297
Aldehyde dehydrogenase 326, 335
Alpha keratins 190, 191
Alzheimer's disease 134, 155, 171, 193, 229, 263
Amyotrophic lateral sclerosis 193, 229
Angiogenesis 332, 346,
Antigen-dependent cellular cytotoxicity (ADCC) 270
Anoikis 256, 257, 258, 259, 260, 261
Apoptotic protease-activating factor 1 (Apaf-1) 252
Aquaporin 10, 126, 151
Aspartylglucosaminuria 115
Autolysosomes 103, 109, 110, 132
Autophagy 15, 30, 47, 52, 106, 109, 110, 111, 112, 118, 119, 120, 124, 131

B

Basal cell carcinoma 341
Biconcave cells 9
Binary fission 3, 10, 29
Biofuel production 20, 30
Bioremediation 22, 26, 31
Bitopic 141, 142
Blastocyst 292, 301
Blood disorders 289, 298, 314, 316
Bone marrow transplantation 93, 305
Brain organoids 300

Breast cancer 95, 98, 156, 325, 328, 329, 334, 388, 389

C

Cajal bodies 37, 38
Calcium-activated potassium channel 149
Callogenesis 307
Callus 296, 306, 307
Cancer 300, 303, 321, 322, 325, 326, 329
   heterogeneity 322, 329
   stem cells 300, 303, 321, 325, 326
Cardiac cells 149
Cardiomyocytes 57, 199, 222, 305
CAR-T cell therapy 138
Caseous necrosis 274, 275
Caspases 247, 252, 278, 280
Cell
   adhesion 24, 134, 256, 258, 325, 346
   communication 209, 210, 212, 251
   cycle 10, 125, 209, 305, 315, 339, 346
   division 10, 29, 35, 189, 207, 290, 308, 315, 346
   fusion 311
   junctions 11
   motility 10, 182, 216
   shrinkage 247, 249, 253, 254, 280
   signaling 8, 24, 165, 194, 206, 207, 09, 212, 213
   surface receptors 208, 215, 216, 218
   therapy 138, 289, 290, 296, 297, 304, 305
Cerebral myopathy 171
Cervical cancer 345, 346
Chemotherapy 87, 156, 230, 303, 314, 324, 329, 333, 335, 340, 341, 346, 347
Cirrhosis 297
Clathrin 64, 65, 110, 119, 157
Coagulative necrosis 272, 273, 274
Colorectal cancer 328, 343, 344
Combination therapy 342, 343

Subject Index

Co-translational translocation 48, 79
CRISPR-Cas9 34, 42, 299, 312

D

Detoxification 123, 124, 125, 126, 135, 265, 297, 308, 309, 312, 313
DNA 37, 39, 40, 42, 96, 142, 195, 249, 253, 254, 267, 268, 270, 280, 338
   fragmentation 249, 253, 254, 267, 268, 270, 280
   methylation 308, 309, 312, 313
   repair 37, 39, 40, 42, 96, 267, 268, 338
Drug delivery systems 122, 138, 157

E

Electron transport chain 166, 167
Embryogenesis 207, 310, 328
Embryonic stem cells 113, 128, 167, 183, 194
Endocytosis 143, 157, 220
Endocrine signaling 208
Endolysosomes 103
Endosymbiosis 163, 173
Enzyme replacement therapy 114, 118
Epigenetics 289, 308, 309, 316, 322
Epithelial-mesenchymal transition 192, 261, 330

F

Ferroptosis 248, 249, 262, 263, 264, 265, 266, 267, 281, 282
Fetal stem cells 292
Fluid mosaic model 138, 140, 142, 158

G

Gangrenous necrosis 275, 276
Gap junctions 210
Gaucher disease 70, 117, 118
Genome editing 161, 170, 172, 178, 312
Glial fibrillary acidic proteins 192, 196
Glycosylation 49, 64, 65, 66, 69, 71, 74, 107, 213
Growth factors 207, 208, 213, 214, 216, 217, 225, 314, 327
Guanine nucleotide–binding proteins 216

H

Hematopoietic stem cells 291, 314
Heterochromatin 37, 310, 313
Heteroplasmy 164, 169
Hedgehog pathway 328
Histone modification 308, 310
Huntington's disease 68, 171, 311
Hydrolytic enzymes 102, 131

I

Immune checkpoint inhibitors 256, 324
Immunotherapy 256, 327, 340, 342, 343, 344, 349
Induced pluripotent stem cells (iPSCs) 292, 294
Insulin 207, 208, 211, 214, 217
Integral proteins 141, 147, 150
Intrinsic pathway 251, 252, 258, 259, 264
Ion 130, 138, 215, 217, 222, 232
   channels 138, 215, 217, 222, 232
   regulation 130
Isotonic solutions 146

J

Juxtacrine signaling 207, 208

K

Keratin 190, 195, 328
Kinesins 197, 198

L

Lamellipodia 189
Laminopathies 39, 42, 43, 194
Leber's hereditary optic neuropathy 171
Ligand-gated ion channels 215, 217, 232
Lipid 139, 140, 142, 143, 145, 147, 153, 154, 155, 215, 232, 248, 263, 264, 265, 266, 281
   bilayer 139, 140, 143, 145, 147, 153, 232
   metabolism 215, 264, 265. 266
   peroxidation 248, 263, 264, 265, 266, 281
   rafts 142, 154, 155
Liquefactive necrosis 273, 274, 275
Liver failure 41, 93, 297

Lysosomal 70, 102, 107, 108, 110, 113, 114, 115, 116
   dysfunction 70, 102, 114
   membrane proteins 107, 108, 110, 113, 114
   storage disorders (LSDs) 115, 116

M

Macrophages 104, 143, 212, 255, 270, 271, 276, 324
Melanoma 337, 341
Membrane blebbing 249, 253, 254, 270
Mesenchymal stem cells 292, 293, 303, 310, 311, 313
Metabolic disorders 54, 55
Metastasis 10, 73, 98, 145, 228, 261, 302, 322, 330, 333, 339, 341, 348, 349
Microfilaments 182, 183, 184, 189, 201
Mitochondrial DNA 161, 168, 170, 171, 178, 269, 301
Molecular chaperones 49, 196
Monotopic 141, 142
Multipotent stem cells 291
Muscular dystrophy 118, 119, 194
Myelodysplastic syndrome 92
Myofibrillar myopathy 192
Myosin 186, 187, 188, 192, 196

N

Nanodelivery 72
Necroptosis 249, 269, 276, 277, 278, 279, 281
Necrosis 246, 247, 252, 262, 267, 272, 273, 277, 279, 281, 282
NETosis 248, 249, 269, 270, 281
Neural cells 296, 305, 309
Neurodegenerative diseases 68, 73, 74, 134, 182, 192, 200, 248, 255, 268, 269, 278, 305, 313
Neurofilament proteins 193
Neurotransmitters 207, 211, 215, 217, 222
Notch pathway 303, 322, 328
Nuclear 34, 37, 39, 142, 194
   pore complexes 34, 37
   lamins 39, 142, 194

O

Oligonucleotide fingerprinting 82

Oligopotent stem cells 291
Onco-ribosomes 95
Oncogenes 278, 304, 348
Oogenesis 163
Organ regeneration 289, 297
Osmosis 146
Osmoregulation 122, 124, 125, 128
Osteoblasts 309, 314
Oxidative phosphorylation 161, 165, 171, 172, 178, 279

P

Paracrine signaling 207
Parthanatos 267, 268, 269, 281
Pauci-molecular theory 140
Phagolysosomes 103, 110, 111
Phagophore 106, 112, 132, 133
PI3K/Akt pathway 98, 259, 261
Pluripotent stem cells 291, 292, 294, 296, 300, 301, 305, 315
Poly (ADP-ribose) polymerase 267, 338
Polytopic 141, 142
Progenitor cells 92, 95, 291, 305, 314, 315, 323
Protein 8, 24, 35, 48, 49, 50, 51, 64, 65, 66, 77, 78, 79, 80, 81, 83, 85, 96, 97, 183
   glycosylation 64, 65
   folding 48, 49, 50, 51, 66, 80
   synthesis 8, 24, 35, 48, 77, 78, 79, 80, 81, 83, 85, 96, 97, 183
Protofilament 195, 196, 197
Proto-lysosomes 109, 110
Pyroptosis 248, 249, 282

Q

Quiescence 322, 338, 346

R

Reactive oxygen species 166, 169, 170, 178, 270, 277
Regenerative medicine 294, 300, 311, 315, 316
Retinitis pigmentosa 114, 222, 310
Ribosomal proteins 77. 78, 175
Ribosome 90, 91, 97, 98, 99
   profiling 97, 98, 99
   recycling 90, 91

Ribosomopathies 91, 95, 99

S

Sarcomere 186, 187, 188
Schistosomiasis 164
Signal transduction 196, 206, 207, 208, 209, 213, 214, 215, 216, 217, 218, 231, 232
Single-cell RNA sequencing 256, 261, 281
Skin cancer 339, 341, 342
Sliding filament model 186
Sphingolipidoses 153, 154
Sphingolipids 142, 153, 154,
Spinal muscular atrophy 39, 43
Squamous cell carcinoma 341
Synaptic 154, 217
  plasticity 154
  signaling 217

T

Tay-Sachs disease 134, 154
Therapeutic potential 47, 77, 232, 289, 298, 301, 303, 314, 316
Thylakoids 174, 176
Tissue homeostasis 119, 212, 249, 252, 254, 255, 313
Tonoplast 123, 126, 127, 128, 135
Trans-golgi network (TGN) 64, 71, 105
Tumor 43, 223, 232, 304, 325, 327, 326, 330, 331, 332, 334, 346, 347
  angiogenesis 330, 331, 332
  microenvironment 326, 331, 334, 346, 347
  progression 223, 232, 325, 327, 330, 332,
  suppressor genes 43, 304
Turgor pressure 122, 123, 125, 128, 130, 135

U

Ulcerative colitis 54
Umbilical cord 291, 292, 294, 305, 314
Unipotent stem cells 292
Uniparental inheritance 164

V

Vacuolar pH 134
Vimentin 191, 192, 195, 196, 331
Voltage-gated potassium channel 150

W

Wnt/β-catenin pathway 328
Wound healing 207, 211, 213, 315

Z

Zygote 30, 291

www.ingramcontent.com/pod-product-compliance
Lightning Source LLC
Chambersburg PA
CBHW041455280526
45792CB00004B/1023